Transition Metal Catalysis in Aerobic Alcohol Oxidation

RSC Green Chemistry

Titles in the Series:

1: The Future of Glycerol: New Uses of a Versatile Raw Material
2: Alternative Solvents for Green Chemistry
3: Eco-Friendly Synthesis of Fine Chemicals
4: Sustainable Solutions for Modern Economies
5: Chemical Reactions and Processes under Flow Conditions
6: Radical Reactions in Aqueous Media
7: Aqueous Microwave Chemistry
8: The Future of Glycerol: 2nd Edition
9: Transportation Biofuels: Novel Pathways for the Production of Ethanol, Biogas and Biodiesel
10: Alternatives to Conventional Food Processing
11: Green Trends in Insect Control
12: A Handbook of Applied Biopolymer Technology: Synthesis, Degradation and Applications
13: Challenges in Green Analytical Chemistry
14: Advanced Oil Crop Biorefineries
15: Enantioselective Homogeneous Supported Catalysis
16: Natural Polymers Volume 1: Composites
17: Natural Polymers Volume 2: Nanocomposites
18: Integrated Forest Biorefineries
19: Sustainable Preparation of Metal Nanoparticles: Methods and Applications
20: Alternative Solvents for Green Chemistry: 2nd Edition
21: Natural Product Extraction: Principles and Applications
22: Element Recovery and Sustainability
23: Green Materials for Sustainable Water Remediation and Treatment
24: The Economic Utilisation of Food Co-Products

25: Biomass for Sustainable Applications: Pollution Remediation and Energy
26: From C-H to C-C Bonds: Cross-Dehydrogenative-Coupling
27: Renewable Resources for Biorefineries
28: Transition Metal Catalysis in Aerobic Alcohol Oxidation

How to obtain future titles on publication:
A standing order plan is available for this series. A standing order will bring delivery of each new volume immediately on publication.

For further information please contact:
Book Sales Department, Royal Society of Chemistry, Thomas Graham House, Science Park, Milton Road, Cambridge, CB4 0WF, UK
Telephone: +44 (0)1223 420066, Fax: +44 (0)1223 420247
Email: booksales@rsc.org
Visit our website at www.rsc.org/books

Transition Metal Catalysis in Aerobic Alcohol Oxidation

Edited by

Francesca Cardona
University of Florence, Italy
Email: francesca.cardona@unifi.it

Camilla Parmeggiani
University of Florence and CNR, Italy
Email: camilla.parmeggiani@unifi.it

ROYAL SOCIETY OF CHEMISTRY

THE QUEEN'S AWARDS FOR ENTERPRISE: INTERNATIONAL TRADE 2013

RSC Green Chemistry No. 28

Print ISBN: 978-1-84973-823-1
PDF eISBN: 978-1-78262-165-2
ISSN: 1757-7039

A catalogue record for this book is available from the British Library

Published by The Royal Society of Chemistry,
Thomas Graham House, Science Park, Milton Road,
Cambridge CB4 0WF, UK

Registered Charity Number 207890

For further information see our web site at www.rsc.org

Printed and bound by CPI Group (UK) Ltd, Croydon, CR0 4YY

Foreword

I welcome this new contribution in the RSC Green Chemistry book series on the use of transition metal catalysis in the aerobic oxidation of alcohols. It is timely and resonates well with all aspects of the Green Chemistry agenda. Safe, selective oxidation of alcohols is a fundamentally and strategically important process for all molecule makers who are involved in serving Society's needs for new functional materials. More than any other reaction, this process provides an opportunity and sets the benchmark for the discovery of new hetero- and homogeneous catalysts under aerobic conditions.

These reactions, in combination with sustainable solvent selection, consideration of costs and recyclability of the systems, provide a step change in how compounds will be prepared on scale in the future. As evidenced by the chapters in this book, the field is evolving rapidly. Clearly, nanoparticulate materials, multi-catalytic systems and novel immobilization methods are becoming increasingly important. Pleasingly, all these developments pay respect to our desire for greatly improved safe and sustainable chemical practices. Many of the chapters emphasize the robustness of the processes to facilitate scaling-up of the reactions while also addressing the issues that arise during downstream processing.

Potentially hazardous components during these oxidation reactions relate to the safe use of oxygen gas. This is addressed by using air or, even more wisely, oxygen gas further diluted with nitrogen. Although some practitioners in the area are also naturally disposed to criticize previous methods or prior art in order to justify incremental advances, the real importance of any new procedure usually derives from the selectivity of the transformations when compared directly with other methods. This is an area that is not always appreciated, yet is the key to the success of any new oxidative strategy. The trends in the use of these methods are to harness synergistic effects that

RSC Green Chemistry No. 28
Transition Metal Catalysis in Aerobic Alcohol Oxidation
Edited by Francesca Cardona and Camilla Parmeggiani

Published by the Royal Society of Chemistry, www.rsc.org

arise by combination with supporting materials or using the special characteristics that arise from tuneable nanoparticle aggregation. Indeed, in general it is the physical aspects of the new reagents that play a greater and increasing role than previously suspected. Likewise, quantitative recovery of expensive transition metal catalysts is becoming a crucial component of any new procedure. It is apparent from the various chapters that the methods by which oxygen gas is introduced into the reaction environment will have a noticeable impact both on future reaction design and on the precise experimental conditions. In this regard, flow chemistry methods are increasingly being touted as providing improved reaction kinetics through better mixing and gas introduction.

These are clearly interesting times, where the pace of change in these new procedures is leading to real innovation in solving these fundamentally and intrinsically important reactions.

While many challenges still need to be faced in terms of selectivity, compatibility, cost and robustness, this new textbook provides an excellent starting point for the discovery of novel aerobic oxidation processes using transition metal catalysis that can even go beyond those of simple alcohol oxidations.

Steven V. Ley
University of Cambridge, UK

Preface

Oxidation reactions are excellent tools for converting a functionality in a lower oxidation state to the desired one. They have been, in the past, among the most useful and commonly used reactions in the industrial processes that employ petroleum-based feedstocks as starting materials. However, in addition to often lacking selectivity, traditional oxidation procedures are among the most polluting and hazardous methods, often occurring with a high E-factor (mass of waste per unit mass of product)[1] and delivering considerable amounts of toxic waste, for instance, metal salts in oxidations with stoichiometric Cr(VI) or Mn(VII) derivatives or nitrogen oxides when carried out with HNO_3. For instance, it was estimated that $\sim$5–8% of the worldwide anthropogenic emissions of N_2O, an inevitable stoichiometric waste in oxidations with HNO_3, was produced during cyclohexanol/cyclohexanone conversion to adipic acid, an important key intermediate in the manufacture of nylon-6,6.[2] The pharmaceutical industry, which deals with complex and sensitive molecules, has begun to restrict the use of oxidation reactions on a preparative scale for reasons related to process sustainability. Indeed, many are the "non-green" aspects of oxidation reactions. The more selective reagents produce undesired waste by-products, which are often highly reactive and thermally unstable. However, the greatest factor that limits the employment of oxidation reactions on a large scale is the safety of the process, since many common procedures are carried out in flammable organic solvents. The inherent safety concerns associated with oxidative processes combined with the disposal of hazardous by-products limit the use of these reactions to no more than 3–4% of the total reactions on a preparative scale in the pharmaceutical industry.[3] Despite the existing challenges, oxidation reactions are routinely run to produce many of the commercial pharmaceuticals available today.[4]

RSC Green Chemistry No. 28
Transition Metal Catalysis in Aerobic Alcohol Oxidation
Edited by Francesca Cardona and Camilla Parmeggiani

Published by the Royal Society of Chemistry, www.rsc.org

Nowadays, the principles of green and sustainable chemistry[5] have to be carefully considered when designing a new oxidation procedure. For all the reasons stated above, the traditional oxidation methods are no longer acceptable and great efforts have recently been devoted by the synthetic chemistry community in both academia and industry to the search for new, more sustainable oxidants.[6]

In particular, the oxidation of alcohols to the corresponding aldehydes and ketones[7] or carboxylic acids[8] is of fundamental importance in organic synthesis, due to occurrence of these moieties in many drugs, vitamins and fragrances (Figure 1).

This pivotal reaction remains one of the most active research areas for the identification of more effective and practical methods. This is partially due to the availability of a plethora of orthogonal protecting groups for alcohols,[9] which often allow chemoselective deprotection prior to oxidation.

A publication by Pfizer's medicinal chemists[10] showed that the three most popular oxidants used at Pfizer for the oxidation of primary alcohols to the corresponding aldehydes are hypervalent iodine reagents[11] such as Dess–Martin periodinane[12] or its precursor IBX, the Swern reagent[13] and the tetrapropylammonium (TPAP)–*N*-methylmorpholine *N*-oxide (NMO)[14] protocol (Figure 2).

All of these methods still have poor atom economies[15] and significant scale-up issues. The Swern reagent produces the evil-smelling dimethyl sulfide as by-product and the Dess–Martin periodinane is shock sensitive and prohibitively expensive for use on a large scale. Stoichiometric TPAP again has very poor atom economy and is too expensive, and since 1980 only

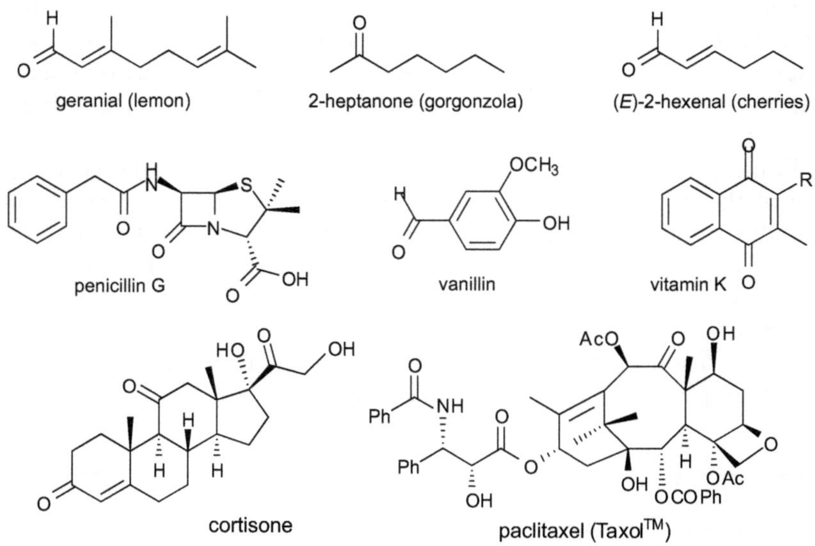

Figure 1 Some drugs, vitamins and fragrances containing a carbonyl or carboxylic moiety.

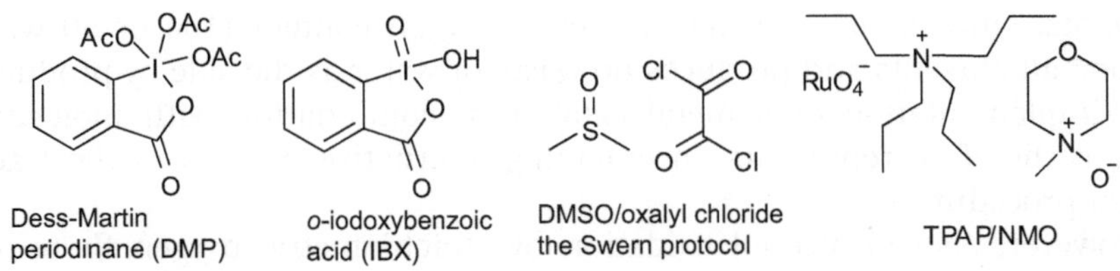

Figure 2 The most popular oxidants used in the pharmaceutical industry for alcohol oxidation.

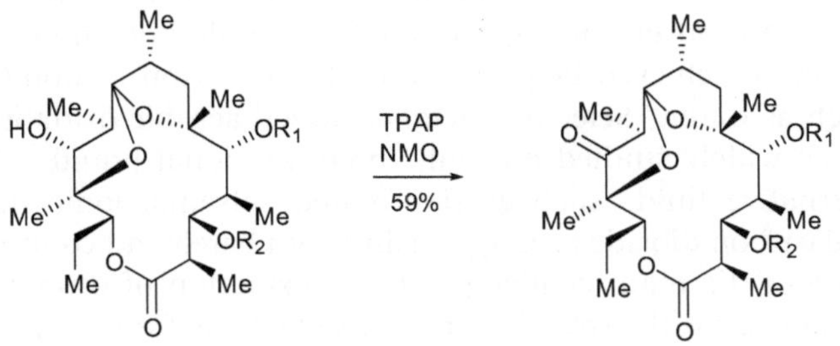

Scheme 1 TPAP oxidation of a macrolide.

one large-scale use of TPAP to catalyze an oxidation in combination with NMO as co-oxidant was reported, namely the oxidation of a secondary alcohol of a very sensitive macrolide (Scheme 1).[16]

As a result, the oxidation of an alcohol to a carbonyl compound, in spite of being a fundamentally important reaction, is actually avoided by the pharmaceutical industry. Therefore, there is still an increasing need in the fine chemical and pharmaceutical industries for systems that are green, scalable and have broad synthetic utilities.

An ideal oxidizing reagent should: (i) oxidize in good yield a broad variety of alcohols bearing different functionalities, (ii) have the ability to be used for scale-up procedures and (iii) be as green as possible, considering worker safety, ecotoxicity and atom economy.[10]

The use of NaOCl as a stoichiometric oxidant certainly represents a dramatic improvement compared with carcinogenic chromium(VI) salts and other reagents referred to above, since it produces one equivalent of acceptable NaCl (environmental quotient $EQ = 1$)[5] as by-product. However, the possibility of forming chlorinated impurities cannot be completely excluded in oxidations carried out with NaOCl.

Therefore, there is definitely a pressing need for more catalysis in alcohol oxidation processes. Catalysis allows reactions to occur under mild conditions in order to save the overall implied energy, and for this reason is strongly encouraged.[17] Catalytic methodologies employing greener oxidants such as molecular oxygen and hydrogen peroxide would seem to represent a further improvement in this respect. Oxygen in particular (or even better air) is among the cheapest and less polluting stoichiometric oxidants, since it has the highest active oxygen content (100 or 50 wt%) among all reagents and produces no waste or water as the sole by-product.[6] The implementation of a metal catalyst in combination with molecular oxygen therefore represents an emerging alternative process to the traditional procedures.

However, also in alcohol oxidation by stoichiometric oxygen there are many shades of green. As Dunn and co-workers pointed out,[10] the use of molecular oxygen presents significant safety issues related to the flammability of mixtures of oxygen with volatile organic solvents. These concerns, however, can be reduced by using air or even oxygen diluted to 10% with nitrogen. Moreover, an improved safety profile and more acceptable scalability can be achieved by performing the oxidation in non-flammable solvents such as water.[5] Selective metal-catalyzed aerobic oxidation of alcohols has been widely studied not only in conventional liquid solvents but also in alternative fluids such as fluorinated solvents, ionic liquids and supercritical carbon dioxide ($scCO_2$)[18] using both batch and continuous-flow reactors. One of the main challenges in the oxidation of primary alcohols with metal catalysts is the selectivity towards aldehyde formation. Aldehydes are prone to hydration and the hydrated aldehyde is considered to be oxidized faster than the non-hydrate compound to carboxylic acids. Therefore, the key to achieving high selectivity in aldehyde formation is to suppress the formation of a carboxylic acid, which can further react with the starting alcohol to yield the ester, which contributes to lowering the aldehyde selectivity.

In this sense, $scCO_2$ seems more suitable for the oxidation of primary alcohols than an aqueous medium, in which aldehyde hydration cannot be avoided (see Chapter 1, Scheme 1.20). However, the addition of catalytic TEMPO (2,2,6,6-tetramethyl-1-piperidine-*N*-oxyl) in the palladium-catalyzed aerobic oxidation of alcohols in water proved to be effective in the complete suppression of carboxylic acid formation (Scheme 2).[19] For details on Pd catalysis see Chapter 4.

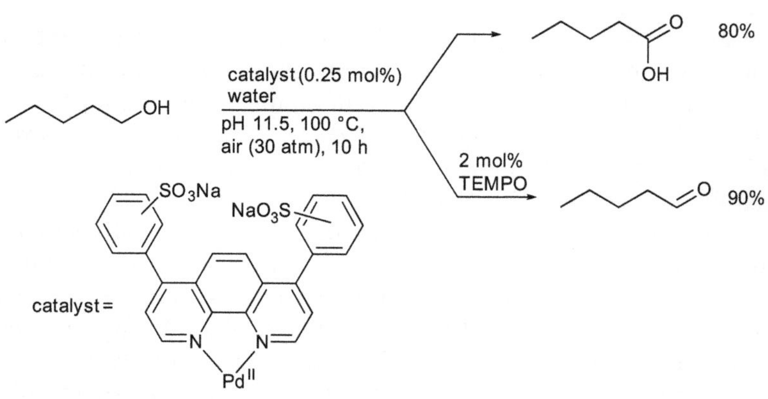

Scheme 2 Palladium-catalyzed aerobic oxidation of alcohols in water.

Selectivity in aldehyde or acid formation from primary alcohols is only one of the many and different challenges of this transformation. A good procedure should need low pressures of O_2, especially in flammable organic solvents, mild reaction conditions, low catalyst loadings and avoidance of expensive or toxic additives. Moreover, it should be able to convert the alcohol moiety selectively in the presence of other groups prone to oxidation (chemoselectivity) and other sensitive functionalities. A further goal is the development of methods able to oxidize one class of alcohols in the presence of another. Finally, an ultimate goal is the development of diastereo- and/or enantioselective alcohol oxidations. In this regard, the groups of Sigman and Stoltz independently discovered[20] that in the presence of the chiral diamine (−)-sparteine the scope of the palladium-catalyzed aerobic oxidations can be extended to asymmetric catalysis, as for example the oxidative kinetic resolution (OKR) of racemic secondary alcohols (Scheme 3)[21] or the oxidative desymmetrization of *meso*-diols (Scheme 4).[22] Whereas the former method (also in cases of perfect selectivity) produces an undesired ketone as by-product, which can eventually be reduced and recycled, the selective oxidation of one of two alcohols in a *meso* substrate can in principle give a product with 100% conversion and enantiomeric excess (*ee*).[23]

This excellent procedure, which leads to remarkably high *ee* values under optimized conditions, was recently applied to the enantioselective total synthesis of various alkaloids such as (−)-aurantioclavine, (+)-amurensinine, (−)-lobeline and (−)- and (+)-sedamine,[24] and to the kinetic resolution of key pharmaceutical building blocks relevant to the enantioselective preparation of Prozac, Singulair and the promising hNK-1 receptor antagonist from Merck (Figure 3).[25]

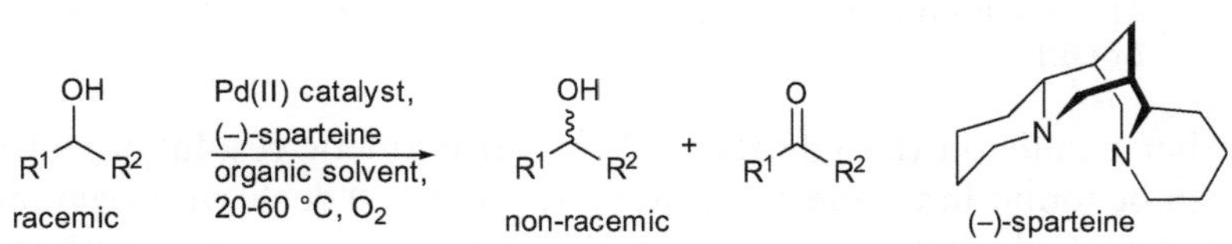

Scheme 3 Palladium-catalyzed oxidative kinetic resolution (OKR) of secondary alcohols.

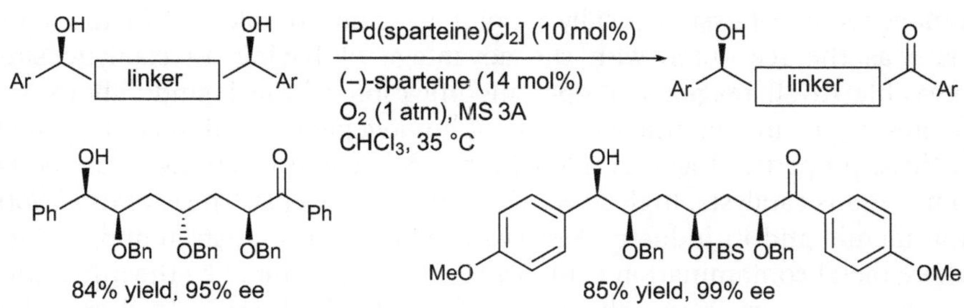

Scheme 4 Oxidative desymmetrization of *meso*-diols.

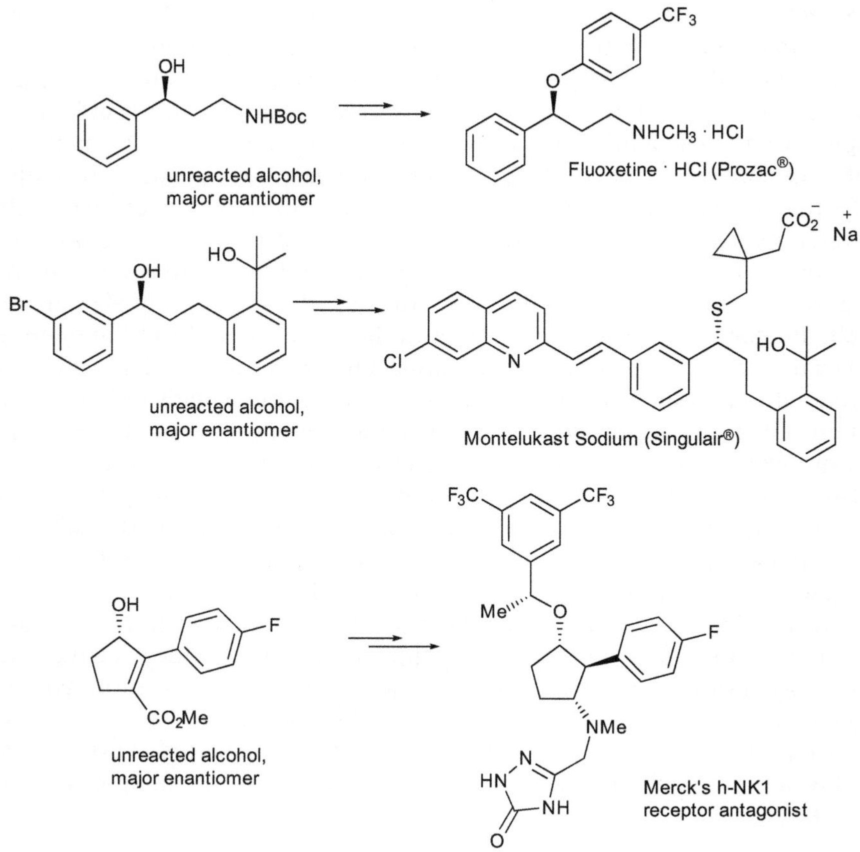

Figure 3 The resolution of important pharmaceutical building blocks by means of OKR.

Further details on the Pd-catalyzed oxidative kinetic resolutions of alcohols can be found in Chapter 4. Moreover, Chapter 9 deals with asymmetric oxidations of alcohols with other transition metals, namely ruthenium, iridium and iron complexes.

Both homogeneous and heterogeneous catalytic systems have been developed[26] and, more recently, metals in the form of nanoparticles.[27] A homogeneous catalyst (typically a soluble metal complex) is in the same phase as the reactants, with the advantage of having all catalytic sites accessible to all reagents. Proper modification of the ligands allows the chemo-, regio- and enantioselectivity of homogeneous catalysts to be tuned. All these properties together allow high efficiency in reactions catalyzed by homogenous catalysts, high selectivities and high yields. They are used both in academia and in industry. However, their use in industrial applications (where metal contamination is highly regulated) is limited by the difficulties encountered in separating the catalyst from the final products. Removal of trace amounts of catalyst from the target product is of crucial importance and still remains a key challenge in homogeneous catalysis. To overcome the separation problems encountered in homogeneous catalysis, heterogeneous catalysts have been introduced. Whereas the first attempts at heterogenization were made with polymeric materials as solid supports, most novel heterogenized catalysts are now based on silica supports, since silica has excellent chemical and thermal stability and good accessibility and porosity, which make it well suited to commercialization. Oxidative degradation of the support is a major problem in particular in oxidation reactions. As an alternative, hybrid systems made of a silica scaffold modified with an organic linker have been proposed. The organic moieties can be strongly anchored to the surface to provide catalytic centers (or ligands) for metal-based catalysis. These hybrid organic inorganic catalysts can anchor the catalytic metal through covalent binding or through simple absorption. However, some issues still remain, such as the accessibility of all active sites to the reagents, which renders heterogeneous catalysts often less efficient than homogeneous catalysts, and the leaching of metals from solid supports, which again needs the separation of traces of metals from the final product. For an interesting overview on emerging *in situ* techniques developed to address the mechanistic understanding of heterogeneous Pd catalysts, see Chapter 4.

Nanoparticles are emerging as excellent sustainable alternatives to conventional solid supports, since they increase the exposed surface area of the active component of the catalyst, thus enhancing the contact between the reagents and the catalytic center, as occurs in homogeneous catalysis. However, their insolubility in the reaction solvent allows easy separation from the reaction mixture, which is the main advantage of heterogeneous catalysis. Thus, nanoparticle systems can be considered an interesting compromise between heterogeneous and homogeneous catalytic systems.[27] Particle size is an important parameter influencing both activity and selectivity; for details on this topic, see Chapters 1 and 4. Chapter 5 addresses the preparation and characterization of gold-based catalysts. This metal, which for a long time was regarded as poorly active, has demonstrated surprisingly high activity when in the form of nanoparticles, which has initiated intense research efforts into its use for aerobic oxidations. However, we must always take into account that there is often an overlap between homogeneous and heterogeneous systems, as pointed out by Sheldon and co-workers, who demonstrated that stabilized nanoparticles are the active catalysts in the Pd–neocuproine system (see Chapter 4 for further details).[28]

Various metals have been discovered that can be activated by oxygen to form catalytically active species for alcohol oxidation. Berzelius reported the stoichiometric oxidation of ethanol with K_2PdCl_4 as early as 1828.[29] However, the first synthetically useful procedure dates back to 1984, when Semmelhack *et al.* reported the first practical Cu-catalyzed aerobic oxidation of primary alcohols in the presence of the stable nitroxyl radical TEMPO.[30]

The coupling of more than one catalysts has also been investigated. In the heterogeneous phase, the synergistic activity of bimetallic nanoclusters is thoroughly discussed in Chapter 4 and 5. Concerning the homogeneous phase, some efficient biomimetic catalytic oxidations of alcohols were reported by Piera and Bäckvall in which transition metal catalysts were coupled with electron-transfer mediators (ETMs). ETMs facilitate the reaction by transporting the electrons from the catalyst to the oxidant along a low-energy pathway, thus mimicking what happens in Nature.[31] This topic is reviewed in Chapter 7, together with the possibility of performing multistep reactions by coupling an oxidative reaction to a different reaction. These tandem reactions allow the creation of several new chemical bonds in a one-pot manner without the isolation of sensitive intermediates, thus also simplifying the work-up procedures.

In addition, the gas-phase oxidation of alcohols is a process of industrial relevance. The basic principles of and recent developments in this topic are reviewed in Chapter 8, which also describes the most innovative materials and methods of operation.

Several oxidations of particular industrial relevance are described in this book, such as the oxidation of methanol to produce formaldehyde or dimethoxymethane and the oxidation of ethanol to produce acetaldehyde, acetic acid, ethylene oxide, acetonitrile or 1,2-dichloroethane (an intermediate for the synthesis of vinyl chloride monomer) (Chapter 8). Moreover, the oxidation of allyl alcohol to produce industrially relevant building blocks such as glyceric acid, acrylic acid and 3-hydroxypropionic acid is discussed in Chapter 5.

Finally, the growing production of biodiesel has led to the availability of large volumes of glycerol, the main by-product of the transesterification reaction of triglycerides. Therefore, methods to convert glycerol into different chemicals are highly desirable. Some aspects related to these industrially relevant transformations (in both the liquid and gas phase) are described in Chapters 1, 5 and 8.

This book aims to give an overview of the most significant metal catalysts (homogeneous, heterogeneous or in the form of nanoparticles) that have been developed in this well investigated field of the research. Owing to the huge amount of literature available on this topic, a choice of the metals was made, trying to analyze and discuss the most versatile and best studied ones (copper in Chapter 2, ruthenium in Chapter 3, palladium in Chapter 4, gold in Chapter 5, iron and vanadium in Chapter 6), highlighting their synthetic potential and always taking into account the previously mentioned synthetic challenges. The use of alternative solvents to conventional fluids is also analyzed.

Francesca Cardona
Camilla Parmeggiani

References

1. R. A. Sheldon, *Chem. Ind.*, 1992, 903.
2. K. Sato, M. Aoki and R. Noyori, *Science*, 1998, **281**, 1646.

3. (a) R. W. Dugger, J. A. Ragan and D. H. Brown Ripin, *Org. Process Res. Dev.*, 2005, **9**, 253; (b) J. S. Carey, D. Laffan, C. Thomson and M. T. Williams, *Org. Biomol. Chem.*, 2006, **4**, 2337.
4. S. Caron, R. W. Dugger, S. G. Ruggeri, J. A. Ragan and D. H. Brown Ripin, *Chem. Rev.*, 2006, **106**, 2943.
5. R. A. Sheldon, *Chem. Soc. Rev.*, 2012, **41**, 1437.
6. *Modern Oxidation Methods*, ed. J.-E. Bäckvall, Wiley-VCH, Weinhein, 2010.
7. *Oxidation of Alcohols to Aldeydes and Ketones*, ed. G. Tojo and M. Fernández, Springer, New York, 2010.
8. *Oxidation of Primary Alcohols to Carboxylic Acids*, ed. G. Tojo and M. Fernández, Springer, New York, 2010.
9. *Protective Groups in Organic Synthesis*, ed. T. W. Green and P. G. M. Wuts, J. Wiley, New York, 1991.
10. K. Alfonsi, J. Colberg, P. J. Dunn, T. Fevig, S. Jennings, T. A. Johnson, H. P. Kleine, C. Knight, M. A. Nagy, D. A. Perry and M. Stefaniak, *Green Chem.*, 2008, **10**, 31.
11. M. Uyanik and K. Ishihara, *Chem. Commun.*, 2009, 2086.
12. D. B. Dess and J. C. Martin, *J. Org. Chem.*, 1983, **48**, 4155.
13. A. J. Mancuso, S. L. Huang and D. Swern, *J. Org. Chem.*, 1978, **12**, 2480.
14. S. V. Ley, J. Norman, W. P. Griffith and S. P. Marsden, *Synthesis*, 1984, 639.
15. B. M. Trost, *Science*, 1991, **254**, 1471.
16. A. B. Jones, *J. Org. Chem.*, 1992, **57**, 4361.
17. *Green Chemistry and Catalysis*, ed. R. A. Sheldon, I. Arends and U. Hanefeld, Wiley-VCH, Weinheim, 2007.
18. T. Seki and A. Baiker, *Chem. Rev.*, 2009, **109**, 2409.
19. G.-J. ten Brink, I. W. C. E. Arends and R. A. Sheldon, *Science*, 2000, **287**, 1636.
20. (a) D. R. Jensen, J. S. Pugsley and M. S. Sigman, *J. Am. Chem. Soc.*, 2001, **123**, 7475; (b) E. M. Ferreira and B. M. Stoltz, *J. Am. Chem. Soc.*, 2001, **123**, 7725.
21. M. S. Sigman and D. R. Jensen, *Acc. Chem. Res.*, 2006, **39**, 221.
22. D. C. Ebner, J. T. Bagdanoff, E. M. Ferreira, R. M. McFadden, D. D. Caspi, R. M. Trend and B. M. Stoltz, *Chem. Eur. J.*, 2009, **15**, 12978.
23. M. Wills, *Angew. Chem. Int. Ed.*, 2008, **47**, 4264.
24. S. Krishnan, J. T. Bagdanoff, D. C. Ebner, Y. K. Ramtohul, U. K. Tambar and B. M. Stoltz, *J. Am. Chem. Soc.*, 2008, **130**, 13745.
25. D. D. Caspi, D. C. Ebner, J. T. Bagdanoff and B. M. Stoltz, *Adv. Synth. Catal.*, 2004, **346**, 185, and references cited therein.
26. For some reviews, see: (a) R. A. Sheldon, I. W. C. E. Arends, G.-J. ten Brink and A. Dijksman, *Acc. Chem. Res.*, 2002, **35**, 774; (b) T. Mallat and A. Baiker, *Chem. Rev.*, 2004, **104**, 3037; (c) I. E. Marko, P. R. Giles, M. Tsukazaki, I. Chellé-Regnaut, A. Gautier, R. Dumeunier, F. Philippart, K. Doda, J.-L. Mutonkole, S. M. Brown and C. J. Urch, *Adv. Inorg. Chem.*, 2004, **56**, 211; (d) S. S. Stahl, *Angew. Chem. Int. Ed.*, 2004, **43**, 3400;

(e) M. J. Schultz and M. S. Sigman, *Tetrahedron*, 2006, **62**, 8227; (f) M. S. Sigman and D. R. Jensen, *Acc. Chem. Res.*, 2006, **62**, 8227; (g) T. Matsumoto, M. Ueno, N. Wang and S. Kobayashi, *Chem. Asian J.*, 2008, **3**, 196; (h) B. Karimi and A. Zamani, *J. Iran. Chem. Soc.*, 2008, **5**, S1; (i) K. M. Gligorich and M. S. Sigman, *Chem. Commun.*, 2009, 3854; (j) C. P. Vinod, K. Wilson and A. F. Lee, *J. Chem. Technol. Biotechnol.*, 2011, **86**, 161; (k) C. Parmeggiani and F. Cardona, *Green Chem.*, 2012, **14**, 547.

27. For some reviews, see: (a) A. T. Bell, *Science*, 2003, **299**, 1688; (b) R. Schlögl and S. B. A. Hamid, *Angew. Chem. Int. Ed.*, 2004, **43**, 1628; (c) D. Astruc, F. Lu and J. R. Aranzaes, *Angew. Chem. Int. Ed.*, 2005, **44**, 7852; (d) V. Polshettiwar and R. Varma, *Green Chem.*, 2010, **12**, 743; (e) S. E. Davis, M. S. Ide and R. J. Davis, *Green Chem.*, 2013, **15**, 17.
28. M. Mifsud, K. V. Parkhomenko, I. C. W. E. Arends and R. A. Sheldon, *Tetrahedron*, 2010, **66**, 1040.
29. J. J. Berzelius, *Liebigs Ann. Chem.*, 1828, **13**, 435.
30. M. F. Semmelhack, C. R. Schmid, D. A. Cortés and C. S. Chou, *J. Am. Chem. Soc.*, 1984, **106**, 3374.
31. J. Piera and J.-E. Bäckvall, *Angew. Chem. Int. Ed.*, 2008, **47**, 3506.

Contents

RSC Green Chemistry No. 28
Transition Metal Catalysis in Aerobic Alcohol Oxidation
Edited by Francesca Cardona and Camilla Parmeggiani

Published by the Royal Society of Chemistry, www.rsc.org

CHAPTER 1

Homogeneous, Heterogeneous and Nanocatalysis

STEFANIA ALBONETTI, RITA MAZZONI AND FABRIZIO CAVANI*

Dipartimento di Chimica Industriale 'Toso Montanari', Università di Bologna, Viale Risorgimento 4, 40136 Bologna, Italy
*Email: fabrizio.cavani@unibo.it

1.1 Introduction

For the conversion of raw materials into fine chemicals and high-value building blocks, reliable oxidation methods have been found to be indispensable tools in modern organic synthesis.[1–6] Various methods with remarkable efficiencies have been developed for a wide array of oxidative transformations which are now well established. However, since most of these standard protocols rely on the use of hazardous terminal oxidants or cause the generation of significant amounts of waste products, there is still an evident need for methodological improvements with respect to environmental and economic issues. Especially in terms of waste product minimization and the use of renewable materials, catalytic aerobic transformations offer some ideal features of a 'green' process[7] towards sustainability. In particular, the oxidation of alcohols is one of the most important synthetic operations in both the organic chemistry laboratory and the chemical industry. Although classic oxidation reactions can be very efficient and selective, they often involve the use of stoichiometric reagents and halogenated solvents, resulting in the generation of large quantities of wastes.[8] The urgent need for more sustainable chemical processes has

RSC Green Chemistry No. 28
Transition Metal Catalysis in Aerobic Alcohol Oxidation
Edited by Francesca Cardona and Camilla Parmeggiani

Published by the Royal Society of Chemistry, www.rsc.org

prompted the development of mild and selective oxidation methods based on the use of green reagents and solvents. In this context, the direct use of oxygen as an oxidizing reagent is a very desirable feature for modern synthetic methods. Hence new catalysts for the aerobic oxidation of alcohols have attracted much attention in recent years. Transition metal-based systems have been successfully used for the aerobic oxidation of alcohols to the corresponding carbonyl compounds with excellent performances of several transition metal catalysts.[5,9]

Homogeneous, heterogeneous and nanocatalysis are the subject of intense research in this field and, rather than aspiring to be exhaustive, this chapter deals with the comparison of those macro-areas within the context of efficiency in reaction design, by following the principles of green and sustainable chemistry for oxidation processes.

1.2 Homogeneous Catalysis for the Aerobic Oxidation of Alcohols

The latest advances in homogeneously catalyzed aerobic oxidation have recently been reviewed for oxovanadium,[10] palladium,[11] copper complexes[12] and bifunctional molecular catalysts.[13] Moreover, cooperative catalysis, combining transition metals with either 2,2,6,6-tetramethylpiperidine-1-oxyl (TEMPO) or other nitroxyl radicals has been gaining in importance.[5,14] At the same time, as a result of their redox–amphoteric properties, significant progress has been made in the development of metal-free aerobic oxidation processes based on TEMPO and related nitroxides as transition metal surrogates.[15] The wealth of papers (more than 140 within the past 3 years) on tailored homogeneous catalysts for the aerobic oxidation of alcohols demonstrates the great efforts being undertaken in this field. The complete history of the selection of metals such as copper, ruthenium, palladium, gold and vanadium is exhaustively explored in later chapters; here, however, we summarize the recent advances in this field by describing some representative examples that follow the principles of rational design of homogeneous catalysts along with the fundamentals of the green and sustainable chemistry already mentioned in the introductory section: (i) oxidizing in good yield with broad functional group compatibility; (ii) achieving high selectivity, if possible without the use of protecting groups; (iii) using non-halogenated solvents, with water being best; (iv) considering safety and using air instead of pure oxygen as the oxidant; and (v) taking into account the ecotoxicity and atom economy of the process.

1.2.1 Selective Oxidation of Alcohols: Cu/TEMPO System Case Study

Copper[16] and, more recently, iron salts[17] in combination with particularly persistent nitroxide as the active oxidation catalyst proved to be superior systems for homogeneous catalyzed alcohol oxidation under an O_2

atmosphere.[5] Extensive descriptions of the history and development of this subject can be found in Chapters 2 and 6. In any case, copper/TEMPO and related systems deserve a brief introduction in order to understand their important role in sustainable developments in homogeneous catalysis. Recent advances reported by Stahl and co-workers[16] demonstrated that the (bpy)Cu(I)/TEMPO catalyst system with NMI (bpy = 2,2′-bipyridine, NMI = *N*-methylimidazole) overcomes nearly all of the limitations associated with Pd(II) catalysts. This catalyst system permits the chemoselective oxidation of benzylic, allylic and aliphatic primary alcohols to the corresponding aldehydes, with rates at least one order of magnitude higher than those observed with Pd(II) catalysts. Moreover, this method is compatible with substrates bearing diverse functional groups and uses ambient air as the source of the O_2 oxidant. A significant rate enhancement has been found with the replacement of Cu(II) with a Cu(I) source and the catalyst system in Scheme 1.1 is notable for its efficiency in the oxidation of aliphatic alcohols.

Mechanistic investigations[16f] support a two-stage catalytic mechanism consisting of (1) 'catalyst oxidation' in which Cu(I) and TEMPO–H are oxidized by O_2 *via* a binuclear Cu_2O_2 intermediate and (2) 'substrate oxidation' mediated by Cu(II) and the nitroxyl radical of TEMPO *via* a Cu(II)–alkoxide intermediate (Scheme 1.2). Catalytic rate laws, kinetic isotope effects and spectroscopic data show that reactions of benzylic and aliphatic alcohols have different turnover-limiting steps. Catalyst oxidation by O_2 is turnover limiting with benzylic alcohols, whereas numerous steps contribute to the turnover rate in the oxidation of aliphatic alcohols.

Another excellent study by Stahl *et al.* explored in depth the substrate scope and the applicability of different nitroxyl co-catalysts, particularly evidencing the complementary role of TEMPO *versus* ABNO (9-azabicyclo-[3.3.1]nonane *N*-oxyl).[18] A catalyst system consisting of (MeObpy)Cu^I(OTf) and ABNO (MeObpy = 4,4′-dimethoxy-2,2′-bipyridine) mediates the aerobic oxidation of all classes of alcohols, including primary and secondary allylic, benzylic and aliphatic alcohols, with nearly equal efficiency. The catalyst exhibits broad functional group compatibility and most reactions are complete within 1 h at room temperature, using ambient air as the source of oxidants. Whereas Cu(I)/TEMPO shows high chemoselectivity for primary alcohols, allowing excellent steric discrimination between unprotected primary and secondary alcohols and also primary alcohols in different steric environments, Cu(I)/ABNO is equally effective with all classes

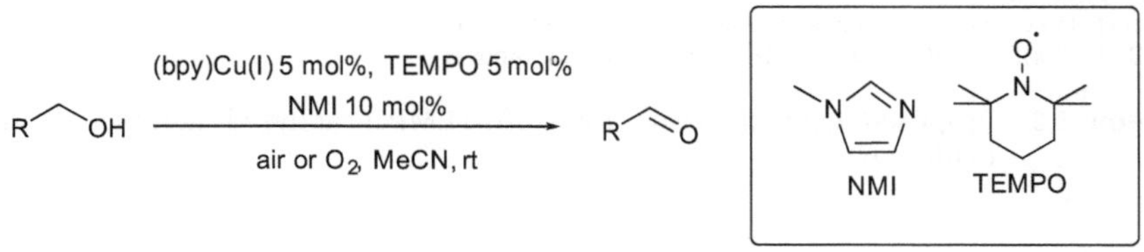

Scheme 1.1 (bpy)Cu(I)/TEMPO/NMI alcohol oxidation system.

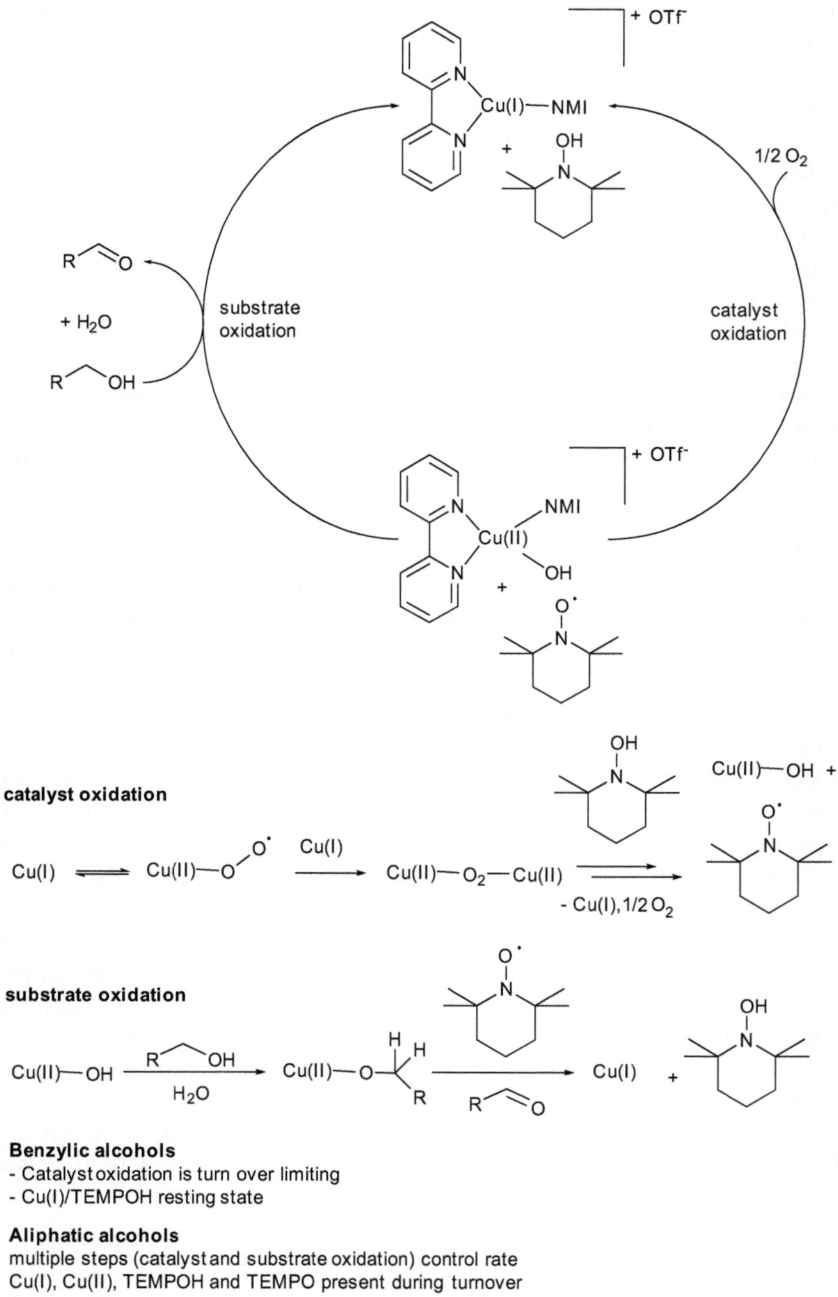

Scheme 1.2 Proposed catalytic cycle for Cu(I)/TEMPO-catalyzed aerobic alcohol oxidation.

of alcohol substrates. Together, these catalytic systems provide compelling aerobic alternatives to traditional alcohol oxidation methods. This Cu(I)/ABNO protocol is also suitable for larger-scale applications; 10–50 mmol (1–9 g) of selected alcohols produced the desired products in ≥90% yield in an open-reaction flask at room temperature within 1–3 h. On the other hand, several drawbacks due to catalyst deactivation require further attention (reactions with a 1 mol% MeObpy/Cu loading suggest that the catalyst deactivates before reaching full conversion of the substrate).

Following the route indicated by the rational design well defined in the work of Stahl and co-workers,[16] in this first section devoted to homogeneous catalysis, the most recent examples of aerobic oxidation are described in two subsections. The first concerns the ligand design and the second focuses on the use of fine-tuned homogeneous catalytic systems for the one-pot synthesis of high-value products such as heterocyclics or nitrogen-containing compounds which are suitable either as building blocks or for fine-chemical synthesis.

1.2.2 Tailored Catalysts for the Selective Oxidation of Alcohols: Ligand Design

Many different ligands – such as pyridine, *N*-heterocyclic carbene chelated *N-O-*, *N-N-*, tridentate, pincer and tetradentate ligands – are used for the preparation of copper, palladium and iron complexes which are active in the aerobic oxidation of alcohols.

1.2.2.1 N-O-*Ligands*

N-O-dentate ligands for copper-catalyzed alcohol oxidation under air or oxygen conditions were first investigated by Punniyamurthy and co-workers, who used a salen analog *N-O*-ligand to achieve the efficient oxidation of primary alcohols to the corresponding aldehydes under oxygen conditions.[19] In this field, Ding and co-workers recently reported highly efficient performances – in particular for secondary alcohols – of the commercially available and inexpensive *N-O*-bidentate ligand L-proline under mild conditions (Scheme 1.3).[20] When using 5 mol% of Cu(I) as the metal precursor, in the presence of 5 mol% of TEMPO as the co-catalyst and air as the oxidant, a wide range of primary and secondary benzylic alcohols are transformed smoothly into corresponding aldehydes and ketones with high yields and selectivities in DMF at room temperature.

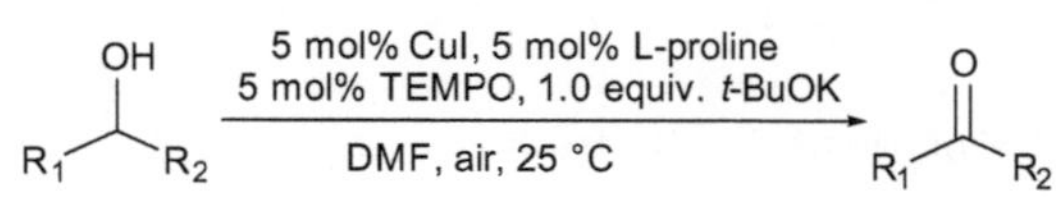

Scheme 1.3 Reaction conditions for the aerobic alcohol oxidation of secondary alcohols to ketones.

Well-defined palladium complexes of the type [Pd(N–O)(X)(L)], in which N–O is an anionic chelate, L is a monodentate base and X is a generic anionic ligand, are attractive catalysts for aerobic alcohol oxidation because they contain within themselves the essential elements to generate catalytic activity. Cámpora and co-workers[21] developed a versatile synthesis that provides access to a wide variety of neophylpalladium complexes of the type [Pd(CH_2CMe_2Ph)(N–O)(L)], where N–O is picolinate or a related *N-O*-bidentate, monoanionic ligand (6-methylpyridine-2-carboxylate, quinoline-2-carboxylate, 2-pyridylacetate or pyridine-2-sulfonate) and L is pyridine or a pyridine derivative. These complexes promote the aerobic oxidation of benzylic, allylic and aliphatic alcohols by oxygen. The chelating *N-O*-ligand is the key element in ensuring the stability of the catalyst and controls the stereoelectronic properties of the active center. Furthermore, the carboxylate group imparts bifunctional character to this ligand, facilitating proton transfer from the substrate (alcohol) to the final electron acceptor (oxygen). The co-ligand L also plays an important role. It must be labile enough to generate the coordinative unsaturation required to allow catalytic activity, but at the same time it contributes to the stability of the system, preventing excessively rapid catalyst decay. Among the different co-ligands tested, pyridine itself showed the best balance of these two properties. Whereas the oxidation of benzyl and secondary aliphatic alcohols with pyridine–carboxylate catalysts is highly selective, carboxylic acids are produced in the case of the primary aliphatic and allyl alcohols. Since the activity and selectivity of these catalysts are ligand controlled, it is expected that the catalyst design may be tuned to improve activity, selectivity and resistance to aggressive oxidation conditions or to generate desirable properties such as compatibility with water or other environmentally friendly solvents. Another useful property of this system is that catalysts perform without additives, facilitating product separation and purification.

With regard to *N-O*-vanadium(v) complexes, 8-hydroxyquinolinates have recently been extensively investigated by Hanson and co-workers for the vanadium-catalyzed aerobic oxidation of lignin model compounds, including benzylic, allylic and propargylic alcohols (Scheme 1.4).[22] The vanadium complex $(HQ)_2V(O)(OiPr)$ (2 mol%, HQ = 8-quinolinate) and NEt_3 (10 mol%) catalyze the oxidation of benzylic, allylic and propargylic alcohols

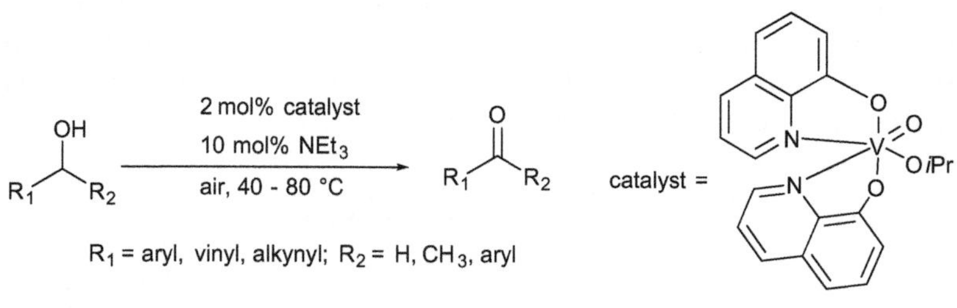

Scheme 1.4 Catalytic oxidation of benzylic, allylic and propargylic alcohols by [$(HQ)_2V(v)(O)OiPr$].

with air. The catalyst can be easily prepared under air using commercially available reagents and is effective for a wide range of primary and secondary alcohols. Reactions proceed under mild conditions (air, 40–80 °C) and in a variety of solvents.

1.2.2.2 *Pyridine Ligands*

Neutral and cationic pyridine-based palladium compounds have been developed by Oberhauser and co-workers as efficient catalysts for the atom-economic aerobic oxidation of unprotected diols to yield chemoselectively the corresponding hydroxy ketone.[23] A comparative catalytic study showed that the bis-cationic precursor (Figure 1.1), in combination with an external base (*i.e.*, K_2CO_3) in a 19 : 1 v/v toluene–DMSO solvent mixture, outperforms neutral complexes. The efficiency of bis-cationic catalyst precursors has been found to depend on (i) the coordination properties of the diol employed with respect to Pd(II) (*e.g.*, 1,2-diols show higher conversion than 1,3-diols) and (ii) the bulkiness and coordination properties of the counterion (*e.g.*, the OTs precursor showed the highest catalytic activity and the BAr_4 counterpart the lowest).

1.2.2.3 *Palladium Pincer Complexes*

The use of palladium NCN and CNC pincer complexes have been found to be effective in the aerobic oxidation of secondary benzyl alcohols at atmospheric pressure in PEG-400, a sustainable reaction medium with excellent yields and functional group tolerability[24] (Scheme 1.5). The recycling of active catalytic species is performed up to the fifth run, while catalyst loadings decreased to 10^{-8} mol%, thus achieving significant turnover number (TON) and turnover frequency (TOF) values. The same conditions proved to be effective for the aerobic oxidation of benzyl methylene compounds, a little-explored process, by palladium catalysts.

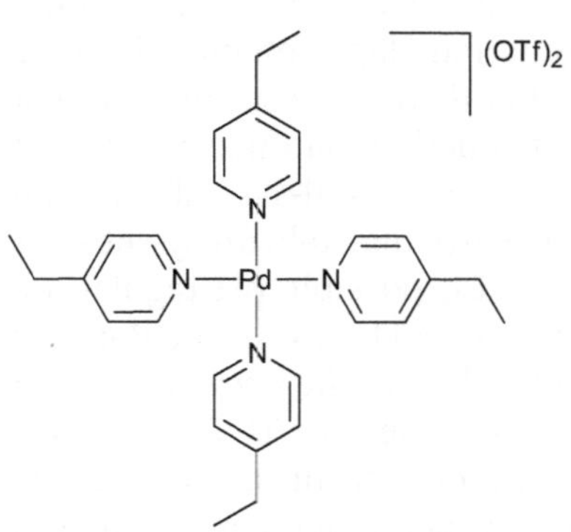

Figure 1.1 Pyridine-based palladium best precursor for the aerobic oxidation of unprotected diols.

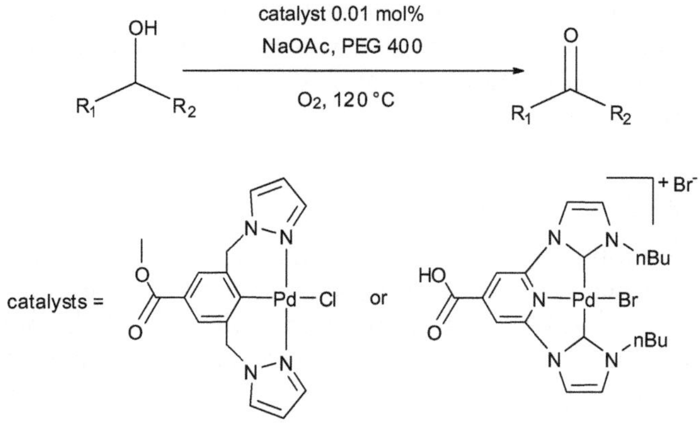

Scheme 1.5 Palladium NCN and CNC pincer complexes effective in the aerobic oxidation of secondary benzyl alcohols.

1.2.2.4 *Water-Soluble Copper Complexes*

Following the pioneering work of the groups of Sheldon,[25] Reedijk[26] and Repo[27] on the use of Cu-based, water-tolerant oxidation catalysts, Pombeiro and co-workers recently developed several copper complexes for the TEMPO-mediated aerobic oxidation of alcohols in water.[28] Catalytic processes in aqueous media may require constrained active particles in the aqueous phase and the design of homogeneous water-soluble catalysts usually involves the use of a suitable ligand with a hydrophilic function.[29] In most cases, such a ligand can be obtained by attaching a water-solubilizing group, *e.g.* a sulfonate or carboxylate group. Another major challenge is the reversible tuning of the acid–base properties of the catalyst, which can enhance the activity/selectivity towards a certain product and/or reversibly change the hydrophilicity. In addition, many processes require a certain pH interval. Therefore, a complex is needed that, in addition to its main catalytic function, can maintain the pH within a certain range (*i.e.* have buffer properties). Hence the synthesis of water-soluble complexes with several pH-tunable sites for protonation–deprotonation is of interest for new green catalytic processes. In this field, the coordination chemistry of copper has been explored with *N*-ethyldiethanolamines for the formation of dicopper(II) alkoxo-bridged complexes using a self-assembly method.[28] Azo derivatives of β-diketones (ADBs) – in particular *o*-hydroxy-substituted ADBs[30] and their congeners – contain sites for possible pH regulation properties, especially sulfo and amino groups such as 3-(5-chloro-2-hydroxy-3-sulfophenylhydrazo)pentane-2,4-dione.[31] All of the above-mentioned systems have shown to be significantly active in the aerobic TEMPO-mediated oxidation of benzyl alcohols to aldehydes. A more recent study investigated the template formation of alkoxy-1,3,5-triazapentadien(e/ato)copper(II) complexes:[32] the latter are active in the oxidative conversion of primary and secondary alcohols to the corresponding carbonyl, making possible molar yields of aldehydes of up to 100% with >99% selectivity. Reactions were carried out in a basic (K_2CO_3) aqueous medium, without any organic solvent and under atmospheric pressure of air (or O_2), with high yields and selectivities. Those mild conditions are of industrial significance.

1.2.2.5 Porphyrins as Photosensitizers

Safari and co-workers introduced a new selective, economical and efficient aerobic pathway for the selective photo-oxidation of a variety of aromatic and aliphatic alcohols to the corresponding aldehyde and ketone derivatives.[33] Their method uses molecular oxygen (1 atm pressure of air) in the presence of free base porphyrins and metalloporphyrins as sensitizers (Sens*) and white light or sunlight in CH_3CN at room temperature. Scheme 1.6 shows the structures of the sensitizers used and the reaction conditions. This method is broad in scope, exhibits chemoselectivity and proceeds under mild reaction conditions. The resulting products are obtained in good conversions within a reasonable time. The best results with both primary and secondary alcohols were obtained with metallated H_2TMP and iron–porphyrins ClFeTMF.

1.2.2.6 Dimethylglyoxime

Transition metal species (ML_x, M = Co, Fe, Cu, Mn, Ni), which are usually used in the form of metalloenzymes, synthetic porphyrins and Schiff base

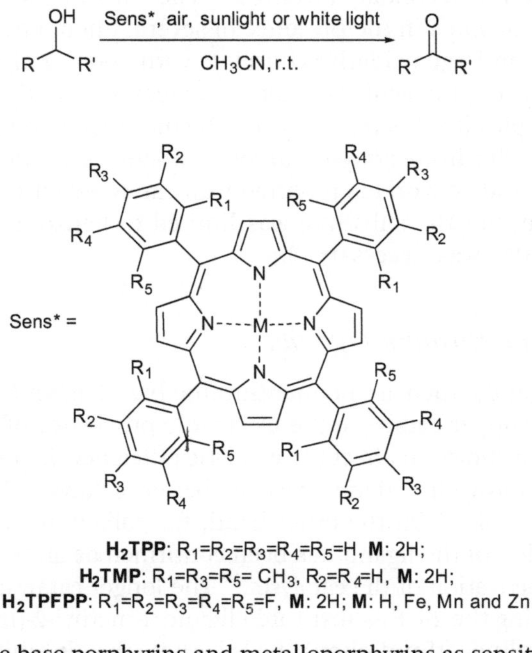

Scheme 1.6 Free base porphyrins and metalloporphyrins as sensitizers for selective photo-oxidation.

complexes, may activate dioxygen for aerobic oxidations.[34] The structure of the majority of these complexes includes cyclic N_4 or open-chain N_2O_2 donor equatorial ligands. Within this field, the simple and inexpensive ligand dimethylglyoxime (DH_2), in combination with cobalt nitrate and TEMPO, efficiently accomplished the aerobic oxidation of various alcohols (including primary and secondary benzylic, allylic and aliphatic alcohols), which were quantitatively converted to the corresponding aldehydes or ketones at 70 °C under 0.4 MPa dioxygen pressure.[35] Several other metal salts (based on Co, Cu, Fe, Mn and Ni) and reaction conditions were screened, but led to poorer results. Mechanistic investigations shed light on the role of the three-component $Co(NO_3)_2/DH_2$/TEMPO catalyst. During oxidation, the *in situ*-generated cobaloxime and NO_x played crucial roles in the activation of dioxygen, thus resulting in two concerted catalytic pathways: cobaloxime-activating-dioxygen TEMPO-catalyzed and NO-activating-dioxygen TEMPO-catalyzed aerobic oxidation of alcohols. Therefore, the present system could efficiently catalyze the aerobic oxidation of aliphatic and secondary alcohols without the need for an additional base.

1.2.2.7 Role of Naphthoxide in Iron-Catalyzed OKR

Within the scope of the oxidative kinetic resolution of secondary alcohols (OKR),[36] a naphthoxide-bound iron(salan) complex has been recently reported as a good example of an efficient iron-catalyzed aerobic oxidative kinetic resolution of secondary alcohols.[37] The substrate scope was studied with secondary alcohols in the presence of several phenol derivatives, which are reluctant to undergo oxidative coupling, with good to high enantiomeric differentiation using molecular oxygen. The ligand plays the main role [the iron(salan) complex itself is inactive] and the modified complex represents a step forward in the iron-catalyzed aerobic kinetic resolution of secondary alcohols (Sekar and co-workers reported iron-catalyzed OKR using O_2 as the terminal oxygen, but the substrate was limited to benzoins and a catalytic amount of TEMPO was necessary[38]).

1.2.2.8 Phenanthroline Ligands

Various *N-N*-ligands such as phenanthroline-based ligands have been investigated. For copper catalysts, the electronic properties of the supporting ligand have been shown to affect the catalytic efficiency, identifying electron-rich 1,10-phenanthroline derivatives as better catalysts for the aerobic oxidation of alcohols.[39] On the other hand, for palladium catalysts, the oxidative degradation of the ligand has been confirmed as an ‘Achilles heel’ due to the rapid deactivation of the catalysts.[40] The longer catalyst lifetimes of Pd complexes bearing the 2-CF_3-substituted ligand 4-methyl-2-(trifluoromethyl)-1,10-phenanthroline (tfmm-phen) reveal that the inhibition of ligand oxidation can lead to stronger catalysts for aerobic alcohol oxidation.

Nevertheless, the modest TON observed (22 mol ketone per mol of Pd) suggests that other decomposition pathways may limit the lifetime of catalysts derived from tfmm-phen.

1.2.2.9 N-Heterocyclic Carbene Complexes

As potential alternatives to traditional nitrogen and phosphine ligands, *N*-heterocyclic carbenes (NHCs) have attracted great interest owing to their unique properties.[41] The performance of NHCs is easily tuned through the introduction of various functional groups at the *N*-positions.[42] Chen and co-workers reported TEMPO-functionalized imidazolium salts which react with commercially available copper powder to produce Cu–NHC complexes.[43] The *in situ*-generated Cu/NHC/TEMPO complexes are fairly efficient catalysts for the aerobic oxidation of primary alcohols to aldehydes (Scheme 1.7). The catalyst is easily available and various primary alcohols were selectively converted to aldehydes in excellent yields. Structural changes to the catalyst in the catalytic cycle are unclear. It was difficult to isolate the catalyst after oxidation was complete because of the low catalyst loading and good solubility. To understand fully the catalytic performance, further work is needed.

Cazin and co-workers[44] reported on a series of Pd(0) complexes bearing mixed ligand systems, NHC/PR_3, which react cleanly with O_2, forming peroxo complexes [Pd(η^2-O_2)(NHC)(PR_3)]. [Pd(NHC)(PR_3)] complexes are used as pre-catalysts in the oxidation of alcohols at a low catalyst loading (0.25 mol%) under mild conditions (60 °C and 1 atm of O_2). Additives such as acetic acid and molecular sieves have a major effect on the reaction outcome. In the case of the most reactive substrates, oxygen can be replaced by air.

1.2.3 Tandem Catalysis for the Multistep Synthesis of Fine Chemicals

The one-pot synthesis of valuable chemicals such as heterocyclics, imines and nitriles from alcohols and a nitrogen source is another challenging perspective of green homogeneous catalysis. Various single-step or tandem

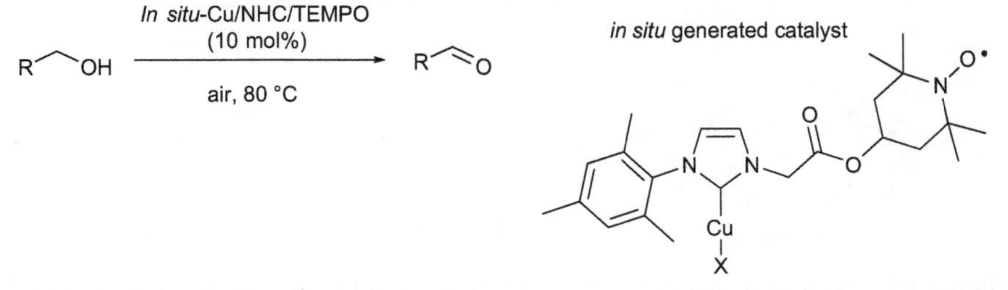

Scheme 1.7 *In situ*-generated Cu/NHC/TEMPO-catalyzed aerobic alcohol oxidation.

reactions with mono- or multicomponent catalysts which exploit the aerobic oxidation of alcohols are steps forward in the concept of 'greener' homogeneous processes.

1.2.3.1 *Biomimetic Approach in the Synthesis of Imines, Lactones, Lactams and Benzoxazoles*

By exploiting the biomimetic approach developed by Bäckwall and co-workers – which combines a ruthenium catalyst (Shvo complex), a cobalt–salen complex and a benzoquinone – the ruthenium catalyst can efficiently catalyze the aerobic oxidation reactions of alcohols[45] and amines[46] to give the corresponding ketones and imines, respectively. Recently, the scope of this biomimetic catalytic system has been expanded, demonstrating that diols and amino alcohols can be oxidized to lactones[47] and lactams,[48] and that the oxidative coupling of benzylamines and 2-aminophenols makes possible the synthesis of benzoxazole structural elements into natural products and building blocks for pharmaceuticals and organic materials (Scheme 1.8).[49]

The proposed mechanism is reported in Scheme 1.9.

This biomimetic system can be applied to the synthesis of benzimidazoles and benzothiazoles and was applied to the synthesis of benzoxazoles in the same way. This biomimetic approach should be a useful method for the

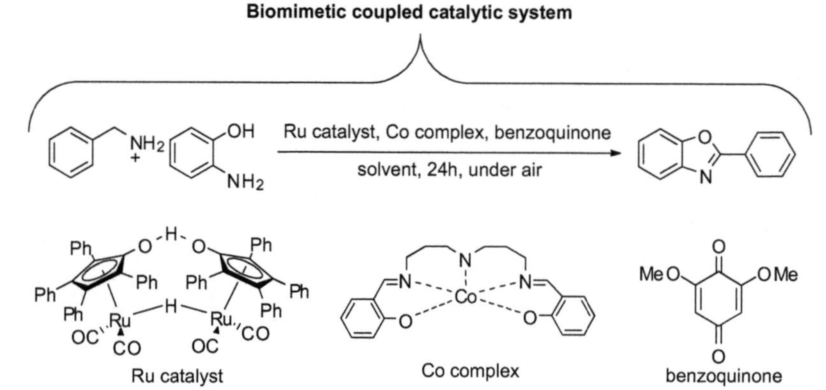

Scheme 1.8 Oxidative coupling to give benzoxazoles by using a biomimetic oxidation system.

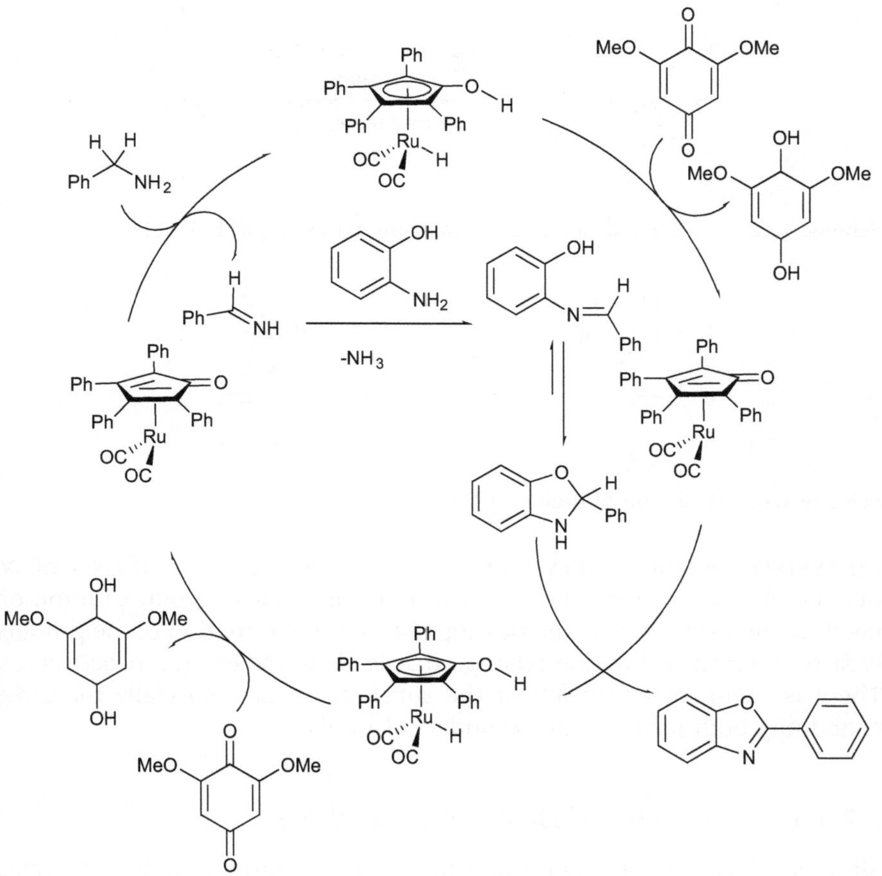

Scheme 1.9 Proposed mechanism for the oxidative coupling that gives benzoxazole.

synthesis of pharmaceutically important heterocyclics under environmentally benign reaction conditions.

1.2.3.2 Copper-Mediated Heterocycle Synthesis

Another recent report on the preparation of benzimidazoles, benzoxazoles and benzothiazoles directly from aromatic alcohols and *o*-phenylenediamine, *o*-aminophenol and *o*-aminothiophenol exploits a simple but efficient Cu(I)/TEMPO/Bpy catalytic system under ambient conditions (Scheme 1.10).[50]

N-Heterocycles can also be prepared using alcohol oxidation as a key synthesis step. The potential of the Cu/TEMPO/O_2 catalytic system as an aerobic oxidation catalyst was explored by Cook and co-workers for the synthesis of substituted indoles and quinolones.[51] Different Cu(I) and Cu(II) salts, *N*-*N*-ligands (such as 2,2′-bipyridine and 1,10-phenanthroline), bases and TEMPO combinations were screened. The study indicated that the

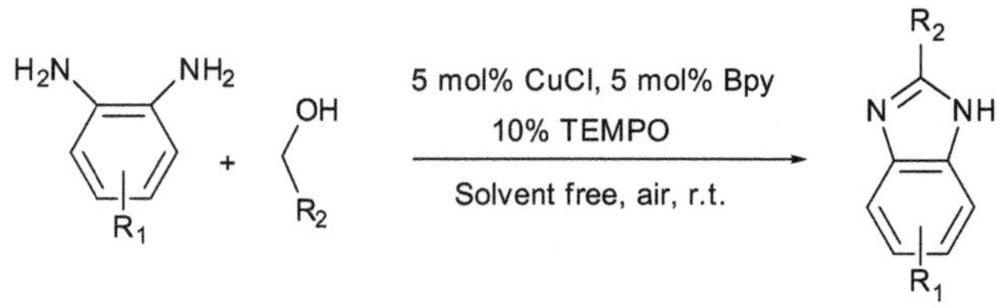

Scheme 1.10 Benzimidazole preparation catalyzed by Cu(I)/TEMPO/Bpy.

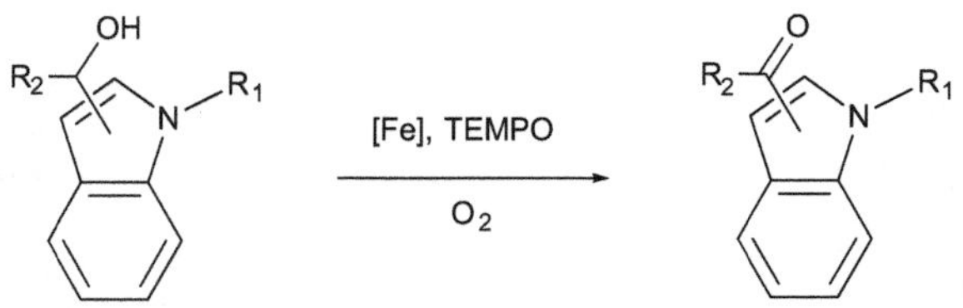

Scheme 1.11 Oxidation of indole carbinols.

Cu/TEMPO oxidation catalyst has the potential for the synthesis of *N*-heterocyclics. In the case of indole synthesis, the issue of product inhibition needs to be resolved. The kinetics suggest that the formation of the product is faster than the inhibition reaction that leads to the catalyst deactivation. There is room for expansion of the substrate scope, especially for those containing both primary and secondary alcohols.

1.2.3.3 Iron-Mediated Heterocycle Synthesis

Ma *et al.* developed a general and practical iron nitrate/TEMPO-catalyzed aerobic oxidation of the alcohols most commonly used in academic and industrial laboratories, such as primary and secondary alkanols, benzylic, allylic and propargylic alcohols and allenols, efficiently expedited by a catalytic amount of sodium chloride.[52] Subsequently, the same $Fe(NO_3)_3$/TEMPO/NaCl system was efficiently used in the aerobic oxidation of (i) propargylic alcohol to α,β-unsaturated alkynals and alkynones,[53] (ii) allylic alcohols with retention of the C=C double-bond configuration[54] and (iii) indole carbinols (an important unit in organic synthesis, since a large number of indole carbonyls with bioactivity can be found in natural products) (Scheme 1.11).[55] The practical use of these preparations was also verified on a large scale (0.1 mol or 1 mol of selected substrates) by using toluene as solvent and purifying the products by distillation or recrystallization, with 90% isolated yield.

1.2.3.4 Imine Synthesis

Imines are useful compounds for organic synthesis of pharmaceuticals and agricultural chemicals. Traditionally, imines are synthesized by the condensation of aldehydes (or ketones) with primary amines. The required carbonyl compound is generally prepared by the oxidation of the corresponding alcohol. With a view to creating environmentally benign processes, the development of catalytic systems for tandem reactions has attracted considerable attention. From this standpoint, oxidative coupling of alcohols and amines should be the most direct way to produce imines. The dicopper complex $[Cu_2(bpnp)(\mu\text{-}OH)(HCOO)_3]$ (Scheme 1.12), stabilized by the bpnp ligand, is an example of an efficient catalyst for the oxidative coupling of alcohols with amines to yield the corresponding imines in the presence of oxygen under solvent-free conditions.[56] Even though more detailed studies are needed to study the effect of the metal centers, this method holds the promise of general applicability, producing good yields of products with various benzyl alcohols and amines, low catalyst loading and solvent-free reaction conditions.

Yu and co-workers reported on the palladium-catalyzed aerobic oxidative synthesis of imines from alcohols and amines under ambient conditions (Scheme 1.13).[57] Ligand, TEMPO and base proved to be key factors for the reaction and, from a detailed screening, the best results were obtained with $Pd(OAc)_2$, TEMPO and Et_3N as the base. The imine synthesis occurs in one pot from alcohols and amines *via* a low-loading palladium-catalyzed tandem aerobic alcohol oxidation–dehydrative condensation reaction that can be readily carried out in open air at room temperature. The same group also reported findings in the area of air-promoted metal-catalyzed aerobic *N*-alkylation.[58] Their method is rather general in catalyst and substrate scope and uses many simpler, cheaper, readily available ligand-free metal catalysts and a wide range of amines, amides and alcohols which behave with respect to activities more effectively than those under conventional anaerobic conditions. Although a few N-alkylation reactions have been carried out under aerobic conditions and found to be more efficient than the anaerobic reactions, the role played by air in such reactions remains to be fully elucidated.

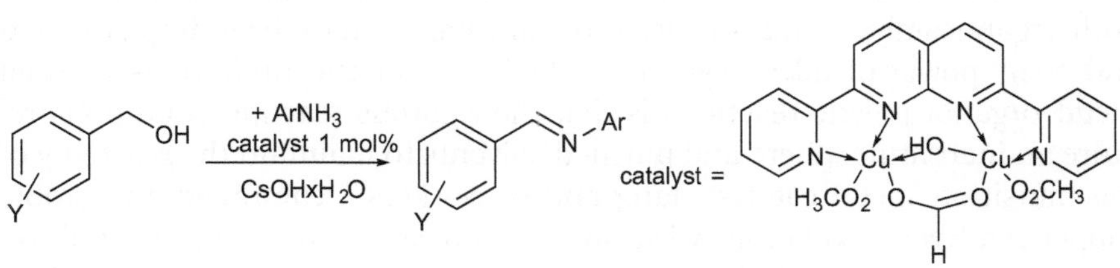

Scheme 1.12 Imine formation from arylamines and benzyl alcohol catalyzed by the dicopper complex $[Cu_2(bpnp)(\mu\text{-}OH)(HCOO)_3]$.

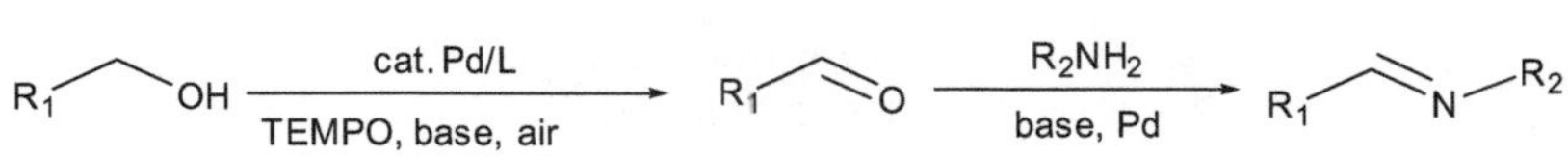

Scheme 1.13 Proposed reaction path for the Pd-catalyzed aerobic oxidative preparation of imines from alcohols and amines.

1.2.3.5 *Nitrile Synthesis*

Nitriles are widely used in both chemical and biological applications as they can be important building blocks in the pharmaceutical, fine chemical, dye and agrochemical industries. However, despite the potential usefulness of their available synthetic protocols,[59] the use of toxic solvents and expensive reagents along with the production of large amounts of inorganic waste and tedious work-up procedures limit their application from a green chemistry perspective. The oxidative conversion of alcohols to nitriles is an attractive alternative and the development of a highly efficient and environmentally benign solvent-free protocol is still a challenging task. As an example, Bhanage and co-workers exploited a copper catalyst for the aerobic oxidative reaction of benzylic and allylic alcohols, using ammonium formate as the nitrogen source, for the synthesis of nitrile compounds under an air atmosphere and solvent-free conditions (Scheme 1.14). A wide range of substrates were well tolerated in the reaction, which produced water as a by-product.[60] From a green perspective, the developed protocol uses an inexpensive and readily available $CuCl_2 \cdot 6H_2O$ catalyst; the system circumvents the use of oxidants such as 2,3-dichloro-5,6-dicyano-1,4-benzoquinone (DDQ), I_2, H_5IO_6 and 1,3-diiodo-5,5-dimethylhydantoin (DIH) and the use of a handy and inexpensive nitrogen source such as ammonium formate.

1.2.3.6 *Asymmetric Synthesis*

Transition metal catalysis, organocatalysis and biocatalysis have attracted much attention in the field of organic synthesis because they can provide highly complex molecules in addition to good chemo-, regio- and stereoselectivity. In recent years, dual catalytic systems, which combine different catalytic species among transition metal complexes, organic catalysts and biocatalysts, have been reported for multifunctionalization and multistep, one-pot reactions.[61] Although a number of multicatalytic reactions using both organocatalysts and transition metal catalysts have been implemented and their powerful advantages described, one of the limitations of dual metal–organocatalytic reactions is that the rigorous reaction conditions require an inert atmosphere and purified solvents to maintain the reactivity of the transition metal catalyst. Jang and co-workers[62] combined an aerobic copper-catalyzed oxidation with an organo-catalyzed Michael addition reaction (iminium catalysis). By controlling the amount of TEMPO, allylic

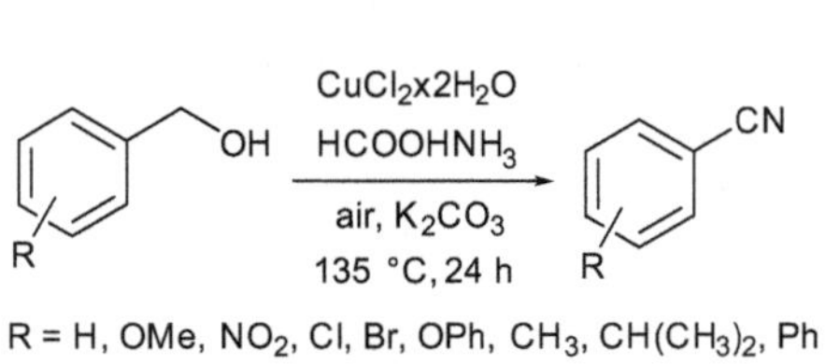

Scheme 1.14 Oxidative synthesis of nitriles from alcohols and ammonium formate.

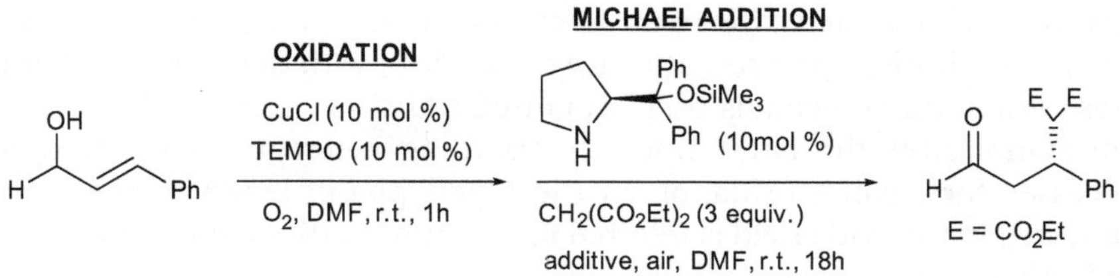

Scheme 1.15 Tandem oxidation–Michael addition.

alcohols were converted into β-substituted aldehydes and α,β-substituted aldehydes in good yields and with excellent levels of stereoselectivity, extending the substrate scope of α,β-unsaturated aldehydes to low molecular weight aliphatic alcohols in addition to aromatic alcohols (Scheme 1.15). Furthermore, merging iminium and enamine catalysis provides a highly stereoselective protocol for the formation of α,β-substituted aldehydes as a single diastereomer with remarkably high enantioselectivity. Large amounts of solvents and silica gel to purify the resulting aldehydes are not required.

1.2.3.7 Esterification of Alcohols

Traditionally, alcohols could be converted into esters by multiple steps. However, the direct conversion of alcohols to esters in the presence of catalysts may represent a step forward towards green, economic and sustainable processes. Despite the fact that great attention has been paid to the palladium-catalyzed selective oxidation of alcohols to aldehydes in recent decades, reports on the oxidative esterification of alcohols are scanty. The first palladium-catalyzed direct aerobic oxidative esterification of benzylic alcohols with methanol and various long-chain aliphatic alcohols was reported by Lei *et al.*[63] Benzylic alcohol and methanol were used in the model reaction and molecular oxygen was used as oxidant. The applicability of this method, where $[PdCl_2(CH_3CN)_2]$ with NaO*t*Bu gave the best results in MeOH as solvent, was shown for a range of different substrates to give their corresponding esters in moderate to high yields. The challenging esterification reactions of long-chain aliphatic alcohols were accomplished by using a *P*-olefin ligand to control the selectivity. The direct nature of this route and the use of O_2 as oxidant represent a step towards an environmentally benign and sustainable process.

1.3 Heterogeneous Catalysis for the Aerobic Oxidation of Alcohols

Even through heterogeneous catalysts are preferable to homogeneous ones owing to their easier recovery and reuse, there are still some concerns and practical problems in the use of heterogeneous catalysts in the aerobic oxidation of alcohols owing to lower catalytic activity and deactivation.[64,65]

Nevertheless, the heterogeneous approach is highly desirable for commercial applications and recent reviews have dealt with the subject of liquid-phase oxidation of alcohols over supported transition metals.[65,66] This section summarizes the latest research regarding heterogeneously catalyzed processes for alcohol oxidation in the liquid phase; however, the reader interested in this wide field is referred to the authoritative papers cited above for further details.

1.3.1 Oxidation of Alcohols Using Single-Metal and Alloy Catalysts

Heterogeneous, late transition metal catalysts – notably ruthenium, gold, platinum and palladium supported catalysts – are particularly active for the liquid-phase aerobic oxidation of alcohols, but other metals and metal compounds have also been studied recently, including Ag[67] and Co_3O_4.[68]

Cinnamyl alcohol (CA) and benzyl alcohol (BA) oxidation are often used as model reactions. As with other alcoholic substrates, CA and BA oxidations generally proceed through an aldehyde intermediate to the final acid products. These two molecules are both highly reactive and the C=C bond present in the CA molecule may undergo side reactions, forming 3-phenylpropan-1-ol, 3-phenylpropionaldehyde and 3-phenylpropanoic acid (Scheme 1.16). In fact, apart from the oxidation pathway, side reactions due to hydrogen transfer and hydrogenolysis, depending on the reaction conditions and the catalyst used, can be observed. As an example, the side reaction forming 3-phenyl-1-propanol is typical for allylic alcohols and may be considered a transfer dehydrogenation reaction in which the reactant plays the role of hydrogen acceptor.[69]

Benzyl alcohol was also reported to be subjected to different reactions depending on the catalyst used and the reaction conditions (Scheme 1.17). Indeed, in addition to the oxidation of BA to form benzaldehyde and benzoic acid and benzoate, reactions such as (i) disproportionation to form toluene, benzaldehyde and H_2O and (ii) dehydration to form dibenzyl ether have been reported.[70]

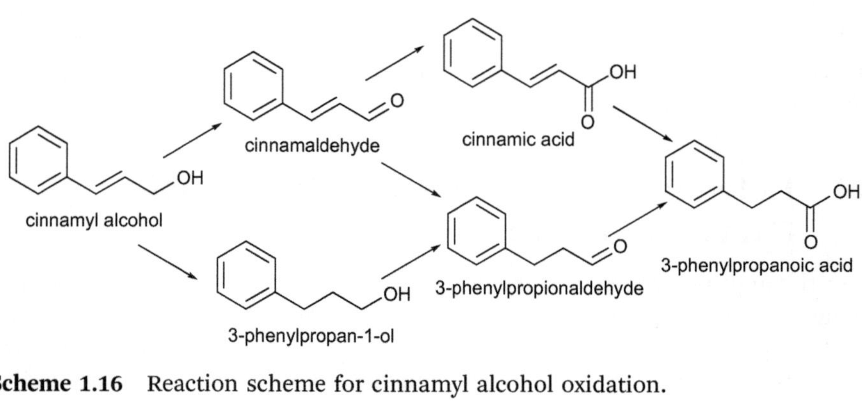

Scheme 1.16 Reaction scheme for cinnamyl alcohol oxidation.

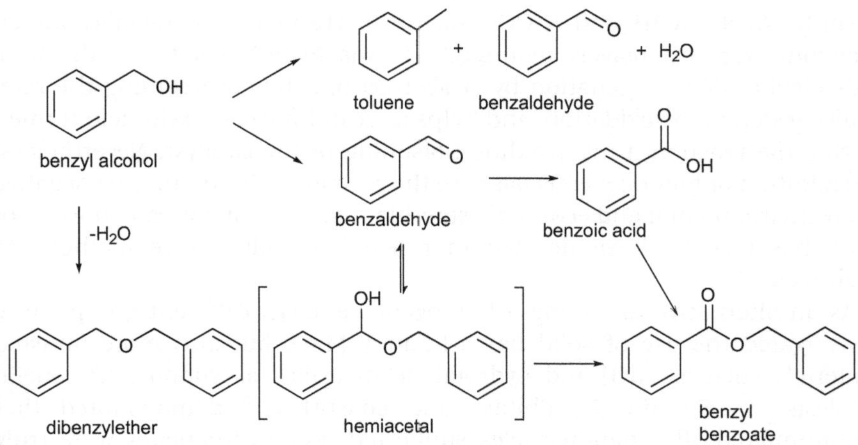

Scheme 1.17 Reaction scheme for benzyl alcohol oxidation.

Particle size-dependent catalytic activity has been reported for BA and CA oxidation over metals such as Pd, Pt and Au. The optimum particle size was identified as around 3–5 nm, indicating that the reaction is structure sensitive.[71,72] This significant size-dependent reactivity was confirmed with unsupported gold nanoparticles, showing optimum performance for particles with diameters <5 nm. Nevertheless, unsupported bulk Au itself was also reported to catalyze the aerobic oxidation of π-activated alcohols, such as BA and CA,[73] indicating that the favorable interaction between the gold surface and the π-system of the aromatic ring is able to enhance the overall efficiency of the active site even in the presence of bulk metal. In fact, the authors found that the π-activation for the $-CH_2OH$ group by the phenyl group in benzyl alcohol was critical for oxidation using bulk Au.

Micro- and mesoporous oxide architectures with high surface area and regular channels have been widely used as a framework for metals, limiting the agglomeration of nanoparticles and relative deactivation. Parlett and co-workers[74–76] recently reported that Pd and Pt dispersed over mesoporous SiO_2 are very active in the oxidation of crotyl alcohol and cinnamyl alcohol. Moreover, they demonstrated that extremely low palladium loadings generate atomically dispersed Pd^{2+} surface species that impart unusual activity in the oxidation of allylic alcohols. The same group demonstrated the usefulness of hierarchically ordered nanoporous Pd/SBA-15 in the oxidation of sterically hindered allylic alcohols – such as farnesol and phytol – to their aldehydes.[75] The results data indicated that, in order to obtain Pd catalysts that are efficient for the aerobic oxidation of these large molecules, support materials capable of stabilizing high amounts of PdO nanoclusters (<2 nm) but with pores large enough to minimize mass-transport limitations are required.

The rate and selectivity of alcohol oxidation depend strongly on the reaction conditions and the positive effects of a homogeneous added base (mainly NaOH, KOH and K_2CO_3) on the performance of metal-catalyzed reactions were extensively reported.[66] It was hypothesized that the base aids alcohol dehydrogenation by H abstraction, thus overcoming the rate-limiting step in the oxidation, and helps in desorbing carboxylic acid formed during the reaction, thus avoiding poisoning of the catalyst. Nevertheless, the addition of homogeneous bases to the reaction solution may be negative from environmental and economic standpoints, and a significant amount of work has recently been devoted to research on alternative methods to their use.

As an alternative to adding a homogeneous base, different groups have investigated the use of solid bases (mainly MgO, but also more unusual materials such as NiO) and hydrotalcite materials as supports for metal catalysts.[77,78] Specifically, Ebitani and co-workers[78] demonstrated that platinum/gold alloy nanoparticles supported over hydrotalcites were truly effective catalysts for the selective aerobic oxidation of glycerol and 1,2-propanediol in base-free aqueous solution when using molecular oxygen at room temperature and atmospheric pressure. The high activity and selectivity of these catalysts were explained in terms of alterations of geometric and electronic states of the catalytically active surface Pt sites by Au atoms and starch ligands used for nanoparticle synthesis.

On the other hand, it appears that, using basic supports, the acid formed in the oxidation reaction can either be absorbed on the solid surface or react with the support, thus leading to catalyst deactivation or leaching of the material into the reaction medium. Accordingly, basic supports seem to serve more as a stoichiometric replacement for the homogeneous base rather than as catalytic materials.

A major topic in the aerobic oxidation of alcohols is the performance of bimetallic catalysts. The alloying of different metals has indeed proven to have the potential for preventing catalyst deactivation, enhancing reaction rates and improving product selectivity.[79]

As an example, the preparation of Au/Pd- and Au/Pt-supported materials and their use in the oxidation of alcohols has attracted high levels of research.[80,81] Supported Au/Pd nanoparticles proved to be highly effective as catalysts for the solvent-free oxidation of BA.[82] Nevertheless, in some cases, the coproduction of large amounts of toluene was observed, due to the presence of a disproportionation reaction of benzyl alcohol.[70] This latter reaction seems to be especially promoted by Pd. The use of supports such as MgO or ZnO was reported to stop toluene formation completely, while the thermal treatment of prepared catalysts at different temperatures led to significant changes in the ratio between main products and by-products.[83]

Recently, in order to elucidate the structure–activity relationships in the reaction of BA oxidation, Hutchings and co-workers reported the catalytic performance of Pt/Au and Pt/Pd nanoparticles supported over TiO_2 and active carbon using *tert*-butyl hydroperoxide (TBHP) as the oxidant. The results demonstrated that when Au/Pd catalysts were used, a notable amount of toluene was identified,[84] whereas when Pt was used instead of Pd in the alloy, toluene was not formed and the main products were benzaldehyde and molecules deriving from aldehyde.[85]

As alternatives to noble metals, low-cost metals and metal oxides are currently being explored as catalysts for the reaction of alcohol oxidation.[68,86]

The use of a manganese-doped cobalt mixed oxide catalyst prepared by a solvothermal method was proposed in the oxidation of vanillyl alcohol to vanillin without using a base.[87] Three different types of metal oxides were observed in the prepared catalysts, which could be identified as Co_3O_4, Mn_3O_4 and $CoMn_2O_4$. Among these, the tetragonal phase of $CoMn_2O_4$ was found to be the most active and selective for vanillyl alcohol oxidation. The successful recycling of the catalyst was also achieved in this oxidation reaction.

A recent study by Zhu *et al.*,[68] using various metal oxides supported on activated carbon (AC), indicated that cobalt and nickel oxides are the most promising catalysts for the aerobic oxidation of benzyl alcohol, whereas manganese, iron and copper oxides showed lower conversion and selectivity. The Co_3O_4/AC catalyst showed high activity under mild conditions in the absence of an added base and the activity was ascribed to the synergistic effect of Co_3O_4 and AC. The proposed reaction mechanism indicated that Co_3O_4 is responsible for the alcohol dehydrogenation step, whereas carbon provides sites for molecular oxygen activation. However, a high load of cobalt oxide over the support was necessary to obtain significant activity and the catalyst could be reused only after thermal treatment at high temperature (350 °C).

1.3.2 Selective Oxidation of Alcohols Derived from Biomass

The selective oxidation of oxygenate compounds derived from renewable feedstocks has been extensively explored over the past 10 years.[65,66,88] The use of metals, mainly Au-, Pd- and Pt-supported catalysts, for the liquid-phase oxidation of a variety of molecules of interest in biomass conversion (polyols, carbohydrates, glycerol and furans) has attracted a great deal of attention from researchers. The main example of this class of reactions is glycerol oxidation.[88–91] Glycerol is a major by-product of biodiesel synthesis and its oxidation to high-value chemicals – such as tartronic acid, glyceric acid and dihydroxyacetone (Scheme 1.18) – can help biodiesel economics become more competitive.

The use of gold catalysts for the selective oxidation of glycerol has increased significantly over the past decade.[92,93] Davis and co-workers demonstrated that Au was more active than Pd and Pt with the use of high base concentration, whereas the activity of bimetallic Pd/Au systems was significantly lower than that of monometallic Au but led to greater formation of glyceric acid.[94] It was also demonstrated that hydrogen peroxide was formed during glycerol oxidation and its formation was associated with an increased presence of base in solution;[95] the H_2O_2 formed can cause carbon–carbon

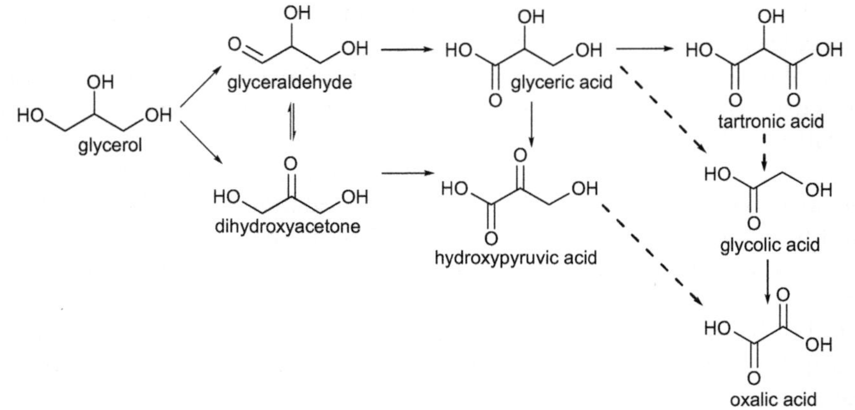

Scheme 1.18 Reaction scheme for glycerol oxidation.

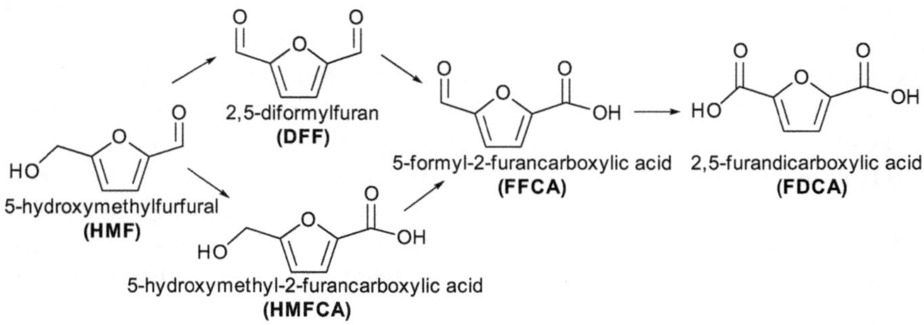

Scheme 1.19 Reaction scheme for 5-hydroxymethyl-2-furfural oxidation.

bond cleavage and increase the formation of by-products during oxidation reactions.

The selective oxidation of glycerol by bimetallic Au/Pd and Au/Pt catalysts supported over MgO at ambient temperature and under base-free conditions was reported by Hutchings and co-workers.[96] The reactivity at low temperature under base-free conditions highlighted the activity of PtAu/MgO materials. In particular, the alloying of Au with Pt leads to the enhancement of both glycerol conversion and selectivity towards C_3 products compared with Pd.

One of the most important sources of biomass is sugars, which are widely available and easily transformed. The homogeneous dehydration of glucose and/or fructose leads to the formation of 5-hydroxymethyl-2-furfural (HMF), which is a key precursor for the synthesis of chemicals that have applications in the polymer and pharmaceutical industries.[97–99] In particular, HMF can be oxidized to 2,5-furandicarboxylic acid (FDCA), which has recently been proposed as a possible surrogate for terephthalic acid,[100] the monomer used for the production of terephthalate plastics.

The synthesis of FDCA from HMF has been widely studied in the last two decades by using different catalysts and reaction conditions. Scheme 1.19 shows the general HMF oxidation pattern. FDCA is generally produced in two stages: the aldehyde functional group is first oxidized to a carboxylic acid, producing 5-hydroxymethyl-2-furancarboxylic acid (HMFCA); then, typically, the oxidation of the hydroxymethyl group produces FDCA through 5-formyl-2-furancarboxylic acid (FFCA) intermediates. Furthermore, sometimes the formation of 2,5-diformylfuran (DFF) has been observed, mainly in the absence of an added base.

Recent representative results reported in the literature are summarized in Table 1.1.

Supported Au nanoparticles have been found to be very active catalysts for 2,5-furandicarboxylic acid synthesis and many researchers have focused their attention on searching for the best supports and reaction conditions for improving the product yield.[101–106] In particular, Davis *et al.*[104] reported that gold-based materials were more active than other metals in the first step of the HMF oxidation, leading to HMFCA very quickly, even though they showed less activity for the subsequent conversion of HFCA to FDCA. The mechanism of HMF oxidation over Au and Pt catalysts in the presence of high amounts of NaOH was recently investigated through the use of isotopically labeled dioxygen and water.[105] The source of oxygen insertion was shown to be water rather than oxygen in all cases. It was suggested that the role of O_2 was that of an electron scavenger, closing the catalytic cycle and allowing the reaction to proceed. However, generally, process efficiency and, in particular, catalyst stability over gold-based samples proved to be rather low.

As also reported for other alcohols, the addition of a second metal to gold was shown to improve strongly the catalytic activity and stability in the reaction of HMF oxidation. In particular, Au–Cu-supported nanoparticles have been shown to produce active and stable catalysts for this reaction.[107,108] A strong synergistic effect was evident with the addition of Cu to Au up to a Cu : Au ratio of 1 : 1, especially in terms of sample stability and resistance to poisoning. Au–Cu alloy nanoparticles were considerably more active and selective than their monometallic counterparts. Moreover, reusability tests showed that the Au–Cu-based catalysts were significantly more stable than their monometallic counterparts.

Table 1.1 Oxidation of HMF using monometallic- and bimetallic-supported catalysts.

Catalyst[a]	T (°C)	Solvent	Time (h)	Conversion (%)	Selectivity (%)		Ref.
					Aldehyde	Acid	
Au/CeO_2 (150)	65	Water–NaOH	8	>99	<1	>99	102
Au/TiO_2 (100)	22	Water–NaOH	22	>99	35	65	105
Pt/C (150)	22	Water–NaOH	6	>99	33	67	105
Au/HT (40)	95	Water	7	>99	<1	>99	106
Au–Cu (1 : 1)/ TiO_2 (100)	95	Water–NaOH	4.5	>99	<1	>99	107
Au–Pd (8 : 2)/ C (200)	60	Water–NaOH	4	>99	3	97	109

[a]In parentheses: ratio HMF : metal (mol/mol).

Villa *et al.*[109] demonstrated that the modification of Au-based catalysts with Pt or Pd metal also produced stable and recyclable catalysts. In particular, they reported that bimetallic Au_8Pd_2 species supported over active carbon – where Au and Pd metal are present in an 8 : 2 molar ratio – have the highest activity and stability for the production of FDCA.

Another solution to the problem of catalyst durability was the addition of bismuth to Pt-containing catalysts.[110] In this case, the oxidation of HMF proceeded *via* the HMFCA and DFF intermediates; both of which were very reactive and rapidly oxidized to 5-formylfurancarboxylic acid (FFCA). The *ex situ* or *in situ* addition of a Bi promoter prevented the deactivation of the Pt catalysts and accelerated the reaction. The highest activity was observed for a Bi/Pt molar ratio of approximately 0.2.

Similar results were obtained by Villa *et al.*[111] when studying the modification of Au–Pd catalysts with bismuth and using the prepared systems in alcohol oxidation. Bi-containing materials were demonstrated to increase selectivity by suppressing parallel reactions in both benzyl alcohol and glycerol oxidation. However, the selectivity of the reactions notably varied only when Bi was deposited on the surface of metal nanoparticles.

1.3.3 Selective Oxidation of Alcohols in Non-Conventional Media

One of the greatest challenges in liquid-phase oxidation catalysis is the development of clean technologies that can operate in water; many of the studies reported so far used this solvent, even though reactants and products in alcohol oxidation are often insoluble in water, so other alternatives to standard organic solvents have also been proposed, such as ionic liquids[112,113] and supercritical fluids.[114–116]

The properties of supercritical carbon dioxide ($scCO_2$) are a combination of those associated with gases or liquids[117] and are particularly useful in the reactions involving gaseous reagents such as hydrogenation with H_2 and oxidation with O_2. Moreover, $scCO_2$ is totally non-flammable and its properties can be manipulated by varying the pressure applied.

The selective aerobic oxidation of alcohols has been studied in depth in $scCO_2$ using both batch and continuous-flow reactors.[114,118] The products usually observed in the oxidation of primary alcohols with metal catalysts and their formation routes are depicted in Scheme 1.20.

Since $scCO_2$ can dry wet material by eliminating the water formed during alcohol oxidation, its application is convenient for achieving high selectivity to aldehyde, while suppressing the formation of carboxylic acid *via* the favored hydration of aldehyde.[119]

Apart from reaction selectivity, a strong dependence of the reaction rate on pressure has also been found in some cases where CO_2 has been used as the solvent. As an example, Caravati *et al.*[120] demonstrated that the conversion of benzyl alcohol to benzaldehyde in CO_2 increased from 25 to 75% when the pressure was increased from 140 to 150 bar. Moreover, the same group[121]

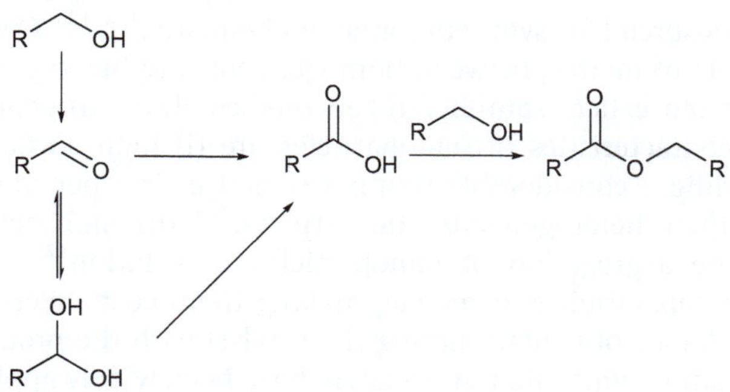

Scheme 1.20 Reaction pathways for the aerobic oxidation of primary alcohols.

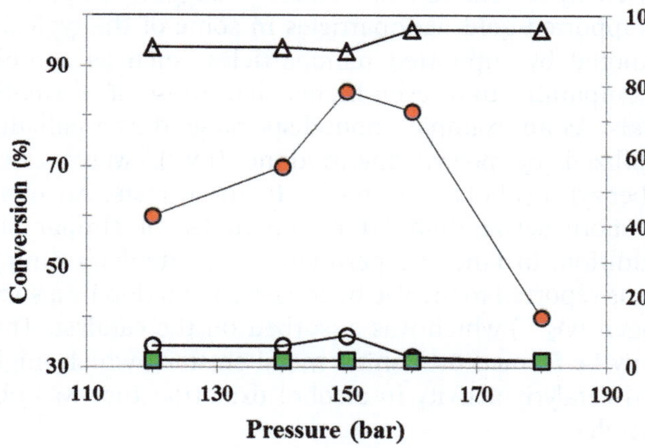

Figure 1.2 Effect of pressure on the oxidation of benzyl alcohol using CO_2 as solvent. Reaction conditions: 80 °C, 1.25 g of catalyst (0.5% Pd/Al_2O_3); feed composition: 0.9 mol% benzyl alcohol, 0.45 mol% O_2, rest CO_2. (●) Conversion; (Δ) benzaldehyde selectivity; (○) toluene selectivity; (■) selectivity to over-oxidation products.

showed that, depending on the pressure, substrate and products are distributed differently in the organic and supercritical phases, thus affecting both the reaction rate and the selectivity (Figure 1.2). Biphasic conditions at pressures close to the dew point resulted in the highest reaction rate, with formation of significant amounts of toluene (25%) as a side product, whereas working in a single phase led to the formation of over-oxidation products as the dominant side products.

1.4 Nanocatalysis for the Aerobic Oxidation of Alcohols

Over recent decades, the study of metal nanoparticles – generally so-called when their size is between 1 and 10 nm – with high specific catalytic activity has been widespread in synthetic organic chemistry.[122–124] Owing to their particular state of matter, between homogeneous and heterogeneous, these species are often called 'semi-heterogeneous' catalysts. In general, the advantageous characteristics of nanoparticles are (i) high surface-to-volume ratios, providing a considerable number of active sites per unit area compared with their heterogeneous counterparts,[125] (ii) high zeta potential, preventing the aggregation of nanoparticles in solution[126] and (iii) the possibility of separation and recycle, making them cost-effective and minimizing the chance of contaminating the catalyst with the product.[127]

Recently, nanometals and nanoclusters have been widely applied in some important reactions,[128] such as the carbon–carbon coupling,[129] carbon–heteroatom bond formation[130] and hydrogenation.[131]

A recent review by Garcia and co-workers[132] suggested that the intrinsic activity of unsupported gold nanoparticles in some of the typical reactions generally promoted by supported nanoparticles, such as aerobic alcohol oxidation, is comparable to or even higher than those of conventional supported materials. As an example, monodisperse gold and palladium nanoparticles, stabilized by polyvinylpyrrolidone (PVP), were tested in the oxidation of benzyl alcohols in water.[133] In these tests, Au nanoparticles proved to be more active than Pd nanoparticles of similar size (about 1.5 nm). In addition, in kinetic experiments, the rate-determining step for BA oxidation was reported to involve hydrogen abstraction by a superoxo-like molecular oxygen ($O_2^{\bullet -}$) which was adsorbed on the catalyst. This specific species can only be formed over small metal clusters, which might explain the size-related catalytic activity in alcohol oxidation that was observed in gold-based samples.

Dumeignil and co-workers[134] recently proposed a complex reaction scheme for the liquid-phase oxidation of glycerol using a quasi-homogeneous solution of Au nanoparticles. The highest conversion of glycerol obtained with this system was 100% after 3 h of reaction at 100 °C and 6 bar oxygen pressure, and the main products formed were glyceric, glycolic, formic, tartronic and oxalic acid with selectivities of 28, 36, 25, 9 and 2%, respectively. The mechanism reported indicated that the direct glycerol oxidation was facilitated in the presence of Au nanoparticles, whereas high temperatures and strongly basic conditions facilitated oxidative cleavage and straight hydrolytic transformations, such as retro-aldol, dicarbonyl cleavage, oxidative cleavage with carbon dioxide evolution and Cannizzaro cross-reactions.

So far, the role of the ligand generally used to stabilize nanometals from grain growth and agglomeration during catalytic tests has not been sufficiently investigated. In this regard, it has been commonly assumed that the ligand plays a negative role, by significantly decreasing the catalytic activity of nanoparticles.[135] Nevertheless, using PVP, it was observed that the interaction of this polymer with metal nanocrystals is accompanied by charge transfer from PVP to the nanometal.[136] Thus, in some cases, this interaction can modify the structure of metal colloids in ways that are sometimes beneficial to their catalytic performance.[137]

Also with nanoparticles, owing to the limited reserves of noble metals, one of the most interesting challenges is that of reducing their content in the active phase. Both intermetallic noble–non-noble alloys and non-noble metals have been demonstrated to be effective in reducing the use of noble metals for catalytic applications. As with the case of supported systems, in many cases the alloyed nanosystems not only combined the properties of monometallic nanoparticles, but also showed a significant improvement because of synergistic effects and different compositions.[81]

Scott and co-workers[138] studied the oxidation of α,β-unsaturated alcohols in water and an ionic liquid (tetraalkylphosphonium chloride) catalyzed by monometallic and bimetallic Pd and Au nanoparticles. In water, bimetallic Au–Pd nanoparticles were demonstrated to oxidize most of the unsaturated alcohols; however, the particle size growth due to Ostwald ripening was problematic. Conversely, the monometallic Pd system showed significant catalytic activity when using tetraalkylphosphonium chloride as a solvent, likely due to the easy oxidation of Pd in the high-chloride environment, whereas gold was inactive.

If heterometals with magnetic properties are alloyed into active phases, it is possible to overcome the drawback of nanoparticle separation and reuse.[139] Thus, magnetically separable nanocatalysts may have potential applications in catalysis.

Definite supports can also be chosen to enhance the recyclability of nanoparticles. Microgels, for example, have been used for this purpose.[140,141] For instance, Prati and co-workers synthesized microgel-stabilized Au nanoparticles by using tailor-made soluble cross-linked polymers as exotemplates and stabilizers.[141] The resulting stabilized system could be conveniently isolated by precipitation, stored in the solid state and re-dispersed in polar organic solvents and water. The prepared catalytic system exhibited remarkable activity in the oxidation of benzylic acid and polyols under mild conditions.

Looking at both the examples presented and the large number of papers on catalysis by nanoparticles which have recently appeared in the literature, we can conclude that significantly increased research in the area of unsupported nanoparticles may be expected in the coming years. In fact, metal colloids can actually be prepared following some accurate and reliable procedures,[142] thus ensuring the synthesis of materials with controlled characteristics and well-defined crystallographic planes, very close to theoretical models. Hence these tuned materials could be used to validate the preferred reaction for specific plane orientation and also for testing specific reaction mechanisms.

1.5 Microwave-Assisted Reactions

The field of microwave (MW)-assisted chemistry is relatively young. Nevertheless, since the first published reports on the use of MW irradiation to carry out chemical reaction in 1986,[143,144] the heating of the chemical

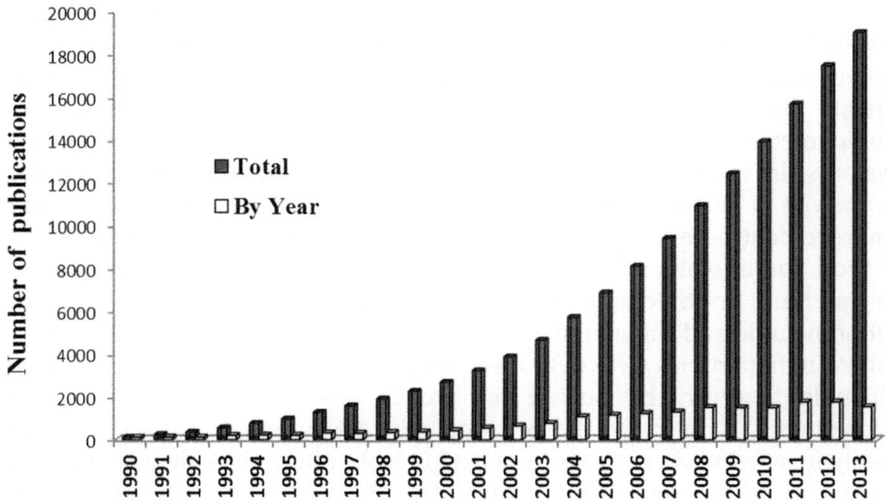

Figure 1.3 Publications on MW-assisted reactions (1990–2013). Number of articles involving MW-assisted reactions. Scopus keyword search on 'microwave reaction'.

reactions by MW energy has been an increasingly popular subject and a large number of papers have been published in this field over the past 20 years (Figure 1.3).

In many of the published studies, MW heating strongly decreased processing times, increased product yields and increased product purity, compared with conventional heating methods.[145–149] The nature of the MW enhancement is usually attributed to several specific thermal effects that cannot be emulated by conventional heating methods, such as the rapid heating rate, the prevention of thermal wall effects and the selective heating of particular reaction components and/or intermediates.

More recently, the advantages of this empowering technology have been exploited in the fields of nanomaterial synthesis[150–153] and catalysis.[154]

As an example, MW effects in heterogeneous catalysis were investigated in the dehydrogenation reaction of decalin and tetralin.[155,156] The use of MW heating proved both to improve the reagent conversion and to lower the deactivation rate of the catalyst. An observed beneficial effect of MW use was the large temperature gradient between the catalyst and surrounding species (direction of heat transfer reversed compared with the thermal mode). This led to the acceleration of mass desorption and species transport in the system, reduction of coke deposition and enhancement of tetralin dehydrogenation. Nevertheless, evidence of the MW effect was only observed in reactions controlled by mass transfer as the rate-limiting step.

Supported iron oxide nanoparticles on aluminosilicate catalysts were found to be efficient and easily recoverable materials in the aqueous selective oxidation of alcohols to their corresponding carbonyl compounds under both conventional and microwave heating.[157] The use of catalysis in conjunction with microwave heating resulted in dramatic reductions in reaction time, which decreased from hours to minutes. Similar results were obtained by García-Martínez and co-workers when using iron oxide nanoparticles (0.5–1.2 wt%) on MCM-41 type silica materials.[158]

1.6 Photochemistry in Oxidation Reactions

Photocatalysis is a branch of chemistry that exploits light radiation to overcome the energy barriers of chemical reactions.[159] Solar light is a renewable source of energy but only in the last century has it been analyzed as a potential motor for chemistry. Photochemical processes between light and matter are now being studied to become both supplementary and complementary to traditional reactions. So far, the most studied applications for photocatalysis have been the destruction of pollutants in water[160–162] and water splitting.[163]

Nevertheless, photocatalysis has recently been used for the oxidation reactions of both aliphatic and aromatic alcohols and a review on the transformation of alcohols by heterogeneous photocatalysts was published fairly recently.[164]

Pristine and doped titania have been mainly used as photocatalysts, owing to their reliability, stability under irradiation and low cost, although different semiconductor and insulator solids have also been used. Recently, Augugliaro and co-workers[165] investigated the UV photo-oxidation of piperonyl alcohol in water using a TiO_2 suspension. All the TiO_2 samples, prepared by the hydrolysis of $TiCl_4$ or from $TiOSO_4$ precursor – less crystalline and more hydroxylated than commercial titania – exhibited higher selectivity, but their reaction rate was higher in the presence of commercial Degussa P25. This behavior was in accordance with previous results obtained for the selective oxidation of other alcohols to the corresponding aldehydes.[166,167] Other products formed, in addition to piperonal, were carbonic anhydride and 1,3-bis[3,4-(methylenedioxy)benzyl] ether (Scheme 1.21); traces of this last compound were found only starting from the highest concentration of alcohol.

The use of heterogeneous photocatalysis for selective chemical transformations of biomass-derived compounds was reviewed recently.[168] The catalytic selective photo-oxidation of biomass was reported to provide various platform molecules such as levulinic, 2,5-furandicarboxylic, succinic,

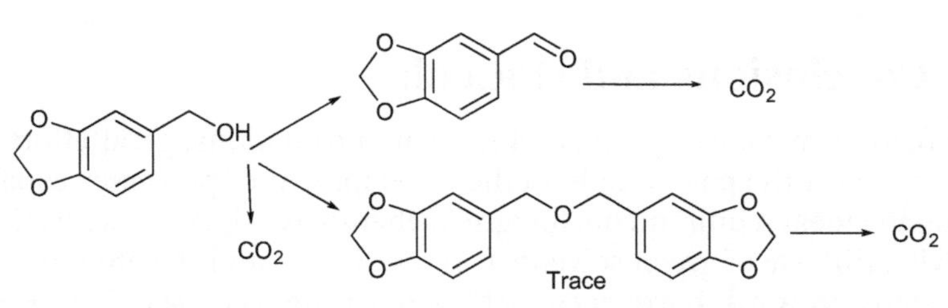

Scheme 1.21 Main reaction pathways for piperonyl alcohol oxidation.

gluconic and glucaric acids.[169] In general, the observed oxidation of alcohols is highly dependent on the class of molecules. The conversion of primary alcohols was usually low but with high selectivity. With regard to the reaction mechanism, it has been suggested that the first step corresponds to the interaction of a surface hole with the hydroxyl group of the alcohol, thus forming metal–oxo species with proton removal. This step becomes easier with carbon branching and as the chain length increases.

The efficiency of TiO_2 catalysts in the selective photocatalytic oxidation of glucose has also been reported recently,[170] and the reaction was found to be very selective towards glucaric and gluconic acids. Similarly, the mechanism of the selective photooxidation of glycerol over TiO_2 was studied.[171] In particular, glycerol was used as a probe molecule to study the mechanism on the catalyst surface, because the determination of the produced intermediates made it possible to discriminate between direct electron transfer and a radical-mediated oxidation mechanism.

A significant improvement in the photoactivity of nanooxides, such as TiO_2, by the use of MW irradiation was described by Serpone and co-workers.[172–174] In particular, they reported that, even though the idea of irradiating TiO_2 with microwaves may appear strange – since the MW photon energy (1×10^{-5} eV) is several orders of magnitude lower than the bandgap energy (3.0–3.2 eV) of TiO_2 – thermal and non-thermal factors can contribute significantly to the enhancement of a TiO_2 photo-assisted reaction, because MW heating may affect both crystalline and surface structures of the metal oxide. In fact, when studying the impact of MW iradiation on the degradation of Rhodamine B (RhB) dye and 1,4-dioxane, the latter demonstrated an important increase in the molecule degradation rate using MW-assisted photo-oxidation.[172] When using this procedure, in fact, the increase in the hydrophobic nature of titania under MW irradiation was reported to facilitate the adsorption of RhB through the aromatic rings with the aid of the three oxygen atoms in RhB. In other words, RhB lies flat on the particle surface. A similar MW effect was observed in the photo-degradation of many organic pollutants in aqueous media. In these studies, the possible advantages of MW heating in obtaining photo-degradations and photo-mineralizations of various organic pollutants was described; nevertheless, the reported research seems to indicate that the coupling of MW irradiation with a UV light source may lead to interesting results for partial oxidation reactions also.

1.7 Conclusions and Outlook

In the field of metal-catalyzed aerobic alcohol oxidations, great efforts have been made over the past decade in the development of processes capable of meeting the ideal requirements for 'green chemistry': (i) good yields, (ii) high selectivity, (iii) use of green solvents (such as water) and (iv) atom economy. Many steps forward have been achieved within the last few years. In particular, with regard to homogeneous catalysis, the case study of Stahl's

Cu/nitroxide system is a good example of the crucial points to be followed in the rational design of new homogeneous catalysts with the aim of broadening the substrate scope and improving reaction conditions. The choice of the ligand and co-catalyst proved crucial, and in-depth mechanistic investigations made it possible to shed more light on the properties of the catalyst to be tailored. Several mono- to tetradentate *N,O*- or carbene-based ligands, recently used for the preparation of Cu, Pd, V and Fe catalysts, proved that, by varying the structure of the ligands, better selectivity and the possibility of working with green solvents such as water can be tailored by design. Great effort has also been devoted to the development of one-pot syntheses toward the preparation of fine chemicals. Elegant tandem approaches have been reported for the direct synthesis of heterocyclics, imines and nitriles from alcohols and a nitrogen source.

New heterogeneous catalysts for the aerobic oxidation of alcohols have attracted much attention in recent years. Monometallic and bimetallic transition metal-based systems have been successfully applied to the aerobic oxidation of alcohols to the corresponding carbonyl compounds using O_2; several of them exhibited excellent performance. In particular, the use of gold-based bimetallic alloys led to significant results for the oxidation of a variety of molecules, especially significant in the production of chemicals from biomass sources. Tuning the product selectivity to acid or aldehyde may be most easily achieved by selecting the solvent. Water facilitates the formation of acid, while unusual solvents such as $scCO_2$ may facilitate high selectivity to aldehyde.

Nanocatalysis is the subject of intense research, and an appreciable increase in the achievement of results in this area may be expected in the coming years. The synthesis of tuned nanomaterials with controlled chemical–physical properties may in fact be used to validate theoretical calculations and to test specific reaction mechanisms while avoiding the interference of support materials on the active phase.

The use of photochemistry under mild conditions to convert alcohols selectively to chemicals is an innovative approach. In the near future, integrated systems of MW and photo-catalyst systems may be available; ideal photo-catalysts for commercial applications, however, should work under solar light irradiation in aqueous solution and only very few studies on sunlight use have been reported in the literature.

Although recent advances have shed more light on the mechanisms of the aerobic oxidation of alcohol and broadened the scope of substrates, selectivity and synthetically significant compounds, there is still plenty of room for further improvements of catalyst performance in the exploration of the potential usefulness of this system in industrial processes.

References

1. H. Adolfsson, I. W. C. E. Arends, J.-E. Bäckvall, M. Beller, A. Beressel, J. Brinksma, W. R. Browne, J. W. de Boer, B. L. Feringa, M. Girhard,

R. Hage, Y. Ishii, E. V. Johnston, N. Komiya, S.-I. Murahashi, R. Neumann, Y. Obora, D. Pijper, S. Sakaguchi, K. Schröder, R. A. Sheldon, Y. Shi, M. K. Tse and V. B. Urlacher, in *Modern Oxidation Methods*, ed. J.-E. Bäckvall, Wiley-VCH, Weinheim, 2nd edn, 2010, p. 1–481.

2. R. A. Sheldon, I. Arends, G. J. ten Brink and A. Dijksman, *Acc. Chem. Res.*, 2002, **35**, 774.
3. T. Mallat and A. Baiker, *Chem. Rev.*, 2004, **104**, 3037.
4. M. Besson and P. Gallezot, *Catal. Today*, 2000, **57**, 127.
5. C. Parmeggiani and F. Cardona, *Green Chem.*, 2012, **14**, 547.
6. F. Cavani and J. H. Teles, *ChemSusChem*, 2009, **2**, 508.
7. For fundamentals of green chemistry, see: (a) P. T. Anastas and J. C. Warner, *Green Chemistry: Theory and Practice*, Oxford University Press, Oxford, 1998, p. 30; (b) S. L. Y. Tang, R. L. Smith and M. Poliakoff, *Green Chem.*, 2005, **7**, 761; (c) R. A. Sheldon, *Chem. Soc. Rev.*, 2012, **41**, 1437.
8. D. Lenoir, *Angew. Chem. Int. Ed.*, 2006, **45**, 3206.
9. For recent reviews, see: (a) M. J. Schultz and M. S. Sigman, *Tetrahedron*, 2006, **62**, 8227; (b) B.-Z. Zhan and A. Thompson, *Tetrahedron*, 2004, **60**, 2917; (c) non-conventional media: S. Liu and J. Xiao, *J. Mol. Catal. A: Chem.*, 2007, **270**, 1.
10. J. A. L. da Silva, J. J. R. F. da Silva and A. J. L. Pombeiro, *Coord. Chem. Rev.*, 2011, **255**, 2232.
11. For recent reviews on Pd-catalyzed aerobic oxidation, see: (a) K. M. Gligorich and M. S. Sigman, *Chem. Commun.*, 2009, 3854; (b) ligand-modulation: M. S. Sigman and D. R. Jensen, *Acc. Chem. Res.*, 2006, **39**, 221; (c) substrate scope: M. J. Schultz, S. S. Hamilton, D. R. Jensen and M. S. Sigman, *J. Org. Chem.*, 2005, **70**, 3343; (d) Pd-catalyzed enantioselective aerobic oxidation: S. Krishnan, J. T. Bagdanoff, D. C. Ebner, Y. K. Ramtohul, U. K. Tambar and B. M. Stoltz, *J. Am. Chem. Soc.*, 2008, **130**, 13745.
12. T. Punniyamurthy and L. Rout, *Coord. Chem. Rev.*, 2008, **252**, 134.
13. T. Ikariya, S. Kuwata and Y. Kayaki, *Pure Appl. Chem.*, 2010, **82**, 1471.
14. R. A. Sheldon and I. W. C. E. Arends, *J. Mol. Catal. A: Chem.*, 2006, **251**, 200.
15. For recent reviews on nitroxide application as direct oxidation catalysts, see: (a) S. Wertz and A. Studer, *Green Chem.*, 2013, **15**, 3116; (b) L. Tebben and A. Studer, *Angew. Chem. Int. Ed.*, 2011, **50**, 5034; (c) R. A. Sheldon and I. W. C. E. Arends, *Adv. Synth. Catal.*, 2004, **346**, 1051.
16. Recent examples: (a) J. M. Hoover and S. S. Stahl, *J. Am. Chem. Soc.*, 2011, **133**, 16901; (b) J. M. Hoover, J. E. Steves and S. S. Stahl, *Nat. Protoc.*, 2012, **7**, 1161; (c) J. U. Ahmad, M. T. Raisanen, M. Kemell, M. J. Heikkila, M. Leskela and T. Repo, *Appl. Catal. A*, 2012, **449**, 153; (d) L. M. Dornan, Q. Cao, J. C. A. Flanagan, J. J. Crawford, M. J. Cook and M. J. Muldoon, *Chem. Commun.*, 2013, **49**, 6030; (e) C. Tao, F. Liu, Y. Zhu, W. Liu and Z. Cao, *Org. Biomol. Chem.*, 2013, **11**, 3349;

(f) J. M. Hoover, B. L. Ryland and S. S. Stahl, *J. Am. Chem. Soc.*, 2013, **135**, 2357.

17. Recent examples: (a) L. Wang, J. Li, Y. Lv, G. Zhao and S. Gao, *Appl. Organomet. Chem.*, 2012, **26**, 37; (b) N. Tamura, T. Aoyama, T. Takido and M. Kodomari, *Synlett*, 2012, 1397; (c) L. Wang, J. Li, X. Zhao, Y. Lv, H. Zhang and S. Gao, *Tetrahedron*, 2013, **69**, 6041; (d) E. Zhang, H. Tian, S. Xu, X. Yu and Q. Xu, *Org. Lett.*, 2013, **15**, 2704; (g) W. Yin, C. Chu, Q. Lu, J. Tao, X. Liang and R. Liua, *Adv. Synth. Catal.*, 2010, **352**, 113.
18. J. E. Steves and S. S. Stahl, *J. Am. Chem. Soc.*, 2013, **135**, 15742.
19. S. Velusamy, A. Srinivasan and T. Punniyamurthy, *Tetrahedron Lett.*, 2006, **47**, 923.
20. G. Zhang, X. Han, Y. Luan, Y. Wang, X. Wen and C. Ding, *Chem. Commun.*, 2013, **49**, 7908.
21. C. Melero, O. N. Shishilov, E. Álvarez, P. Palma and J. Cámpora, *Dalton Trans.*, 2012, **41**, 14087.
22. (a) S. K. Hanson, R. Wu and L. A. Silks, *Org. Lett.*, 2011, **13**, 1908; for mechanistic investigation, see: (b) B. N. Wigington, M. L. Drummond, T. R. Cundari, D. L. Thorn, S. K. Hanson and S. L. Scott, *Chem. Eur. J.*, 2012, **18**, 14981; (c) S. K. Hanson, R. T. Baker, J. C. Gordon, B. L. Scott, L. A. Silks and D. L. Thorn, *J. Am. Chem. Soc.*, 2010, **132**, 17804.
23. L. Bettucci, C. Bianchini, J. Filippi, A. Lavacchi and W. Oberhauser, *Eur. J. Inorg. Chem.*, 2011, 1797.
24. G. Urgoitia, R. SanMartin, M. T. Herrero and E. Dominguez, *Green Chem.*, 2011, **13**, 2161.
25. P. Gamez, I. W. C. E. Arends, J. Reedijk and R. A. Sheldon, *Chem. Commun.*, 2003, 2414.
26. J. S. Uber, Y. Vogels, D. van den Helder, I. Mutikainen, U. Turpeinen, W. T. Fu, O. Roubeau, P. Gamez and J. Reedijk, *Eur. J. Inorg. Chem.*, 2007, **26**, 4197.
27. P. J. Figiel, A. Sibaouih, J. U. Ahmad, M. Nieger, M. T. Raisanen, M. Leskela and T. Repo, *Adv. Synth. Catal.*, 2009, **351**, 2625.
28. P. J. Figiel, A. M. Kirillov, M. F. C. G. da Silva, J. Lasri and A. J. L. Pombeiro, *Dalton Trans.*, 2010, **39**, 9879.
29. K. H. Shaughnessy, *Chem. Rev.*, 2009, **109**, 643.
30. M. N. Kopylovich, K. T. Mahmudov, M. F. C. G. da Silva, M. L. Kuznetsov, P. J. Figiel, Y. Y. Karabach, K. V. Luzyanin and A. J. L. Pombeiro, *Inorg. Chem.*, 2011, **50**, 918.
31. M. N. Kopylovich, K. T. Mahmudov, M. Haukka, P. J. Figiel, A. Mizar, J. A. L. da Silva and A. J. L. Pombeiro, *Eur. J. Inorg. Chem.*, 2011, 4175.
32. M. N. Kopylovich, Y. Yu, M. F. C. G. Karabach, P. J. da Silva, J. Figiel, Lasri and A. J. L. Pombeiro, *Chem. Eur. J.*, 2012, **18**, 899.
33. M. Hajimohammadi, N. Safari, H. Mofakham and F. Deyhimi, *Green Chem.*, 2011, **13**, 991.
34. For the activation of dioxygen with transition metal complexes, see: (a) S. Caron, R. W. Dugger, S. G. Ruggeri, J. A. Ragan and D. H. B. Ripin, *Chem. Rev.*, 2006, **106**, 2943; (b) M. Costas, M. P. Mehn,

M. P. Jensen and L. Que, *Chem. Rev.*, 2004, **104**, 939; (c) E. A. Lewis and W. B. Tolman, *Chem. Rev.*, 2004, **104**, 1047; (d) V. V. Smirnov, D. W. Brinkley, M. P. Lanci, K. D. Karlin and J. P. Roth, *J. Mol. Catal. A: Chem.*, 2006, **251**, 100.
35. Y. Jing, J. Jiang, B. Yan, S. Lu, J. Jiao, H. Xue, G. Yang and G. Zheng, *Adv. Synth. Catal.*, 2011, **353**, 1146.
36. For a review on OKR, see: H. Pellissier, *Adv. Synth. Catal.*, 2011, **353**, 1613.
37. T. Kunisu, T. Oguma and T. Katsuki, *J. Am. Chem. Soc.*, 2011, **133**, 12937.
38. P. Muthupandi, S. K. Alamsetti and G. Sekar, *Chem. Commun.*, 2009, 3288.
39. O. Das and T. K. Paine, *Dalton Trans.*, 2012, **41**, 11476.
40. D. M. Pearson, N. R. Conley and R. M. Waymouth, *Organometallics*, 2011, **30**, 1445.
41. (a) S. Díez-González, N. Marion and S. P. Nolan, *Chem. Rev.*, 2009, **109**, 3612; (b) G. C. Vougioukalakis and R. H. Grubbs, *Chem. Rev.*, 2010, **110**, 1746; (c) G. C. Fortman and S. P. Nolan, *Chem. Soc. Rev.*, 2011, **40**, 5151.
42. (a) J. C. Y. Lin, R. T. W. Huang, C. S. Lee, A. Bhattacharyya, W. S. Hwang and I. J. B. Lin, *Chem. Rev.*, 2009, **109**, 3561; (b) M. Poyatos, J. A. Mata and E. Peris, *Chem. Rev.*, 2009, **109**, 3677; (c) O. Schuster, L. Yang, H. G. Raubenheimer and M. Albrecht, *Chem. Rev.*, 2009, **109**, 3445.
43. X. Liu, Q. Xia, Y. Zhang, C. Chen and W. Chen, *J. Org. Chem.*, 2013, **78**, 8531.
44. V. Jurčík, T. E. Schmid, Q. Dumont, A. M. Z. Slawin and C. S. J. Cazin, *Dalton Trans.*, 2012, **41**, 12619.
45. G. Csjernyik, A. H. Ell, L. Fadini, B. Pugin and J.-E. Bäckvall, *J. Org. Chem.*, 2002, **67**, 1657.
46. J. S. M. Samec, A. H. Ell and J.-E. Bäckvall, *Chem. Eur. J.*, 2005, **11**, 2327.
47. Y. Endo and J.-E. Bäckvall, *Chem.Eur. J.*, 2011, **17**, 12596.
48. B. P. Babu, Y. Endo and J.-E. Bäckvall, *Chem. Eur. J.*, 2012, **18**, 1152.
49. Y. Endo and J.-E. Bäckvall, *Chem. Eur. J.*, 2012, **18**, 13609.
50. J. Yu, J. Xu and M. Lu, *Appl. Organomet. Chem.*, 2013, **27**, 606.
51. J. C. A. Flanagan, L. M. Dornan, M. G. McLaughlin, N. G. McCreanor, M. J. Cook and M. J. Muldoon, *Green Chem.*, 2012, **14**, 1281.
52. S. Ma, J. Liu, S. Li, B. Chen, J. Cheng, J. Kuang, Y. Liu, B. Wan, Y. Wang, J. Ye, Q. Yu, W. Yuan and S. Yu, *Adv. Synth. Catal.*, 2011, **353**, 1005.
53. J. Liu and S. Ma, *Synthesis*, 2012, **44**, 1569.
54. J. Liu and S. Ma, *Org. Lett.*, 2013, **15**, 5150.
55. J. Liu and S. Ma, *Org. Biomol. Chem.*, 2013, **11**, 4186.
56. Y.-S. Lan, B.-S. Liao, Y.-H. Liu, S.-M. Peng and S.-T. Liu, *Eur. J. Org. Chem.*, 2013, 5160.
57. L. Jiang, L. Jin, H. Tian, X. Yuan, X. Yu and Q. Xu, *Chem. Commun.*, 2011, **47**, 10833.
58. C. Liu, S. Liao, Q. Li, S. Feng, Q. Sun, X. Yu and Q. Xu, *J. Org. Chem.*, 2011, **76**, 5759.

59. (a) A. J. Fatiadi, Preparation and synthetic applications of cyano compounds, in *Triple-Bonded Functional Groups: Volume 2*, eds. S. Patai and Z. Rappaport, John Wiley & Sons, Ltd., Chichester, UK, 1983, chapter 26, p. 1057–1303; (b) R. C. Larock, Nitriles, Carboxylic Acids and Derivatives, in *Comprehensive Organic Transformations*, ed. H. C. Brown, VCH, New York, 1989, p. 860–862; (c) J. S. Miller and J. L. Manson, *Acc. Chem. Res.*, 2001, **34**, 563; (d) P. Magnus, D. A. Scott and M. R. Fielding, *Tetrahedron Lett.*, 2001, **42**, 4127; (e) M. B. Smith and J. March, chapter 17, in *March's Advanced Organic Chemistry: Reactions, Mechanisms and Structure*, eds: M. B. Smith and J. March, Wiley, Hoboken, NJ, 6th edn, 2007, p. 1546–1550.
60. D. Kumar, T. Yadav and B. M. Bhanage, *Eur. J. Org. Chem.*, 2013, 5106.
61. Review articles: (a) Z. Shao and H. Zhang, *Chem. Soc. Rev.*, 2009, **38**, 2745; (b) J. Zhou, *Chem. Asian J.*, 2010, **5**, 422; (c) C. Zhong and X. Shi, *Eur. J. Org. Chem.*, 2010, 2999.
62. X.-H. Ho, H.-J. Oh and H.-Y. Jang, *Eur. J. Org. Chem.*, 2012, 5655.
63. C. Liu, J. Wang, L. Meng, Y. Deng, Y. Li and A. Lei, *Angew. Chem. Int. Ed.*, 2011, **50**, 5144.
64. C. P. Vinod, K. Wilson and A. F. Lee, *J. Chem. Technol. Biotechnol.*, 2011, **86**, 161.
65. N. Dimitratos, J. A. Lopez-Sanchez and G. J. Hutchings, *Chem. Sci.*, 2012, **3**, 20.
66. S. E. Davis, M. S. Ide and R. J. Davis, *Green Chem.*, 2012, **15**, 17.
67. M. J. Beier, T. W. Hansen and J.-D. Grunwaldt, *J. Catal.*, 2009, **266**, 320.
68. J. Zhu, J. L. Faria, J. L. Figueiredo and A. Thomas, *Chem. Eur. J.*, 2011, **17**, 7112.
69. C. Keresszegi, T. Burgi, T. Mallat and A. Baiker, *J. Catal.*, 2002, **211**, 244.
70. M. Sankar, E. Nowicka, R. Tiruvalam, Q. He, S. H. Taylor, C. J. Kiely, D. Bethell, D. W. Knight and G. J. Hutchings, *Chem. Eur. J.*, 2011, **17**, 6524.
71. F. Li, Q. H. Zhang and Y. Wang, *Appl. Catal. A: Gen.*, 2008, **334**, 217.
72. S. E. J. Hackett, R. M. Brydson, M. H. Gass, I. Harvey, A. D. Newman and K. Wilson, *Angew. Chem. Int. Ed.*, 2007, **46**, 8593.
73. H. Guo, A. Al-Hunaiti, M. Kemell, S. Rautiainen, M. Leskela and T. Repo, *ChemCatChem*, 2011, **3**, 1872.
74. C. M. A. Parlett, L. J. Durndell, K. Wilson, D. W. Bruce, N. S. Hondow and A. F. Lee, *Catal. Commun.*, 2014, **44**, 40.
75. C. M. A. Parlett, P. Keshwalla, S. G. Wainwright, D. W. Bruce, N. S. Hondow, K. Wilson and A. F. Lee, *ACS Catal.*, 2013, **3**, 2122.
76. C. M. A. Parlett, D. W. Bruce, N. S. Hondow, M. A. Newton, A. F. Lee and K. Wilson, *ChemCatChem*, 2013, **5**, 939.
77. A. Villa, C. E. Chan-Thaw, G. M. Veith, K. L. More, D. Ferri and L. Prati, *ChemCatChem*, 2011, **3**, 1612.
78. D. Tongsakul, S. Nishimura and K. Ebitani, *ACS Catal.*, 2013, **3**, 2199.
79. M. Sankar, N. Dimitratos, P. J. Miedziak, P. P. Wells, C. J. Kiely and G. J. Hutchings, *Chem. Soc. Rev.*, 2012, **41**, 8099.

80. S. Biella and M. Rossi, *Chem. Commun.*, 2003, 378.
81. P. Paalanen, B. M. Weckhuysen and M. Sankar, *Catal. Sci. Technol.*, 2013, **3**, 2869.
82. M. Sankar, E. Nowicka, P. J. Miedziak, G. L. Brett, R. L. Jenkins, N. Dimitratos, S. H. Taylor, D. W. Knight, D. Bethell and G. J. Hutchings, *Faraday Discuss.*, 2010, **145**, 341.
83. J. Pritchard, M. Piccinini, R. Tiruvalam, Q. He, N. Dimitratos, J. A. Lopez-Sanchez, D. J. Morgan, A. F. Carley, J. K. Edwards, C. J. Kiely and G. J. Hutchings, *Catal. Sci. Technol.*, 2013, **3**, 308.
84. M. I. bin Saiman, G. L. Brett, R. Tiruvalam, M. M. Forde, K. Sharples, A. Thetford, R. L. Jenkins, N. Dimitratos, J. A. Lopez-Sanchez, D. M. Murphy, D. Bethell, D. J. Willock, S. H. Taylor, D. W. Knight, C. J. Kiely and G. J. Hutchings, *Angew. Chem. Int. Ed.*, 2012, **51**, 5981.
85. V. Peneau, Q. He, G. Shaw, S. A. Kondrat, T. E. Davies, P. Miedziak, M. Forde, N. Dimitratos, C. J. Kiely and G. J. Hutchings, *Phys. Chem. Chem. Phys.*, 2013, **15**, 10636.
86. J. J. Zhu, K. Kailasam, A. Fischer and A. Thomas, *ACS Catal.*, 2011, **1**, 342.
87. A. Jha, K. R. Patil and C. V. Rode, *ChemPlusChem*, 2013, **78**, 1384.
88. B. T. Kusema and D. Yu. Murzin, *Catal. Sci. Technol.*, 2013, **3**, 297.
89. C. H. Zhou, H. Zhao, D. S. Tong, L. M. Wu and W. H. Yu, *Catal. Rev. Sci. Eng.*, 2013, **55**, 369.
90. J. Xu, H. Zhang, Y. Zhao, B. Yu, S. Chen, Y. Li, L. Hao and Z. Liu, *Green Chem.*, 2013, **15**, 1520.
91. A. Villa, G. M. Veith and L. Prati, *Angew. Chem. Int. Ed.*, 2010, **49**, 4499.
92. S. Carrettin, P. McMorn, P. Johnston, K. Griffin and G. J. Hutchings, *Chem. Commun.*, 2002, 7, 696.
93. D. Wang, A. Villa, D. Su, L. Prati and R. Schlögl, *ChemCatChem*, 2013, **5**, 2717.
94. W. C. Ketchie, M. Murayama and R. J. Davis, *J. Catal.*, 2007, **250**, 264.
95. B. N. Zope and R. J. Davis, *Top. Catal.*, 2009, **52**, 269.
96. L. Brett, Q. He, C. Hammond, P. J. Miedziak, N. Dimitratos, M. Sankar, A. A. Herzing, M. Conte, J. A. Lopez-Sanchez, C. J. Kiely, D. W. Knight, S. H. Taylor and G. J. Hutchings, *Angew. Chem. Int. Ed.*, 2011, **50**, 10136.
97. J. N. Chheda, G. W. Huber and J. A. Dumesic, *Angew. Chem. Int. Ed.*, 2007, **46**, 7164.
98. A. A. Rosatella, S. P. Simeonov, R. F. M. Frade and C. A. M. Afonso, *Green Chem.*, 2012, **14**, 509.
99. R.-J. van Putten, J. C. van der Waal, E. de Jong, C. B. Rasrendra, H. J. Heeres and J. G. de Vries, *Chem. Rev.*, 2013, **113**, 1499.
100. A. Gandini, *Green Chem.*, 2011, **13**, 1061.
101. O. Casanova, S. Iborra and A. Corma, *J. Catal.*, 2009, **265**, 109.
102. O. Casanova, S. Iborra and A. Corma, *ChemSusChem*, 2009, **2**, 1138.
103. Y. Y. Gorbanev, S. K. Klitgaard, J. M. Woodley, C. H. Christensen and A. Riisager, *ChemSusChem*, 2009, **2**, 672.

104. S. E. Davis, L. R. Houk, E. C. Tamargo, A. K. Datye and R. J. Davis, *Catal. Today*, 2011, **160**, 55.
105. S. E. Davis, B. N. Zope and R. J. Davis, *Green Chem.*, 2012, **14**, 143.
106. N. K. Gupta, S. Nishimura, A. Takagaki and K. Ebitani, *Green Chem.*, 2011, **13**, 824.
107. T. Pasini, M. Piccinini, M. Blosi, R. Bonelli, S. Albonetti, N. Dimitratos, J. A. Lopez-Sanchez, M. Sankar, Q. He, C. J. Kiely, G. J. Hutchings and F. Cavani, *Green Chem.*, 2011, **13**, 2091.
108. S. Albonetti, T. Pasini, A. Lolli, M. Blosi, M. Piccinini, N. Dimitratos, J. A. Lopez-Sanchez, D. J. Morgan, A. F. Carley, G. J. Hutchings and F. Cavani, *Catal. Today*, 2012, **195**, 120.
109. A. Villa, M. Schiavoni, S. Campisi, G. M. Veith and L. Prati, *ChemSusChem*, 2013, **6**, 609.
110. H. A. Rass, N. Essayem and M. Besson, *Green Chem.*, 2013, **15**, 2240.
111. A. Villa, D. Wang, G. M. Veith and L. Prati, *J. Catal.*, 2012, **292**, 73.
112. V. I. Parvulescu and C. Hardacre, *Chem. Rev.*, 2007, **107**, 2615.
113. I. Favier, D. Madec and M. Gómez, Metallic nanoparticles in ionic liquids – applications in catalysis, in *Nanomaterials in Catalysis*, ed. P. Serp and K. Philippot, Wiley-VCH, Weinheim, 2013.
114. T. Seki and A. Baiker, *Chem Rev.*, 2009, **109**, 2409.
115. U. Hintermair, T. Chinnusamy and W. Leitner, Transition-metal complexes in supported liquid phase and supercritical fluids – a beneficial combination for selective continuous-flow catalysis with integrated product separation, in *New Strategies in Chemical Synthesis and Catalysis*, ed. B. Pignataro, Wiley-VCH, Weinheim, 2012.
116. X. Han and M. Poliakoff, *Chem. Soc. Rev.*, 2012, **41**, 1428.
117. M. Skerget, Z. Knez and M. Knez-Hrncic, *J. Chem. Eng. Data*, 2011, **56**, 694.
118. S. E. Dapurkar, H. Kawanami, M. Chatterjee, C. V. Rode, T. Yokoyama and Y. Ikushima, *Appl. Catal. A*, 2011, **394**, 209.
119. M. S. Schneider and A. Baiker, *Catal. Rev. Sci. Eng.*, 1995, **37**, 515.
120. M. Caravati, J. D. Grunwaldt and A. Baiker, *Appl. Catal. A*, 2006, **298**, 50.
121. M. J. Beier, J.-D. Grunwaldt, I. Tsivintzelis, A. D. Jensen, G. M. Kontogeorgis and A. Baiker, *J. Supercrit. Fluids*, 2012, **63**, 199.
122. *Nanoparticles and Catalysis*, ed. D. Astruc, Wiley-VCH, Weinheim, 2008.
123. *Nanomaterials in Catalysis*, ed. P. Serp and K. Philippot, Wiley-VCH, Weinheim, 2013.
124. R. Schlgl and S. B. A. Hamid, *Angew. Chem. Int. Ed.*, 2004, **43**, 1628.
125. D. Astruc, F. Lu and J. R. Aranzaes, *Angew. Chem. Int. Ed.*, 2005, **44**, 7852.
126. F. Thielbeer, K. Donaldson and M. Bradley, *Bioconj. Chem.*, 2011, **22**, 144.
127. V. Polshettiwar and R. S. Varma, *Green Chem.*, 2010, **12**, 743.
128. R. B. N. Baig and R. S. Varma, Nanocatalysis in water, in *Metal-Catalyzed Reactions in Water*, ed. P. H. Dixneuf and V. Cadierno, Wiley-VCH, Weinheim, 2013.

129. D. Astruc, *Inorg. Chem.*, 2007, **46**, 1884.
130. B. C. Ranu, R. Dey, T. Chatterjee and S. Ahammed, *ChemSusChem*, 2012, **5**, 22.
131. M. Blosi, S. Albonetti, S. Ortelli, A. L. Costa, L. Ortolani and M. Dondi, *New J. Chem.*, 2014, **38**, 1401.
132. Y. Mikami, A. Dhakshinamoorthy, M. Alvaro and H. Garcia, *Catal. Sci. Technol.*, 2013, **3**, 58.
133. H. Tsunoyama, N. Ichikuni, H. Sakurai and T. Tsukuda, *J. Am. Chem. Soc.*, 2009, **131**, 11293.
134. E. Skrzynska, J. Ftouni, J.-S. Girardon, M. Capron, L. Jalowiecki-Duhamel, J.-F. Paul and F. Dumeignil, *ChemSusChem*, 2012, **5**, 2065.
135. A. Quintanilla, V. C. L. Butselaar-Orthlie, C. Kwakernaak, W. G. Sloof, M. T. Kreutser and F. Kapteijn, *J. Catal.*, 2010, **271**, 104.
136. Y. Borodko, S. M. Humphrey, T. D. Tilley, H. Frei and G. A. Somorjai, *J. Phys. Chem. C*, 2007, **111**, 6288.
137. K. Chen, H. Wu, Q. Hua, S. Chang and W. Huang, *Phys. Chem. Chem. Phys.*, 2013, **15**, 2273.
138. A. Maclennan, A. Banerjee and R. W. J. Scott, *Catal. Today*, 2013, **207**, 170.
139. C. M. Lim and I. S. Lee, *Nano Today*, 2010, **5**, 412.
140. Y. Lu and M. Ballauff, in *Hybrid Nanomaterials: Synthesis, Characterization and Applications*, ed. B. P. S. Chauhan, Wiley, Hoboken, NJ, 2011, p. 1.
141. A. Biffis, S. Cunial, P. Spontoni and L. Prati, *J. Catal.*, 2007, **251**, 1.
142. K. An, S. Alayoglu, T. Ewers and G. A. Samorjai, *J. Colloid Interface Sci.*, 2012, **373**, 1.
143. R. Gedye, F. Smith, K. Westaway, H. Ali, L. Baldisera, L. Laberge and J. Rousell, *Tetrahedron Lett.*, 1986, **27**, 27982.
144. R. J. Giguere, T. L. Bray, S. M. Duncan and G. Majetich, *Tetrahedron Lett.*, 1986, **27**, 4945.
145. J. D. Moseley and C. O. Kappe, *Green Chem.*, 2011, **13**, 794.
146. C. O. Kappe and D. Dallinger, *Mol. Diversity*, 2009, **13**, 71.
147. *Microwaves in Organic Synthesis*, ed. A. Loupy, Wiley-VCH, Weinheim, 2nd edn, 2006.
148. C. O. Kappe, D. Dallinger and S. S. Murphree, *Practical Microwave Synthesis for Organic Chemists*, Wiley-VCH, Weinheim, 2009.
149. I. Bilecka and M. Niederberger, *Nanoscale*, 2010, **2**, 1358.
150. *Microwaves in Nanoparticle Synthesis: Fundamentals and Applications*, ed. S. Horikoshi and N. Serpone Wiley-VCH, Weinheim, 2013.
151. A. Gondolini, E. Mercadelli, A. Sanson, S. Albonetti, L. Doubova and S. Boldrini, *J. Eur. Ceram. Soc.*, 2013, **33**, 67.
152. M. Blosi, S. Albonetti, M. Dondi, C. Martelli and G. Baldi, *J. Nanopart. Res.*, 2011, **13**, 127.
153. M. Blosi, S. Albonetti, A. L. Costa, N. Sangiorgi and A. Sanson, *Chem. Eng. J.*, 2013, **215–216**, 616.

154. A. Richel, P. Laurent, B Wathelet, J.-P. Wathelet and M. Paquot, *Catal. Today*, 2011, **167**, 141.
155. Y. Suttisawat, S. Horikoshi, H. Sakai, P. Rangsunvigit and M. Abe, *Fuel Process. Technol.*, 2012, **95**, 27.
156. Y. Suttisawat, H. Sakai, M. Abe, P. Rangsunvigit and S. Horikoshi, *Int. J. Hydrogen Energy*, 2012, **37**, 3242.
157. F. Rajabi, A. Pineda, S. Naserian, A. M. Balu, R. Luque and A. A. Romero, *Green Chem.*, 2013, **15**, 1232.
158. A. I. Carrillo, E. Serrano, R. Luque and J. García-Martínez, *Appl. Catal. A: Gen.*, 2013, **453**, 383.
159. A. Primo, A. Corma and H. Garcia, *Phys. Chem. Chem. Phys.*, 2011, **13**, 886.
160. H. X. Fan, G. Li, F. Yang, L. Yang and S. M. Zhang, *J. Chem. Technol. Biotechnol.*, 2011, **86**, 1107.
161. A. L. Costa, S. Ortelli, M. Blosi, S. Albonetti, A. Vaccari and M. Dondi, *Chem. Eng. J.*, 2013, **225**, 880.
162. A. L. Costa, S. Ortelli, M. Blosi, S. Albonetti, A. Vaccari and M. Dondi, *J. Photochem. Photobiol. A: Chem.*, 2013, **276**, 58.
163. G. Li, P. Zhang, Z. Bian, J. Zhu, L. Wu and H. Li, *ChemSusChem*, 2013, **6**, 1461.
164. V. Augugliaro and L. Palmisano, *ChemSusChem*, 2010, **3**, 1135.
165. M. Bellardita, V. Loddo, G. Palmisano, I. Pibiri, L. Palmisano and V. Augugliaro, *Appl. Catal. B: Environ.*, 2014, **144**, 607.
166. G. Palmisano, S. Yurdakal, V. Augugliaro, V. Loddo and L. Palmisano, *Adv. Synth.Catal.*, 2007, **349**, 964.
167. S. Yurdakal, G. Palmisano, V. Loddo, O. Alagöz, V. Augugliaro and L. Palmisano, *Green Chem.*, 2009, **11**, 510.
168. J. C. Colmenares and R. Luque, *Chem. Soc. Rev.*, 2014, **43**, 765.
169. J. J. Bozell and G. R. Petersen, *Green Chem.*, 2010, **12**, 539.
170. J. C. Colmenares, A. Magdziarz and A. Bielejewska, *Bioresour. Technol.*, 2011, **102**, 11254.
171. C. Minero, A. Bedini and V. Maurino, *Appl. Catal. B: Environ.*, 2012, **128**, 135.
172. S. Horikoshi and N. Serpone, *Catal. Today*, 2014, **224**, 225.
173. S. Horikoshi, M. Kajitani, N. Serpone and M. Abe, *J. Jpn. Soc. Color Mater.*, 2008, **82**, 51.
174. S. Horikoshi, Y. Minatodani, H. Tsutsumi, H. Uchida, M. Abe and N. Serpone, *J. Photochem. Photobiol. A: Chem.*, 2013, **265**, 20.

CHAPTER 2

Copper Catalysts for Aerobic Oxidation of Alcohols

OINDRILA DAS AND TAPAN KANTI PAINE*

Department of Inorganic Chemistry, Indian Association for the Cultivation of Science, 2A&2B Raja S. C. Mullick Road, Jadavpur, Kolkata-700032, India
*Email: ictkp@iacs.res.in

2.1 Introduction

The catalytic aerobic oxidation of alcohols has been explored by using different metals and many of these catalytic systems are useful in synthetic chemistry.[1–6] However, there remain several practical problems such as reaction conditions and solvents, catalyst loading, recyclability of catalyst, functional group tolerance and chemo-/stereoselectivity. Therefore, catalysis of alcohol oxidation using aerial oxygen on an industrial scale remains a challenging task.

Nature is the source of inspiration where metal-containing enzymes catalyze many organic transformation reactions with high selectivity and specificity under mild conditions through green processes.[7] Copper is one of the most abundant natural elements in the geosphere and has the ability to cycle between the +1 and +2 oxidation states. Because of its inherent electronic properties and accessible redox potentials, copper takes part in many biological functions such as structural shaping, electron transfer and catalysis. The presence of copper in the active sites of different enzymes makes it an appropriate choice for the development of alcohol oxidation catalysts.

RSC Green Chemistry No. 28
Transition Metal Catalysis in Aerobic Alcohol Oxidation
Edited by Francesca Cardona and Camilla Parmeggiani

Published by the Royal Society of Chemistry, www.rsc.org

In biological systems, a number of copper-containing enzymes are known to catalyze metabolically relevant oxidation reactions using molecular oxygen as the terminal oxidant. One such metalloenzyme is galactose oxidase (GO), a mononuclear copper enzyme, that catalyzes the aerobic oxidation of primary alcohols to aldehydes with concomitant reduction of dioxygen to hydrogen peroxide.[8] The copper(II) center is surrounded by four protein-derived ligands (Tyr272, His495, His496 and His581) and a labile aqua ligand, resulting in a square-pyramidal co-ordination geometry.[9,10] Structural studies on the 'inactive' form of GO revealed an unusual tyrosinate ligand (Tyr272) covalently connected to a cysteine (Cys228) (Figure 2.1).[9] In the 'active' form of GO, Tyr272 becomes one-electron oxidized to form a tyrosyl radical. The copper(II)–phenoxyl radical intermediate is the key feature of the mechanism of the two-electron oxidation of alcohols.[11]

Inspired by the coupling of an organic cofactor and reversible copper(I)/copper(II) redox chemistry in the unique O_2-dependent reaction catalyzed by GO, a variety of synthetic systems have been developed.[12–18] These systems exhibit versatile reactivity towards alcohols and provide a valuable insight into the development of alcohol oxidation catalysts. Over the past several decades, extensive studies have been carried out on copper-based catalysts for the aerobic oxidation of alcohols.[2,3,5,19] This chapter presents an overview of the advances made in this direction.

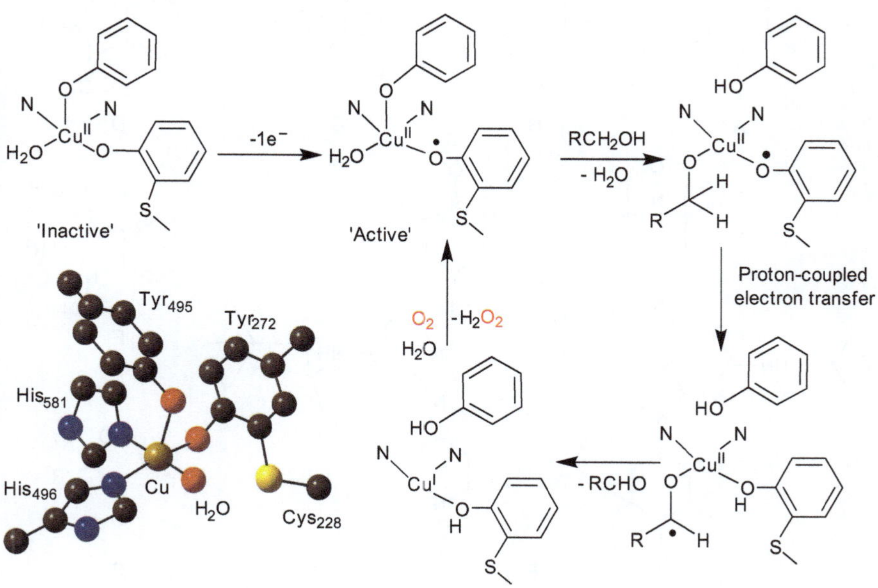

Figure 2.1 Proposed catalytic cycle for the aerobic oxidation of primary alcohol catalyzed by galactose oxidase.

2.2 Alcohol Oxidation Catalysts Based on the Copper–Nitroxyl Radical System

2.2.1 Homogeneous Catalysts

Stable nitroxyl radicals are widely used in oxidation catalysis. 2,2,6,6-Tetramethylpiperidine-1-oxyl (TEMPO) and its derivatives and other nitroxyl radicals (Figure 2.2) have been used in combination with metal-free oxidants for the oxidation of alcohols to the corresponding carbonyl compounds.[20,21] These methods are useful in synthetic chemistry but require stoichiometric oxidants and produce salt wastes. To circumvent these problems, copper-based catalytic systems incorporating organic radicals have been studied for alcohol oxidation. Furthermore, the organic radicals with copper mimic the copper–phenoxyl moiety present in the active site of GO.

In 1984, Semmelhack *et al.* reported the use of copper chloride (10 mol%) and TEMPO (10 mol%) for the aerobic oxidation of alcohols in dimethylformamide (DMF) under ambient conditions.[22] The method was practical and efficient for benzylic and allylic primary alcohols. However, secondary and sterically hindered alcohols were less susceptible to oxidation even with excess catalyst. In the reaction pathway, oxidation of nitroxyl radical to nitrosonium cation by copper(II) ion was proposed (Figure 2.3). The resulting nitrosonium ion has been proposed as the active oxidant in the two-electron

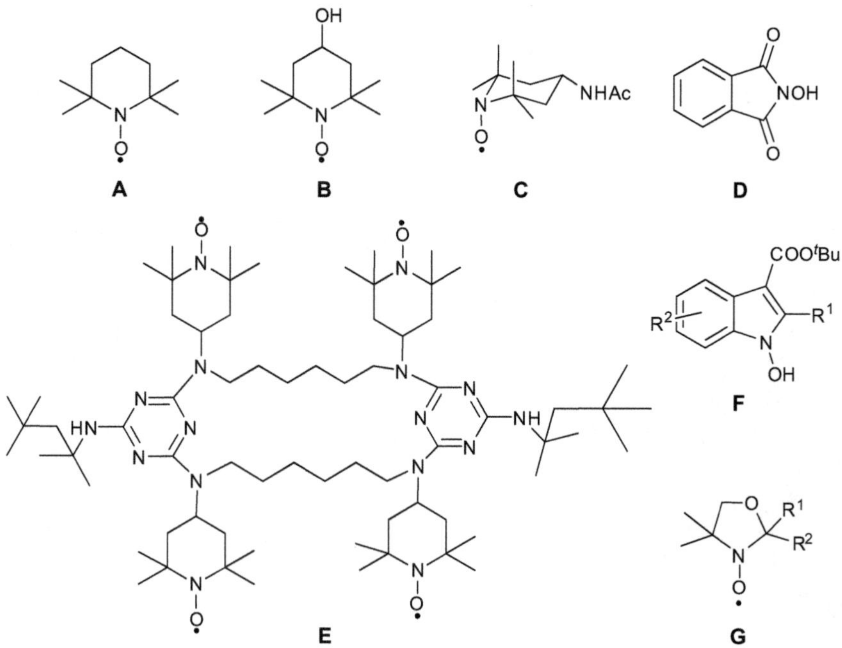

Figure 2.2 Structures of different organic radicals or radical precursors used in combination with copper salts for the aerobic oxidation of alcohols.

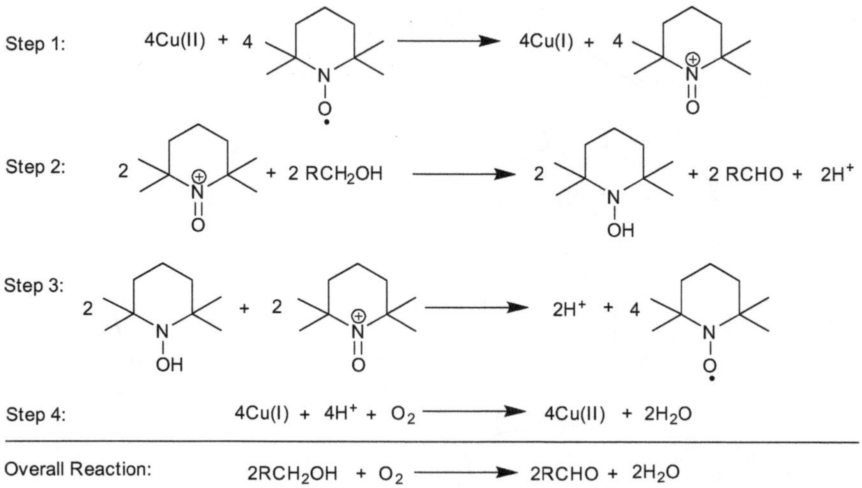

Figure 2.3 Mechanism of the aerobic oxidation of primary alcohols by CuCl–TEMPO catalyst proposed by Semmelhack *et al.*

oxidation of primary alcohols to aldehydes and is reduced to hydroxylamine. Then a rapid comproportionation reaction between hydroxylamine (TEMPO-H) and nitrosonium cation gives rise to nitroxyl radical (TEMPO) species essentially required for the catalytic cycle. The copper(I) ion is simultaneously oxidized to copper(II) by O_2. TEMPO, being a stable radical, prevents the aldehydes from further oxidation to acids.

Following this work, a combination of TEMPO (2 mol%) and mixed Mn(II)–Cu(II) nitrate (2 mol%) catalytic system was reported by Minisci and co-workers for the aerobic oxidation of primary and secondary alcohols in acetic acid under mild conditions with excellent yields of the oxidized products.[23] A macrocyclic tetrafunctional nitroxyl radical (**E** in Figure 2.2) was also employed in the mixed metal nitrate catalysis.[24] Although the exact role of Mn(II) nitrate remained unclear, an oxoammonium cation was proposed as the active oxidant formed *via* disproportionation of TEMPO to oxoammonium and TEMPO-H. Then the metal ion oxidized TEMPO-H to TEMPO in the presence of dioxygen.

Oxoammonium cations have been reported earlier to exhibit the oxidation of simple aliphatic alcohols[25–27] and a Cope-like cyclic elimination of the nitrosonium–alcoholate adduct in the transition state has been proposed.[28] In Semmelhack *et al.*'s procedure, however, aliphatic alcohols were not oxidized to aldehydes. This prompted Sheldon and co-workers to reinvestigate the mechanism. In 2003, they reported a detailed mechanistic study on the CuCl–TEMPO-catalyzed oxidation of benzyl alcohol. On the basis of results from stoichiometric oxidations with TEMPO, kinetic isotope effects and Hammett correlation studies, they proposed a copper-mediated dehydrogenation mechanism rather than the oxoammonium mechanism of Semmelhack *et al.*[29] The formation of a copper(II)–piperidinoxyl complex *via*

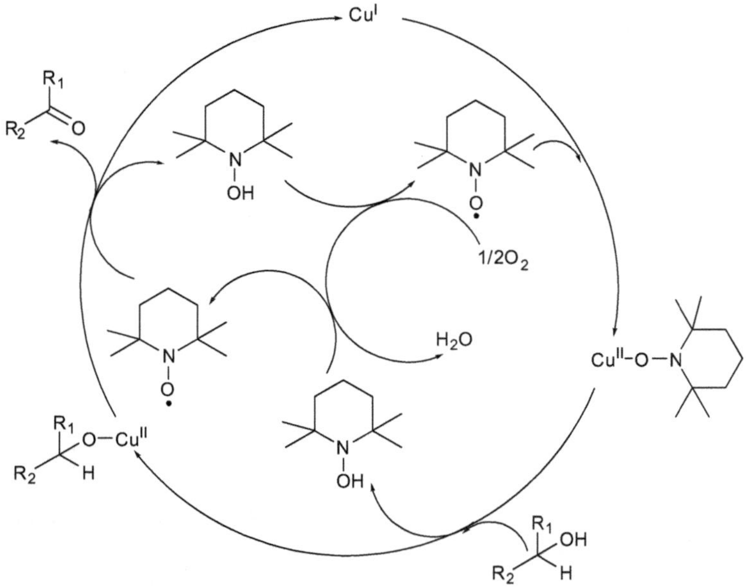

Figure 2.4 Proposed mechanism of the aerobic oxidation of alcohols catalyzed by the CuCl–TEMPO system.

one-electron oxidation of copper(I) by TEMPO was proposed in the catalytic pathway (Figure 2.4). Reaction of this complex with alcohols results in an alcoholate-coordinated copper(II) complex. The latter then reacts with a second molecule of TEMPO to abstract an H-atom, resulting in the formation of carbonyl compound, TEMPO-H and copper(I). TEMPO-H is rapidly oxidized by air to regenerate TEMPO.[29]

Later, Vatèle reported a convenient method for the oxidative rearrangement of tertiary allylic alcohols to β-substituted enones by utilizing a $CuCl_2$–TEMPO catalytic system in the presence of O_2.[30,31] New copper-based catalytic systems using *N*-hydroxyphthalimide[32] and *N*-hydroxyindole[33] (**D** and **F** in Figure 2.2) have been reported for the chemoselective aerobic oxidation of alcohols.

Following the success of Cu–TEMPO systems, a variety of copper salts and organic ligands were screened to improve the chemoselectivity of catalytic alcohol oxidations. In 2003, Sheldon and co-workers reported a simple catalytic system containing copper(II) bromide, TEMPO and 2,2′-bipyridine (bpy) for the aerobic oxidation of alcohols in acetonitrile–water solvent mixture at room temperature (Figure 2.5a).[34] The catalytic system exhibited high chemoselectivity for primary alcohols under mild conditions. Secondary alcohols such as octan-2-ol and 1-phenylethanol were not oxidised. Moreover, no over-oxidation of the aldehydes to the corresponding acids was observed. It was noted that all the components were necessary for efficient transformation.

A mechanism similar to that with GO has been proposed in which TEMPO radical abstracts a hydrogen atom from the metal-coordinated alcoholate.[29]

(a)

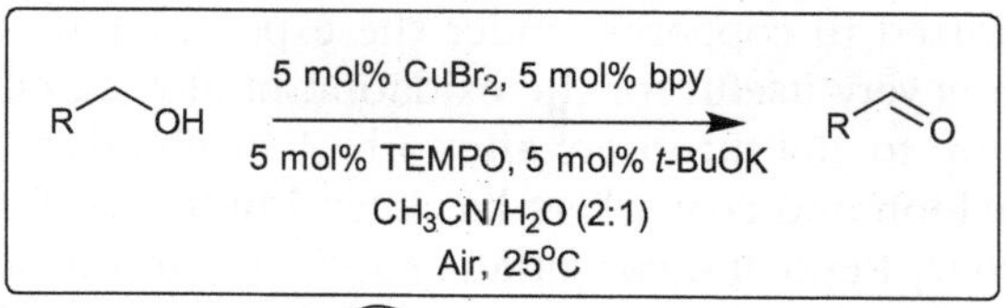

(b)

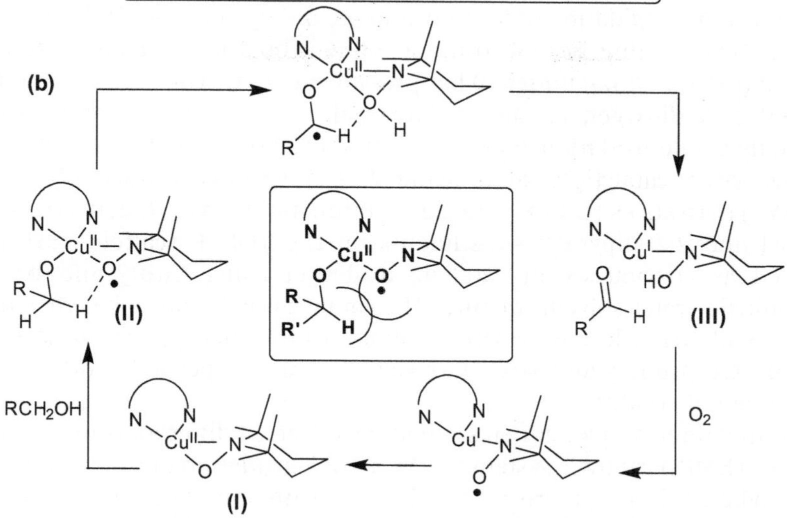

Figure 2.5 (a) Copper–TEMPO catalyst system for the aerobic oxidation of primary alcohols in the presence of bpy reported by Sheldon and co-workers; (b) proposed mechanism for the (bpy)Cu–TEMPO catalyzed chemoselective oxidation of primary alcohols. Inset: possible intermediate responsible for the lack of reactivity of secondary alcohols.

In the proposed mechanism, a ternary TEMPO–copper(II)–alcoholate intermediate (**II**) is formed upon binding of alcoholate to the metal center of a copper(II)–η^2-TEMPO species (**I**) (Figure 2.5b). X-ray structures of the complexes of copper(II) halides with η^2-coordinated TEMPO were reported by Rey and co-workers.[35,36] The rate of the reaction was reported to accelerate in the presence of a base, indicating coordination of the alcoholate to the copper center. Coordinated TEMPO in intermediate **II** abstracts the hydrogen atom to form a copper(I) complex (**III**), aldehyde and TEMPO-H. For secondary alcohols, the coordinated TEMPO is unable to abstract the hydrogen atoms from the alcoholate owing to steric hindrance associated with the methyl groups of TEMPO and the secondary alcohol (Figure 2.5b, inset). TEMPO-H is oxidized to TEMPO by O_2 and the copper(I) is oxidized to regenerate **I** in the catalytic cycle.

Further investigations revealed that a higher catalytic turnover number (TON) could be achieved in the presence of electron-donating substituents at *para* positions of bpy.[37] It was subsequently observed that the catalytic behavior of the Sheldon and co-workers' catalyst is dependent on the pH of the reaction solution. Maximum catalytic activity was found with a smaller amount of elemental copper (0.5 mol%) and 5 mol% TEMPO in the presence of bpy (2.5 mo%) at pH 13–13.5, whereas the activity was lost at pH 8.

Copper(I) is oxidized to copper(II) under the experimental conditions. The catalytic system is very useful for the oxidation of allylic and benzylic alcohols exhibiting up to 250 turnovers after 5 h. A mechanism similar to that proposed by Sheldon and co-workers has been implicated for the oxidation reaction.[38] In 2007, Repo and co-workers succeeded in the selective and efficient catalytic oxidation of primary and secondary benzylic alcohols using a Cu–TEMPO–diimine system (diimine = phen, bpy) in an alkaline aqueous solvent (pH 12.6–13.5) under 10 bar pressure at 80 °C. The use of an aqueous solvent and dioxygen as the oxidant makes it an efficient and environmentally viable oxidation process.[39] Gartshore and Lupton reported a new copper-based catalytic system using 2,2,4,4-tetramethyloxazolidine-*N*-oxyl (TOXYL) derivatives (**G** in Figure 2.2).[40] Bifunctional TOXYL derivatives with pyridyl and 2,2′-bipyridyl substituents on the TOXYL scaffold have been utilized as co-catalysts in catalytic oxidation using $CuBr_2$ and base in acetonitrile–water solvent mixture. The catalyst was found to be efficient for a range of benzylic and allylic alcohols. Unfortunately, the oxidation of unactivated primary and secondary alcohols was not possible under similar experimental conditions.

Kumpulainen and Koskinen studied the effects of different components of the Cu–TEMPO catalytic system on the aerobic oxidation of various aliphatic and allylic alcohols.[41] It was reported that the amount of base was critical for the outcome of the oxidation. A comparative study with a range of bases revealed that amine bases such as *N*-methylimidazole (NMI), *N*,*N*-dimethylaminopyridine (DMAP) and triethylamine (TEA) show moderate catalytic activity whereas high activity is exhibited by 1,8-diazabicyclo[5.4.0]undec-7-ene (DBU) and 1,4-diazabicyclo[2.2.2]octane (DABCO). Kinetic studies with a $CuBr_2$–bpy–base system in acetonitrile showed a first-order kinetic dependence of the reaction on TEMPO and second-order on copper. The oxidation reactions follow a second-order dependence on the alcohol substrate at low concentrations and a first-order dependence on oxygen at high alcohol concentrations. Use of copper(II) trifluoromethanesulfonate instead of copper(II) bromide increased the efficiency of the catalyst for challenging alcohol substrates.

The use of an air-microbubble strategy to increase the gas concentration in the liquid or gas phase in homogeneous catalysis has been experimentally supported to promote catalytic conversion in the aerobic oxidation of alcohols.[42] Although the exact role of the microbubbles is not clear, it provides an efficient method to increase the efficiency of the Cu–TEMPO catalyst.

Recently, Stahl and co-workers reported a highly practical method for the aerobic oxidation of alcohols using (bpy)Cu(I)–TEMPO catalyst and *N*-methylimidazole (NMI) base in acetonitrile solvent with ambient air as the oxidant.[43,44] The catalytic system exhibits high chemoselectivity towards primary alcohols including aliphatic substrates. Moreover, the catalytic method is compatible with alcoholic substrates containing different functional groups. Highly chemoselective oxidations of diols can also be carried out without any protection (Figure 2.6a).

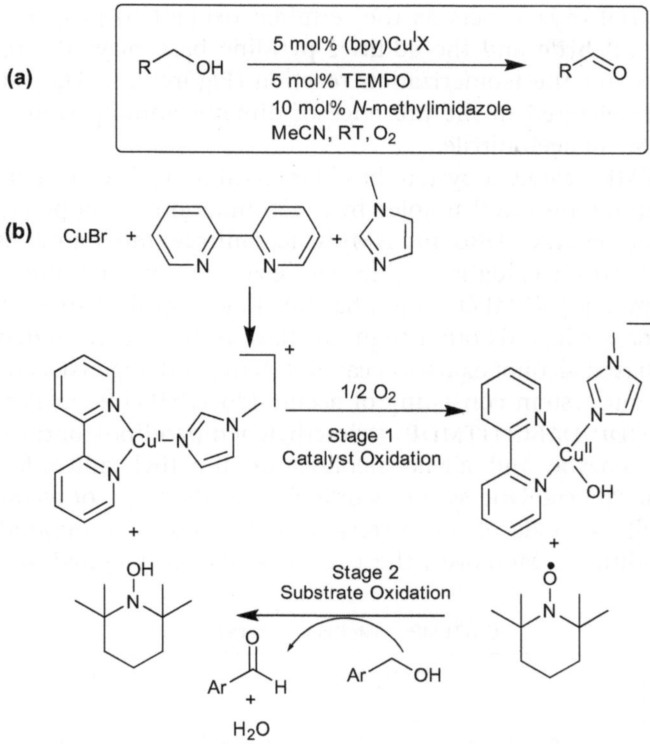

Figure 2.6 (a) Aerobic alcohol oxidation by (bpy)Cu–TEMPO catalyst in the presence of NMI; (b) catalytic cycle for Cu(I)–TEMPO-catalyzed aerobic alcohol oxidation proposed by Stahl and co-workers.

Systematic kinetic and spectroscopic studies provided useful insights into the mechanism of (bpy)Cu(I)–TEMPO-catalyzed aerobic alcohol oxidation. On the basis of detailed mechanistic studies, it has been proposed that the catalytic system involved two copper centers during the reaction with dioxygen.[45] A two-stage catalytic mechanism consisting of (stage 1) 'catalyst oxidation' in which dioxygen oxidizes copper(I) and TEMPO-H *via* a peroxo-bridged dicopper(II) intermediate and (stage 2) 'substrate oxidation' by copper(II) and TEMPO radical *via* a copper(II)–alkoxide intermediate has been proposed (Figure 2.6b). The proposed mechanism can successfully differentiate the reactivity of benzylic and aliphatic alcohols. While the oxidation of catalyst (stage 1) controls the rate of benzyl alcohol oxidation, several steps contribute to the turnover rate in the oxidation of aliphatic alcohols. Studies with substituted benzyl alcohols of variable electronic environment establish a pair of electron-pushing mechanism during the formation of aldehyde in the Cu(I)–bpy–TEMPO–NMI catalytic system in acetone.[46]

Christmann and co-workers used the (bpy)Cu(I)–TEMPO catalytic system to oxidize allylic alcohols *via* a stereoconvergent one-pot oxidation–isomerization process.[47] Aerial oxygen acts as the terminal oxidant to produce the α,β-unsaturated aldehyde and the additive pyridine base plays the role of an organocatalyst for the isomerization reaction (Figure 2.7). The rate of the reaction is accelerated in the presence of dimethylaminopyridine (DMAP) instead of NMI in acetonitrile.

The Cu–TEMPO catalytic system has been used to oxidize aromatic amino alcohols to quinolines and indoles by a combination of copper(ii) triflate, 2,2′-bipyridine, TEMPO, DBU and NMI in acetonitrile under air at 60 °C.[48] A practical method for oxidative imine synthesis from alcohol and amine by using the (bpy)Cu(i)–TEMPO system has been developed (Figure 2.8).[49]

Several *N*-donor ligands other than bpy have been studied to develop the Cu–TEMPO based alcohol oxidation catalyst. Jiang and Ragauskas reported a four-component system consisting of acetamido-TEMPO (**C** in Figure 2.2)–$Cu(ClO_4)_2$–TMDP–DABCO (TMDP = trimethylenedipyridine) for the oxidative catalysis of benzylic and allylic alcohols in dimethyl sulfoxide (DMSO) (Figure 2.9a). The catalytic system works for a wide range of alcohols with good to excellent yields of the corresponding carbonyl compounds under ambient conditions. Moreover, the catalyst could be recycled and reused

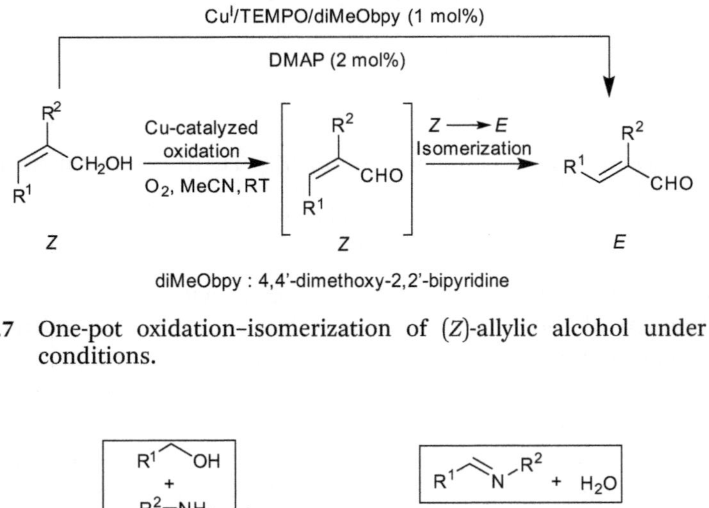

Figure 2.7 One-pot oxidation–isomerization of (*Z*)-allylic alcohol under aerobic conditions.

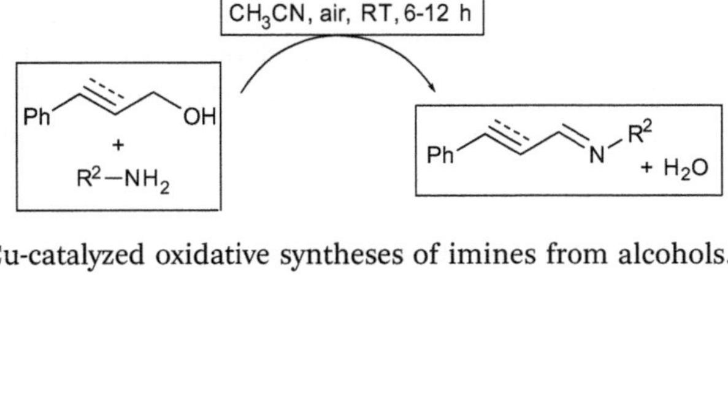

Figure 2.8 Cu-catalyzed oxidative syntheses of imines from alcohols.

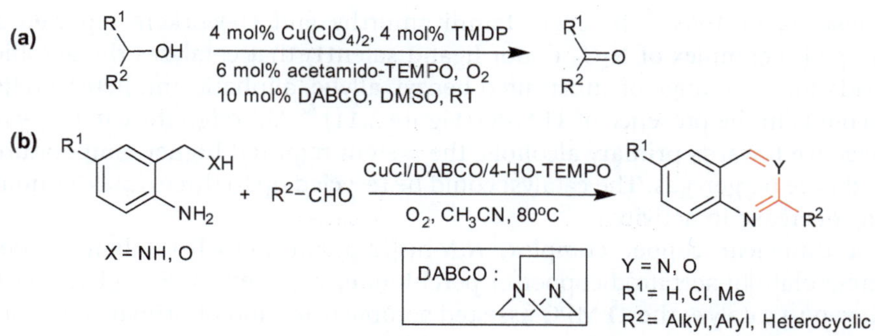

Figure 2.9 (a) Aerobic oxidation of alcohols using acetamido-TEMPO–$Cu(ClO_4)_2$–TMDP–DABCO catalyst system; (b) aerobic oxidative synthesis of 2-substituted quinazolines and 4*H*-3,1-benzoxazines.

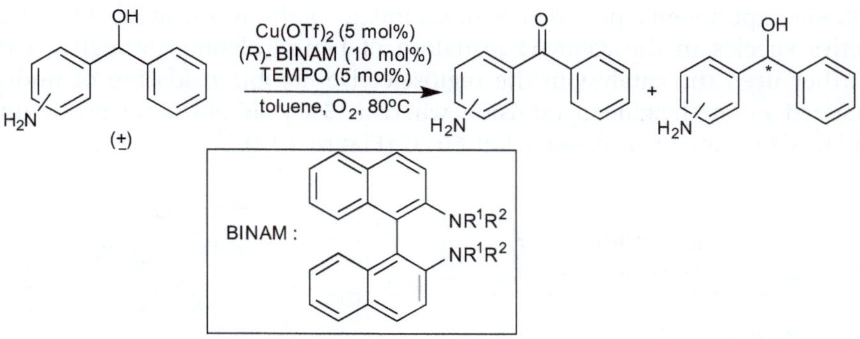

Figure 2.10 Oxidative kinetic resolution of amino alcohols by (BINAM)Cu–TEMPO catalyst.

three times without appreciable loss of catalytic activity, simply extracting the final products with *n*-pentane and adding fresh alcohol to the DMSO phase.[50] A catalyst containing a mixture of DABCO, CuCl and TEMPO (5 mol% each) has been reported that exhibits remarkable conversion of both primary and secondary alcohols to the corresponding carbonyl compounds in toluene at 100 °C.[51] Single-crystal X-ray examination revealed the copper(II) center in a trigonal bipyramidal coordination geometry in the related catalytic complex $[(DABCO)_2CuCl_2 \cdot HCl]$. Recently, this catalytic system in combination with 4-hydroxy-TEMPO (**B** in Figure 2.2) was used to carry out the oxidative synthesis of heteroaromatic compounds such as 2-substituted quinazolines and 4*H*-3,1-benzoxazines in acetonitrile (Figure 2.9b).[52]

An enantiopure bidentate nitrogen donor ligand, (*R*)-BINAM, was used in the $Cu(OTf)_2$–TEMPO-catalyzed oxidative kinetic resolution (OKR) of several benzylic amino alcohols (Figure 2.10).[53,54]

The use of an isolated copper(II) complex of 1,10-phenanthroline (phen) in combination with a stable nitroxyl radical (di-*tert*-butylnitroxyl) for the catalytic aerobic oxidation of alcohols was first reported by Bracksman and

Gaasbeck in 1966.[55] In 2006, Punniyamurthy and co-workers reported a copper(II) complex of N_2O_2 donor ligand salenH_2 that catalyzes the aerobic oxidation of a range of substituted benzyl, allylic, aliphatic and heterocyclic alcohols in the presence of TEMPO (Figure 2.11).[56] Although the catalyst was selective towards primary alcohols, the system required higher temperature and pure oxygen gas. The catalyst could be recycled up to three times without any decrease in activity.

A dinuclear copper complex, *N*,*N*-bis[(2-pyridylmethyl)-1,3-diaminopropan-2-olato](μ-acetato)dicopper(II) perchlorate, reported by Striegler,[57] was shown to catalyze the TEMPO-assisted aerobic oxidation of primary alcohols in aqueous alkaline medium under ambient conditions with a turnover rate of 9 h^{-1}. Moreover, the catalyst is highly regioselective towards primary alcohols. The complex can discriminate various carbohydrates selectively on the basis of the number of hydroxyl groups of the sugar involved in coordination to the metal center. Isothermal titration calorimetry and in-depth kinetic experiments provided a mechanistic pathway involving binuclear active species in the catalytic oxidation of benzyl alcohol. Striegler *et al.* further used the catalyst in the regioselective aerobic oxidation of underivatized α-D-methylglucopyranoside, where a 35% yield of the corresponding aldehyde product was observed at 60 °C (Figure 2.12).[58]

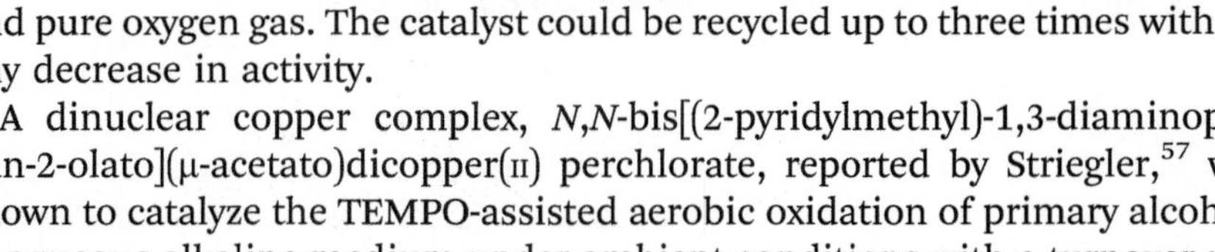

Figure 2.11 Aerobic oxidation of alcohols catalyzed by copper complex of salen ligand.

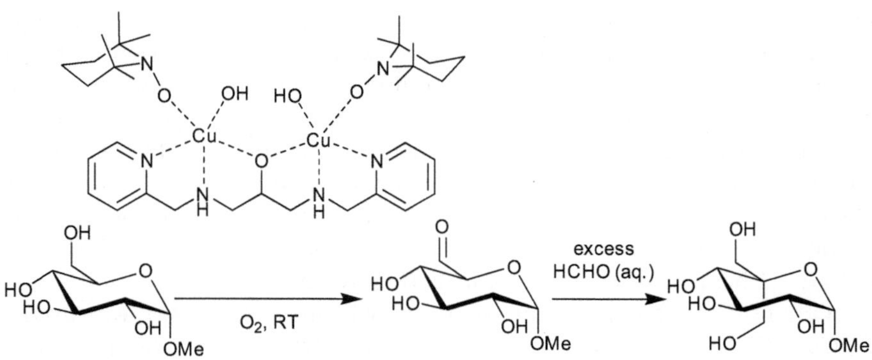

Figure 2.12 Aerobic oxidation of underivatized α-D-methylglucopyranoside into α-D-methylglucopyranosyl-6-carbaldehyde with a binuclear copper(II) catalyst and successive reaction with excess formaldehyde to produce 5-*C*-hydroxymethyl-α-D-methylglucopyranoside.

Reedijk and co-workers used pyrazole–pyridine and pyrazole–naphthol ligands for the Cu–TEMPO catalytic oxidation of benzyl alcohol in the presence of dioxygen under mild conditions. Whereas pyrazole–naphthol ligands inhibit the catalytic ability of the corresponding copper complexes, the copper(II) complex of pyrazole–pyridine ligands can oxidize benzyl alcohol in acetonitrile.[59] Reedijk's group further developed copper complexes of bifunctional triazine-based ligands (Figure 2.13) by combining a pyrazole–pyridine group or substituted bpy and TEMPO on a triazine ring for the oxidative transformation of benzyl alcohol and 1-octanol to the aldehydic products under mild conditions in air.[60,61]

Repo and co-workers reported the catalytic aerobic oxidation of substituted benzyl alcohols by using copper(II) complexes of 2-*N*-arylpyrrolecarbaldimine ligands and TEMPO in the presence of potassium carbonate in aqueous medium.[62] A comparative study of catalytic conversion with electronically and sterically variable aryl-substituted ligands was performed (Figure 2.14). A catalytic system containing *N*-benzylidene-*N*,*N*-dimethylethane-1,2-diamine, $CuBr_2$ and TEMPO has been reported to oxidize various benzylic alcohols in the absence of any base in acetonitrile–water solvent mixture at 80 °C.[63]

Repo and co-workers also developed an efficient catalytic system based on N_2O_2 donor 3,5-di-*tert*-butylsalicylaldimine ligands and TEMPO

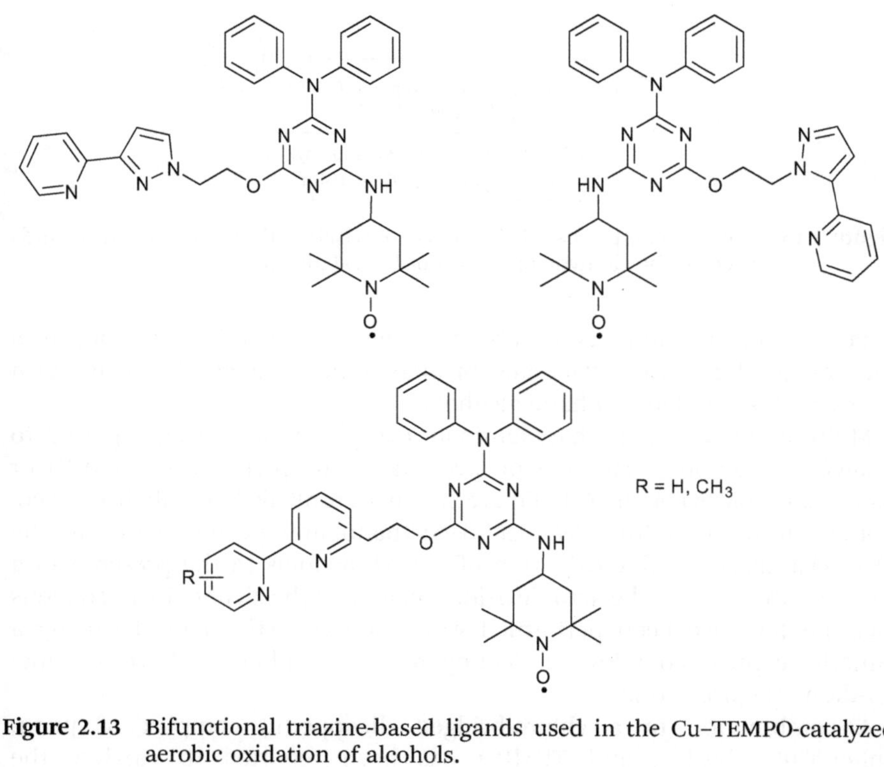

Figure 2.13 Bifunctional triazine-based ligands used in the Cu–TEMPO-catalyzed aerobic oxidation of alcohols.

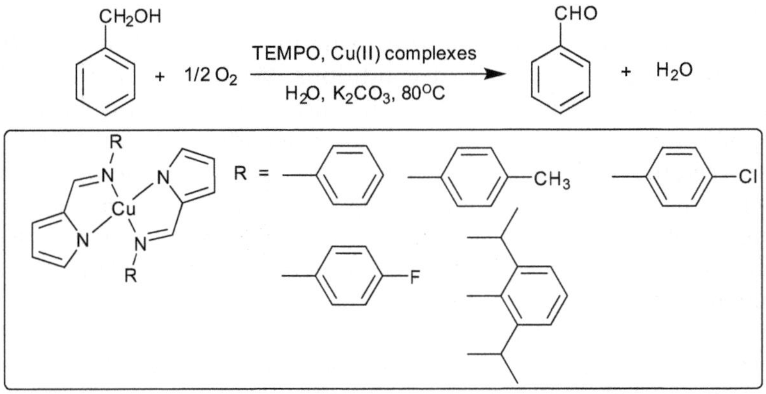

Figure 2.14 Aerobic oxidation of alcohols catalyzed by copper(II) 2-*N*-arylpyrrole-carbaldimine-based catalyst systems.

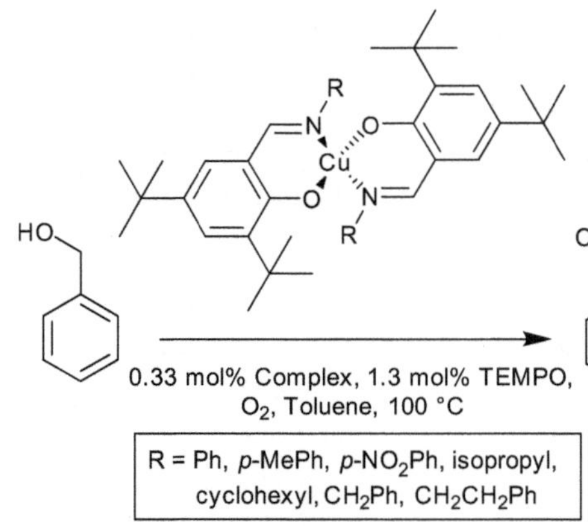

Figure 2.15 Copper complexes of 3,5-di-*tert*-butylsalicylaldimine-derived ligands used in the catalytic oxidation of benzyl alcohol.

(Figure 2.15). The catalysts are able to catalyze the aerobic oxidation of a wide range of benzylic, allylic and primary aliphatic alcohols but perform poorly with secondary aliphatic alcohols.[64,65]

Multinuclear Cu(II)–triethanolaminate complexes have been reported to catalyze the aerobic oxidation of alcohols in the presence of TEMPO or acetamido-TEMPO (**A** and **C** in Figure 2.2) in acetonitrile.[66] Pombeiro and co-workers utilized multinuclear Cu(II)–triethanolaminate complexes in the TEMPO-assisted aerobic oxidation of benzyl alcohols in the presence of a base in pure water.[67] The aerobic oxidation of benzylic alcohol in an aqueous medium has also been performed with good catalytic yields by using a dimeric copper complex of 3-(2-hydroxy-4-nitrophenylhydrazo)pentane-2,4-dione (Figure 2.16).[68]

A combination of a chloro-bridged dicopper(II) complex of phen, $[(phen)Cu(\mu\text{-}Cl)(Cl)]_2$, and TEMPO has been reported to catalyze the

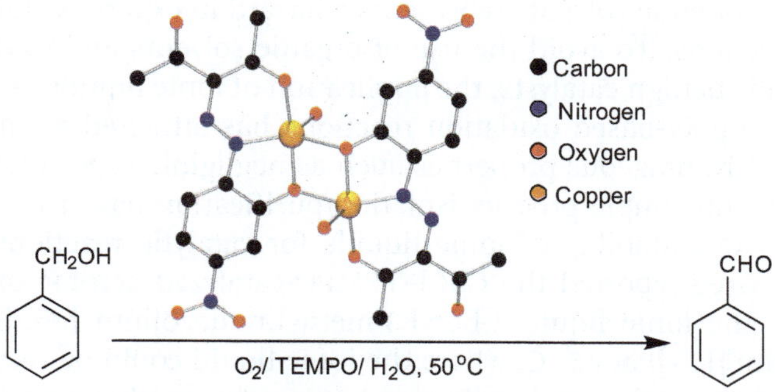

Figure 2.16 Oxidation of benzyl alcohol by a dinuclear copper complex of 3-(2-hydroxy-4-nitrophenylhydrazo)pentane-2,4-dione in aqueous solution.

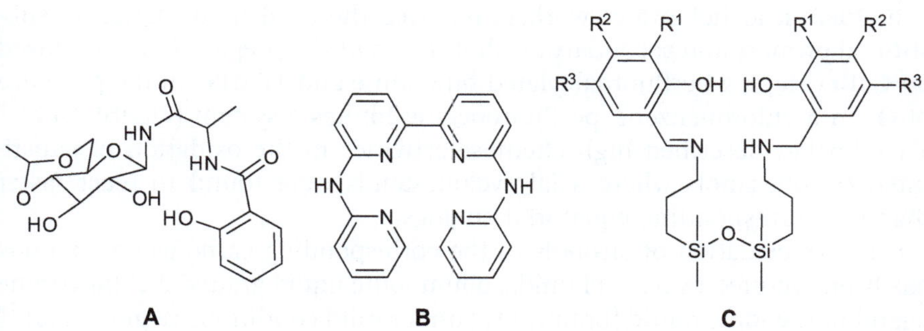

Figure 2.17 Ligands used in the aerobic oxidation of alcohols. **A**, alanyl-4,6-*O*-ethylidene-β-D-glucopyranosylamine; **B**, *N*,*N*-di(pyridine-2-yl)-2,2-bipyridine-6,6-diamine; and **C**, disiloxane-derived Schiff-base ligands.

oxidation of a range of both primary and secondary benzylic and aliphatic alcohols in the presence of a base at room temperature in acetonitrile.[69] TEMPO-mediated aerobic oxidation of benzyl alcohols and its derivatives catalyzed by a copper complex of a β-diketone azo derivative 3-(5-chloro-2-hydroxy-3-sulfophenylhydrazo)pentane-2,4-dione ligand in aqueous medium was reported.[70] The complex builds a three-dimensional supramolecular architecture with hydrogen-bonded water molecules in its structure, as is evident from the X-ray crystal structure. Copper(II) complexes of alkoxy-1,3,5-triazapentadienato have been used as efficient catalysts for the TEMPO-mediated aerobic oxidation of alcohols with high selectivities under mild conditions.[71] Recently, copper complexes of the tetrapyridyl ligand *N*,*N*-di(pyridin-2-yl)-2,2-bipyridine-6,6-diamine (**B** in Figure 2.17) have been reported to serve as a catalyst in the oxidation of benzyl alcohol in the presence of TEMPO and a base.[72] Mononuclear copper complexes of a salen-type ligand containing a disiloxane unit (**C** in Figure 2.17) have been utilized in the aerobic oxidation of alcohols that exhibit very high catalytic TONs of aldehyde products.[73,74]

The use of organic solvents represents a hazard in experimental studies of catalytic reactions. To avoid the use of organic solvents and to develop environmentally benign catalysts, the application of ionic liquids as alternative solvents in copper-based oxidation reactions has attracted attention in recent times. Advantageous properties such as negligible vapor pressure, easy recyclability and simple product isolation/purification have led chemists to investigate the suitability of ionic liquids for catalytic reactions. In 2002, Ansari and Gree reported the CuCl–TEMPO-catalyzed aerobic oxidation of alcohols in the ionic liquid 1-butyl-3-methylimidazolium hexafluorophosphate ([bmim][PF_6]) at 65 °C, where the ionic liquid could be recycled up to eight times but not the catalyst (Figure 2.18).[75] This method was found to be efficient for the oxidation of a series of primary and secondary alcohols with 5 mol% each of CuCl and TEMPO. Aliphatic alcohols, however, were slowly oxidized under the experimental conditions.

In 2000, Knochel and co-workers reported the catalytic oxidation of substituted primary and secondary alcohols by an *in situ*-prepared catalyst based on CuBr · Me_2S, a perfluoroalkylated bipyridine and TEMPO in the presence of O_2 in a chlorobenzene–perfluorooctane biphasic system (Figure 2.19).[76] They further described high chemoselectivities in the oxidation of substituted cyclohexanols where axial cyclohexanols were found to react faster than the corresponding equatorial analogs.[77]

Selective oxidation of alcohols to the corresponding carbonyl compounds has been reported by a novel imidazolium ionic liquid-grafted 2,2′-bipyridine ligand in the ionic liquid [bmim][PF_6] under mild conditions (Figure 2.20).[78]

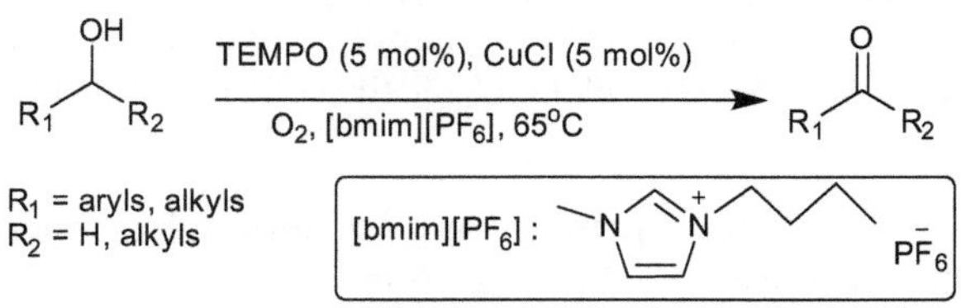

Figure 2.18 Cu–TEMPO-catalyzed aerobic oxidation of alcohols in an ionic liquid.

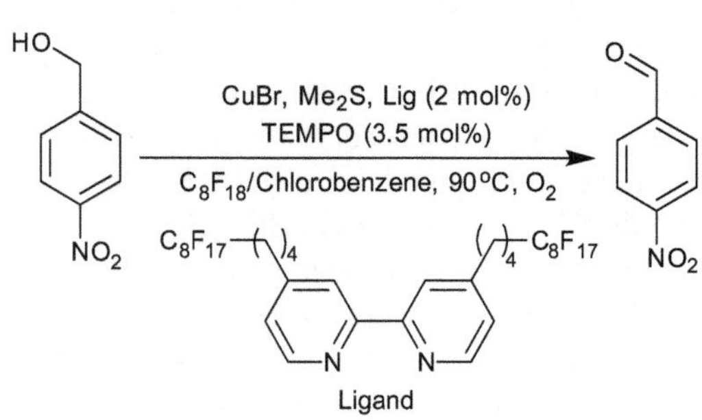

Figure 2.19 Perfluoroalkylated bipyridine ligand used in a copper–TEMPO-catalyzed aerobic oxidation of 4-nitrobenzyl alcohol in a fluorous biphasic system.

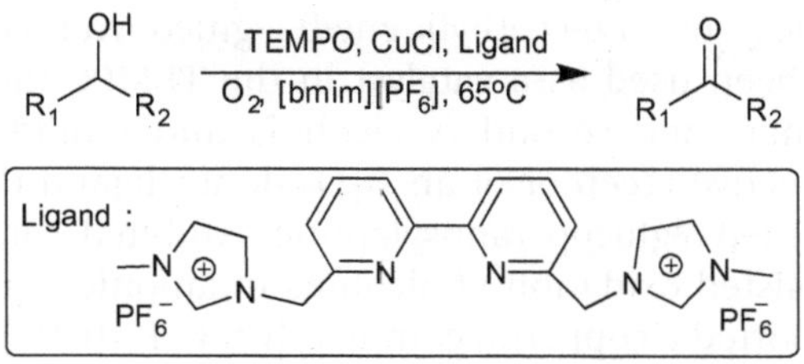

Figure 2.20 Imidazolium ionic liquid-grafted 2,2′-bipyridine ligand in Cu–TEMPO catalyst system.

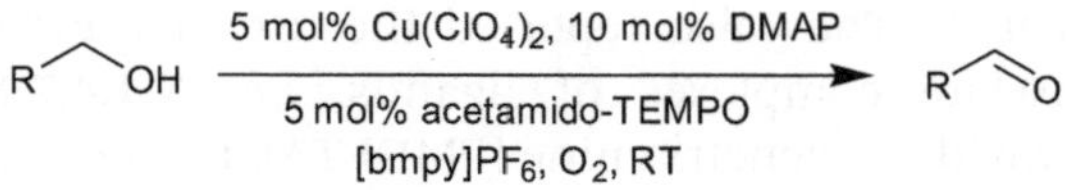

Figure 2.21 Cu–TEMPO-catalyzed oxidation of alcohols in 1-butyl-4-methylpyridinium hexafluorophosphate ([bmpy]PF_6.

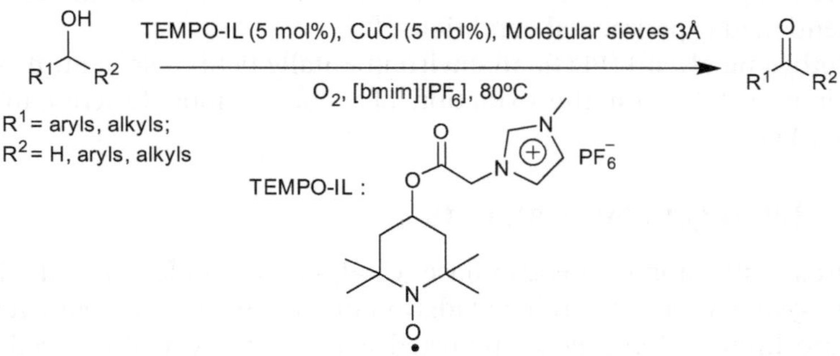

Figure 2.22 Molecular sieve-promoted oxidation of alcohols catalyzed by copper halide and immobilized TEMPO on an ionic liquid.

Jiang and Ragauskas reported a three-component acetamido-TEMPO–Cu catalytic system to catalyze the oxidation of primary alcohols selectively over secondary alcohols using DMAP as a base in ionic liquids (Figure 2.21).[79] Kinetic studies with a range of bases provided mechanistic insights into the catalytic cycle.[80]

Catalytic oxidations of alcohols have been carried out with a copper-based system with TEMPO immobilized on an ionic liquid and a heterogeneous promoter, molecular sieves 3Å, in an ionic liquid (Figure 2.22). The heterogeneous promoter acted as a Brønsted base to increase the reaction rate and the catalyst could easily be recycled and reused without a decrease in activity.[81,82]

A three component catalytic system comprising CuBr–4-pyrrolidinopyridine–acetamido-TEMPO under solvent-free conditions is capable of chemospecific aerobic oxidation of 1-(4-hydroxymethylphenyl)ethanol with high selectivity.[83] The catalyst can be recycled by an antisolvent precipitation method. For solid alcohols, PEG-200 is used as solvent. The dimeric copper(II) complex of 4,6-*O*-ethylidene-β-D-glucopyranosylamine (**A** in Figure 2.17) has been used as a catalyst in the TEMPO-mediated selective oxidation of primary and secondary alcohols under mild conditions.[84] A cupric acetate selective receptor of an N_2O_2 donor ligand derived from alanyl-4,6-*O*-ethylidene-β-D-glucopyranosylamine exhibited catalytic properties in the TEMPO-assisted oxidation of alcohols in anionic liquid.[85]

Contel *et al.* reported a copper(II) complex, $[Cu(C_8F_{17}(CH_2)_2CO_2)_2(R_f\text{-tacn})]$, where $R_f = -(CH_2)_3C_8F_{17}$, as an efficient and recoverable precatalyst for the aerobic oxidation of 4-nitrobenzyl alcohol to 4-nitrobenzaldehyde with TEMPO at 90 °C. The catalyst can be recycled after five runs without a decrease in activity.[86] They also reported the use of fluorocarbon-soluble copper(II) carboxylate complexes of ligands, *i.e.* *N*-1,4,7-Me_3TACN and *N*-1,4,7-pentamethyldiethylenetriamine (PMDETA), in the aerial oxidation of *p*-nitrobenzyl alcohol to *p*-nitrobenzaldehyde in chlorobenzene–toluene using TEMPO at 90 °C in the thermomorphic mode (Figure 2.23).[87] Moreover, a series of copper complexes of fluorinated alkoxide-derived bipyridine ligands have been reported to perform as an efficient catalytic system under the aforementioned experimental conditions.[88]

Poly(ethylene glycol) (PEG), an environmentally benign solvent, produces a recyclable catalyst for the oxidation of alcohols upon functionalization with TEMPO.[89]

2.2.2 Heterogeneous Catalysts

Compared with homogeneous-phase catalysts, examples of Cu–TEMPO catalytic systems for the aerobic oxidation of alcohols in the heterogeneous phase are limited. Yang *et al.* reported a recyclable Cu–Mn mixed oxide supported on active carbon for the aerobic oxidation of various alcohols in the presence of TEMPO in dichloromethane at 80 °C.[90] The catalyst is effective with a low loading of TEMPO (0.5 mol%) and can be recycled after five runs without any detectable decrease in activity and selectivity. An immobilized TEMPO derivative, (TEMPO-4-oxymethyl)polystyrene, has been

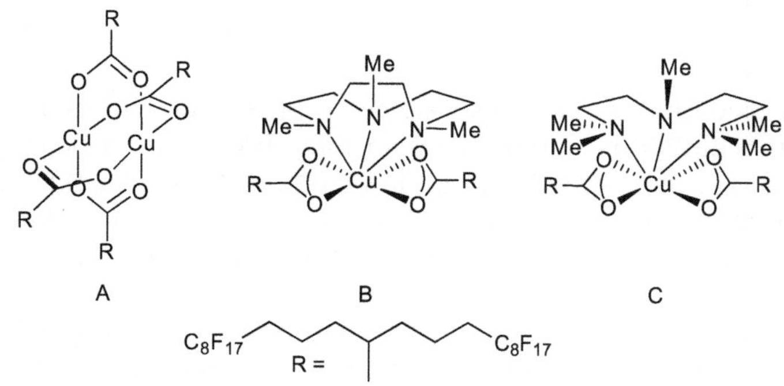

Figure 2.23 Fluorocarbon-soluble copper complexes (**A**, **B** and **C**) of ligands *N*-1,4,7-Me_3TACN (**B**) and *N*-1,4,7-pentamethyldiethylenetriamine (**C**) used as oxidation catalysts.

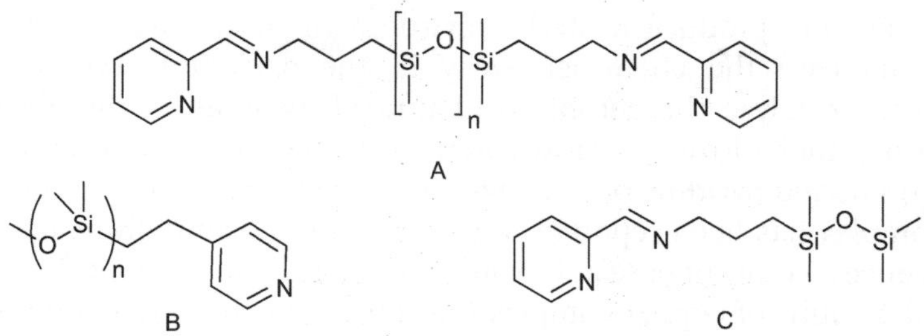

Figure 2.24 Polydimethylsiloxane pyridylimine Schiff-base ligands (**A** and **C**) and functionalized pyridine (**B**) used in the aerobic oxidation of alcohols.

successfully used in combination with Cu(0) and bipy in the oxidation of benzyl alcohol on a larger scale. The immobilized TEMPO catalyst was reused four times without any significant decrease in activity or selectivity.[38] A polymer-supported copper complex of bpy coupled with immobilized TEMPO molecules has served as a catalyst for the aerobic oxidation of primary alcohols.[91] Introduction of a triglycyl peptide to supported TEMPO and pre-adsorption of a Cu complex on resin beads have been reported to enhance the efficiency of the catalyst.[92]

A silica-supported complex of copper(II) acetate and polydimethylsiloxane (PDMS)-functionalized pyridine (**B** in Figure 2.24) with TEMPO was used in the catalytic aerobic oxidation of 4-nitrobenzyl alcohols in supercritical carbon dioxide.[93]

A $CuCl_2$–TEMPO catalyst dissolved in an ionic liquid, dispersed in the form of a film on a solid support consisting of bimodal pore structure silica or an ionogel is capable of producing high TONs in the heterogeneous catalytic oxidation of primary alcohols.[94] TEMPO-coated Fe_3O_4 superparamagnetic nanoparticles in the presence of Mn(II)–Cu(II) nitrates performs as a good alcohol oxidation catalyst.[95] Garcia and co-workers reported a reusable heterogeneous catalyst by exploiting a $Cu_3(BTC)_2$ (BTC = 1,3,5-benzenetricarboxylate) metallo-organic framework and TEMPO in acetonitrile.[96] Recently, a heterogeneous catalyst for the aerobic oxidation of benzylic and allylic alcohols in *n*-octane was developed with immobilized TEMPO on an ionic liquid and a copper salt, all being tethered on a silica support.[97] A catalytic system has been designed with a TEMPO task-specific ionic liquid (TEMPO-IL) with $CuCl_2$ in a silica gel matrix prepared by a sol–gel technique to perform oxidative catalysis of benzylic, allylic and heterocyclic alcohols.[98]

2.3 Copper-Based Alcohol Oxidation Catalysts Without Nitroxyl Radical

2.3.1 Homogeneous Catalysts

In addition to Cu–TEMPO catalytic systems, a large number of copper-based catalysts employing ligands of different denticity have been reported.

Organic ligands produce a steric environment at the metal center and thereby increase the chemoselectivity in the oxidation reaction. These compounds catalyze the aerobic oxidation of alcohols in the absence of TEMPO or other nitroxyl radicals. Aerobic alcohol oxidation employing a copper(I) salt and pyridine or phen was first reported in 1977.[99] The copper-phen complex has been reported to oxidize benzylic and allylic alcohols in the presence 2 equiv. of base at elevated temperature in benzene. The system required 2 equiv. of copper complex for optimum conversion and therefore was not considered catalytic. Sakharov and Skibida reported a novel catalytic system based on copper complexes of phen that exhibited better activity in non-aqueous solvents, as determined by kinetic studies.[100]

Markó and co-workers reinvestigated the copper–phen system and improved the efficiency of the system remarkably by using substituted hydrazines as reducing agents.[101–106] In 1996, Markó and co-workers reported the aerobic oxidation of alcohols using CuCl (5 mol%), phen (5 mol%), di-*tert*-butylazodicarboxylate hydrazine ($DBADH_2$) (5 mol%) and 2 equiv. of K_2CO_3 in toluene at 70–90 °C.[101] The catalytic system was efficient for a number of alcohols with high tolerance of other functional groups. However, the catalytic system required 2 equiv. of K_2CO_3 and was not efficient for the oxidation of primary aliphatic alcohols. Under the basic reaction conditions, alcohols containing α-stereogenic centers were oxidized without racemization. It was proposed that the insoluble K_2CO_3 acted as a heterogeneous base in the catalytic reaction. When fluorobenzene was used instead of toluene, a catalytic amount of base was needed.[103] Further investigation revealed that the use of DMAP or NMI in the catalytic system improved its efficiency to such an extent that conversion of aliphatic alcohols such as 1-decanol was increased markedly in the homogeneous phase (Figure 2.25).[105]

Based on several mechanistic studies and probable roles of the additives, a mechanistic proposal was put forward (Figure 2.25b). Initially, the copper(I)–hydrazide complex (**I**) is oxidized to a peroxo-bridged copper(II)–hydrazide species (**II**). Homolytic O–O bond cleavage of the peroxide intermediate forms a copper(II)–oxyl radical species, which then abstracts a hydrogen atom from the coordinated hydrazide to form intermediate **III**. Reaction of **III** with alcohol affords a copper–alkoxide species (**IV**). Intramolecular hydride transfer to DBAD and subsequent dissociation of the aldehyde regenerate the copper(I) catalyst. The mechanistic pathway of the intramolecular oxidation of coordinated alcoholate bears a similarity to that proposed for GO and Cu–TEMPO system.

Markó and co-workers' copper catalyst in combination with a diazo reagent and triphenylphosphine has been applied to a domino one-pot oxidation–olefination process. The method was applied to a variety of alcohols, including alcohols bearing α-stereogenic centers.[107] A catalytic system composed of both bpy and phen in equimolar proportions in a CuCl–DBAD–K_2CO_3 system was effective in the aerobic oxidation of allylic and propargylic alcohols with high yields.[108] The same strategy has also been employed in the aerobic oxidation of 2,3-allenols to 1,2-allenic ketones.[109]

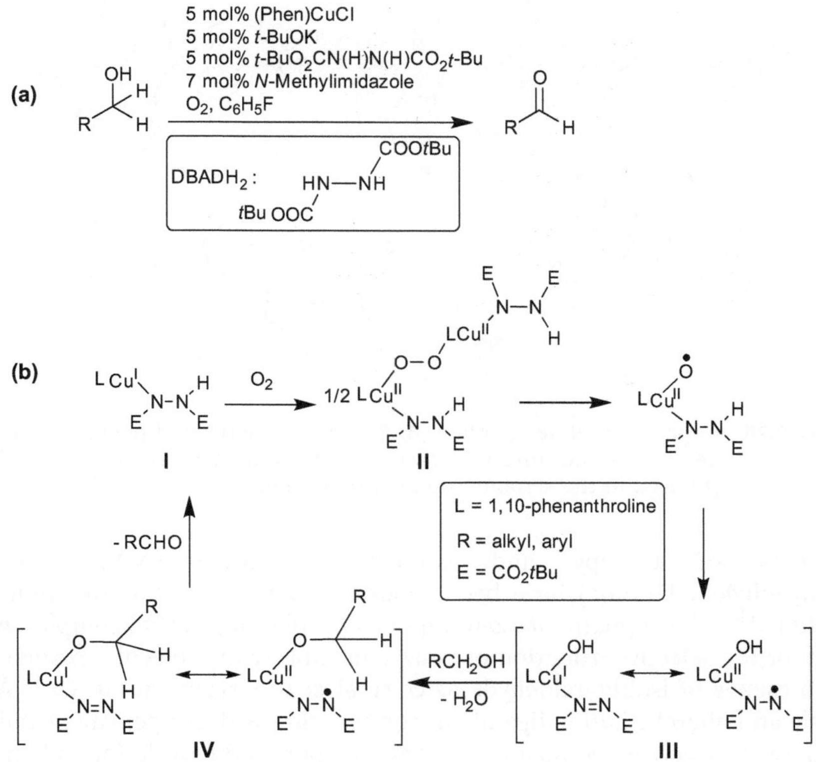

Figure 2.25 (a) Aerobic oxidation of alcohols catalyzed by Cu–phen catalyst in the presence of $DBADH_2$ and (b) the proposed catalytic mechanism.

Copper–phen complexes have also been utilized in the oxidation of veratryl alcohol (3,4-dimethoxybenzyl alcohol) to veratraldehyde with O_2 in alkaline aqueous solutions by Repo and co-workers.[110] Recently, Das and Paine reported a series of copper complexes supported by different phen derivatives for the aerobic oxidation of primary alcohols in the presence of a base (tetrabutylammonium methoxide) under ambient conditions (Figure 2.26).[111] The catalysts exhibit high selectivity towards primary alcohols. The copper complexes of phen (**A** and **B** in Figure 2.26) are found to be better catalysts than other complexes. In the catalytic cycle, a copper(II)–superoxo species is proposed to abstract the hydrogen atom of the metal-bound alcoholate. The effect of structural and electronic factors of the supporting ligand on the catalytic efficiencies was reported.

In addition to copper–phen complexes, numerous catalytic systems with nitrogen donor ligands have been reported. In 1980, Munakata *et al.* used a series of binary copper(I) complexes of 2,2′-bipyridine, 4,4′-dimethyl-2,2′-bipyridine, 2,2′,2″-terpyridine and ethylenediamine and ternary copper(I) complexes of 2,2′-bipyridine and β-diketonates for the oxidation of primary and secondary alcohols in the presence of O_2 at 15 °C.[112] The catalytic

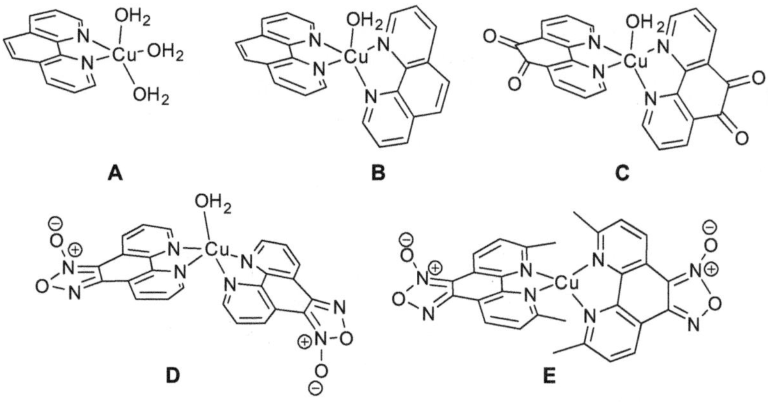

Figure 2.26 Copper complexes of phen (**A**, **B**), phendione (**C**) and phenfuroxan (**D**) derivatives and furoxan derivative of 2,9-dimethyl-1,10-phenanthroline (**E**) used in the aerobic oxidation of alcohols.

activities of Cu–bpy and Cu–TMEDA (TMEDA = *N*,*N*,*N*′,*N*′-tetramethylethylenediamine) have been reported for the efficient oxidation of alcohols.[113,114] A copper(II) *meso*-tetraphenylporphyrin (CuTPP) complex was used for the selective oxidation of benzyl alcohols with molecular oxygen in the presence of isobutyraldehyde as co-catalyst in *o*-xylene at 60 °C.[115] Aliphatic and alicyclic amine ligands in combination with copper salts exhibit good catalytic activity. Repo and co-workers reported the oxidation of benzyl alcohol and 3,4-dimethoxybenzyl alcohol (veratryl alcohol) by using copper complexes of diamine-based ligands.[116] For this purpose, they utilized a range of aliphatic, alicyclic and aromatic amines including ethylenediamine, 1,3-propanediamine, diethylenetriamine, TMEDA, 9,10-diaminophenanthrene (DAPHEN), 1,2-diaminocyclohexane (DACH), *etc.* They also reported a copper complex, bis(pyridine-2-carboxaldehyde oxime)copper(II) sulfate, as the catalyst precursor for the aerobic oxidation of veratryl alcohol.[117] Kinetic study of the oxidation reaction indicated the formation of a binuclear hydroxyl-bridged complex $(LCu\text{-}\mu\text{-}(OH)_2\text{-}CuL)^{2+}$ in catalysis with a Cu–TMEDA complex.[118] Studies on oxidation reactions with a copper–bisisoquinoline catalyst revealed an increase in the catalytic yield of aldehydes and ketones by using ionic liquids [bmim][PF_6], [omim][BF_4] and [hmim][BF_4] (omim = 1-octyl-3-methylimidazolium, hmim = 1-hexyl-3-methylimidazolium).[119]

The catalytic cycle of GO has fuelled interest in developing copper complexes supported by organic ligands, particularly nitrogen- and phenol-containing ligands. Over the years, a number of copper complexes employing multidentate ligands such as mono(pyridyl)-alkylamines, bis(pyridyl)alkylamines, triazacyclononanes, salen and substituted tris(pyrazolyl)borates have been developed as structural and spectroscopic models of GO.[120]

The first catalytic functional model system for GO was reported even before the structure of the enzyme was solved. The model complexes

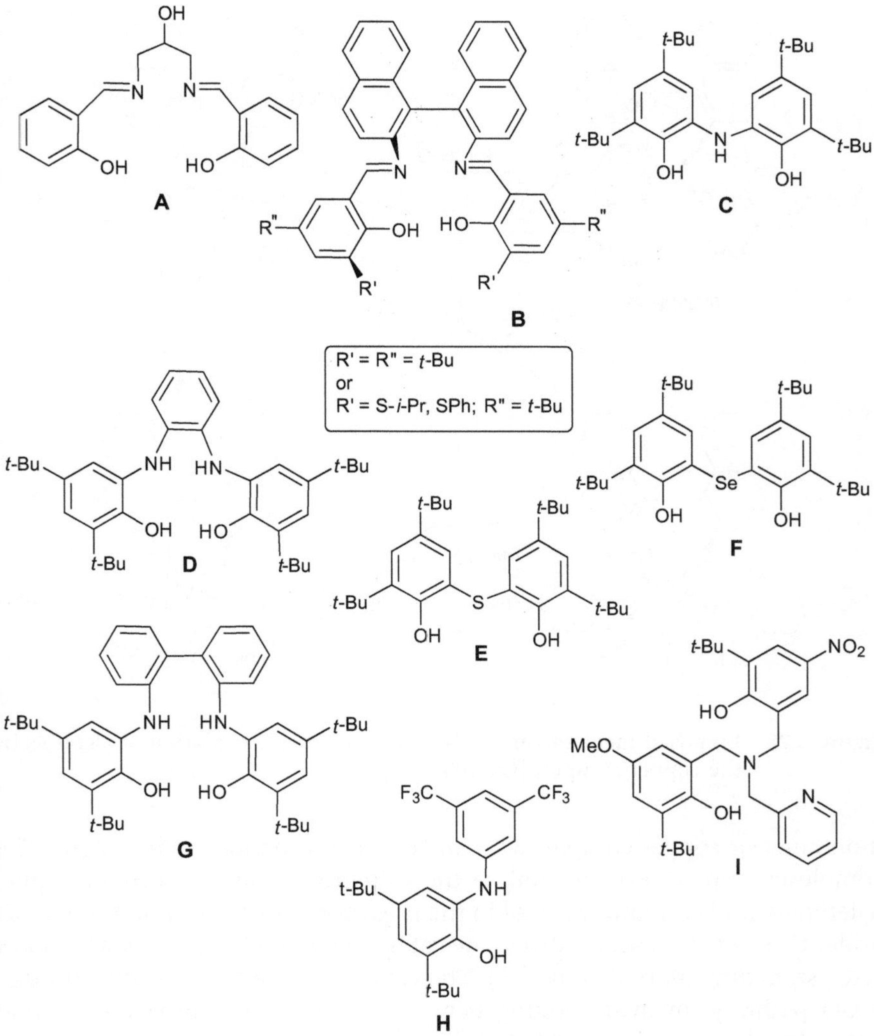

Figure 2.27 Ligands used to prepare functional models of GO.

(bpy)Cu(PR_3)X (R = Me, Et, *n*Bu, Ph; X = Cl, Br, I), have been reported to catalyze the oxidation of ethanol to acetaldehyde under O_2 pressure (30 psi) at 25 °C.[121] Kitajima *et al.* used a copper complex of *N,N'*-(2-hydroxypropane-1,3-diyl)bis(salicylaldiminato) (**A** in Figure 2.27) for the oxidation of ethanol to acetaldehyde using the substrate as solvent under O_2 at 40 °C.[122] A strong base such as KOH was used for catalytic activity. The catalyst could oxidize propanol and hydroxyacetone to their corresponding carbonyl compounds.

In 1998, Stack and co-workers reported the first efficient biomimetic copper catalysts for the aerobic oxidation of alcohols.[123] The ligands used for this study were salen-type compounds with a binaphthyl backbone and thioether functions (**B** in Figure 2.27). The ligands form a non-square-planar

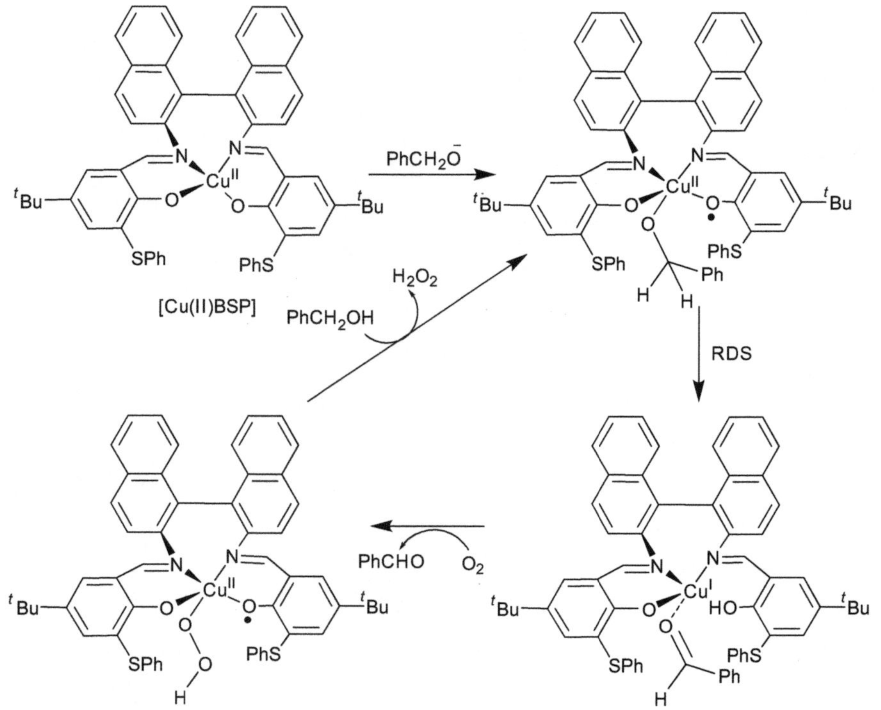

Figure 2.28 Proposed mechanism for the catalytic aerobic oxidation of alcohols by the copper complex [Cu^{II}(BSP)].

mononuclear copper complex with an N_2O_2 coordination environment. The complexes were shown to catalyze the oxidation of benzyl alcohol, 1-phenylethanol and cinnamyl alcohol in the presence of catalytic amount of KOH under O_2 at 22 °C in acetonitrile. The catalyst exhibited high TON with a low catalyst loading (0.01–0.06 mol%). The catalytic oxidation operates through a GO pathway involving cycling between copper(II)–phenoxyl radical and copper(I)–phenolate intermediate (Figure 2.28).

Chaudhuri and co-workers reported a number of excellent synthetic models for GO.[17,124–128] Copper complexes of phenolate ligands containing N, S and Se hetero-donor atoms are capable of catalyzing the aerobic oxidation of alcohols. A range of different primary alcohols could be oxidized using these catalysts. The copper(II) complex with selenobis(2,4-di-*tert*-butylphenol) ligand (**F** in Figure 2.27) selectively oxidizes benzyl alcohol to benzaldehyde with 95 turnovers in 24 h under ambient conditions.[127] The copper complex of the sulfur analog (**E** in Figure 2.27) has been reported to oxidize secondary alcohols and also to effect oxidative C–C bond coupling reactions. The copper complex of the tetradentate ligand *N,N′*-bis(3,5-di-*tert*-butyl-2-hydroxyphenyl)-1,2-phenylenediamine (**D** in Figure 2.27) is capable of oxidizing methanol and ethanol under ambient conditions in organic solvents or in pure substrates.[125] The copper complex derived from

N,*N*-bis(2-hydroxy-3,5-di-*tert*-butylphenyl)amine ligand (**C** in Figure 2.27) has been applied in the aerobic oxidation of benzyl alcohol and ethanol to the corresponding aldehydes.[126] The aerial oxidation of alcohols including methanol and ethanol by a diradical copper(II) complex of the ligand 2-[3,5-bis(trifluoromethyl)aniline]-4,6-di-*tert*-butylphenol (**H** in Figure 2.27) was reported. The copper(II) complex of pyridylbis(2-hydroxy-3-*tert*-butylphenyl)-amine (**I** in Figure 2.27) catalyzes the aerobic oxidation of benzyl alcohols with 220 turnovers in 48 h in the presence of catalytic amount of KOH.[129] Another copper complex of a similar type of ligand, bis(3-*tert*-butyl-5-methyl-2-hydroxybenzyl)(2-pyridylmethyl)amine, has been reported to catalyze efficiently the chemoselective oxidation of primary alcohols to aldehydes under ambient conditions. A maximum of 272 turnovers after 24 h was achieved for the oxidation of benzyl alcohol.[130] A five-coordinate Cu(II) complex of the N_3O_2-donor ligand *cis*-1-(3′,5′-dimethoxybenzylideneamino)-3,5-[2-hydroxy-(3′,5′-di-*tert*-butyl)benzylideneimino]cyclohexane (**H** in Figure 2.27) has been applied in the aerobic oxidation of benzyl alcohol with 160 turnovers in 144 h.[131]

2.3.2 Heterogeneous Catalysts

Copper-based oxidative heterogeneous catalysis has been developing rapidly in recent times. A copper complex of phthalocyanine encapsulated on a polystyrene matrix has been shown to catalyze the aerobic oxidation of alcohols in the presence of KOH in refluxing toluene.[132] The encapsulated reusable catalyst was reported to exhibit better activity than the unencapsulated counterpart. A resin-immobilized CuO and Cu nanocomposite has been shown to be an efficient heterogeneous catalyst for the aerobic oxidation of alcohols under mild conditions.[133]

Layered double hydroxides (LDHs) or hydrotalcite-like compounds (HTLCs) have recently become popular in heterogeneous-phase catalysis for their potential use as adsorbents and anion exchangers. The idea has been exploited in the development of a reusable green catalyst for the aerobic oxidation of a range of benzylic, secondary and heterocyclic alcohols using copper–aluminum hydrotalcite/*rac*-BINOL.[134] Aerobic oxidation of benzyl alcohol using an alumina-supported copper–manganese oxide catalyst in the homogeneous deposition–precipitation method has also been reported.[135] The copper nanoparticle-catalyzed aerobic oxidation of propargylic alcohols in the presence of bpy has been described.[136] Supported Au–CuO catalysts have been reported for the selective oxidation of alcohols by oxygen.[137,138] An eco-friendly heterogeneous catalytic system containing spinel-type chromites MCr_2O_4 (M = Co, Ni and Cu) has been used for the aerial oxidation of benzyl alcohol.[139] However, the catalytic activity of the copper catalyst was found to be the lowest among these chromites.

A nanoporous silica surface is being widely used to develop oxidation catalysts in the heterogeneous phase. Several silica-supported Au–Cu alloys have been used for the aerobic oxidation of alcohols.[140] A heterogeneous hybrid polymer–inorganic copper nanoparticle–polyacrylamide–mesoporous silica SBA-15 has been reported to serve as a good recyclable catalyst in the oxidation of alcohols.[141]

2.4 Conclusion

Investigations on copper complexes have helped to improve the chemo-, regio- and enantioselectivity of catalytic reagents towards the oxidation of alcohols. Although a number of effective copper-based catalysts for alcohol oxidations have been developed, finding catalytic systems with high activity and wide applications needs further effort. The range of substrates needs to be extended so that they can be oxidized under environmentally friendly conditions. The design of green catalysts mandatorily excludes any radical initiator as co-oxidant, and rather includes aerial oxidation reactions in environmentally friendly solvents. Moreover, to increase catalytic activity, attention needs to be paid to the recyclability of the catalyst. Hence the use of polymeric solid supports, nanoporous materials and thermally stable silica as adsorbents has been initiated for heterogeneous-phase catalytic reactions. Immobilized catalysts on solid supports can be easily separated from the reaction medium and offer high selectivity and catalytic yields of the products and are reusable. Further studies in these directions would provide useful information for the development of highly efficient copper-based alcohol oxidation catalysts for practical and industrial applications.

References

1. T. Punniyamurthy, S. Velusamy and J. Iqbal, *Chem. Rev.*, 2005, **105**, 2329.
2. T. Punniyamurthy and L. Rout, *Coord. Chem. Rev.*, 2008, **252**, 134.
3. M. J. Schultz and M. S. Sigman, *Tetrahedron*, 2006, **62**, 8227.
4. M. J. Schultz, S. S. Hamilton, D. R. Jensen and M. S. Sigman, *J. Org. Chem.*, 2005, **70**, 3343.
5. C. Parmeggiani and F. Cardona, *Green Chem.*, 2012, **14**, 547.
6. M. S. Sigman and D. R. Jensen, *Acc. Chem. Res.*, 2006, **39**, 221.
7. J. W. Whittaker, in *Metal Ions in Biological Systems: Metalloenzymes Involving Amino Acid-Residue and Related Radicals*, ed. H. Sigel and A. Sigel, Marcel Dekker, New York, 1994, vol. 30, pp. 315–360.
8. J. W. Whittaker, *Chem. Rev.*, 2003, **103**, 2347.
9. N. Ito, S. E. V. Phillips, C. Stevens, Z. B. Ogel, M. J. McPherson, J. N. Keen, K. D. S. Yadav and P. F. Knowles, *Nature*, 1991, **350**, 87.
10. N. Ito, S. E. V. Phillips, K. D. S. Yadav and P. F. Knowles, *J. Mol. Biol.*, 1994, **238**, 794.
11. J. W. Whittaker, *Arch. Biochem. Biophys.*, 2005, **433**, 227.
12. F. Thomas, *Eur. J. Inorg. Chem.*, 2007, 2379.
13. B. A. Jazdzewski and W. B. Tolman, *Coord. Chem. Rev.*, 2000, **200–202**, 633.

14. I. W. C. E. Arends, P. Gamez and R. A. Sheldon, *Adv. Inorg. Chem.*, 2006, **58**, 235.
15. V. Mahadevan, R. J. M. K. Gebbink and T. D. P. Stack, *Curr. Opin. Chem. Biol.*, 2000, **4**, 228.
16. R. F. Brissos, S. García, A. Presa and P. Gamez, *Comments Inorg. Chem.*, 2011, **32**, 219.
17. P. Chaudhuri, K. Wieghardt, T. Weyhermüller, T. K. Paine, S. Mukherjee and C. Mukherjee, *Biol. Chem.*, 2005, **386**, 1023.
18. L. Que Jr. and W. B. Tolman, *Nature*, 2008, **455**, 333.
19. R. A. Sheldon and I. W. C. E. Arends, *J. Mol. Catal. A*, 2006, **251**, 200.
20. R. A. Sheldon and I. W. C. E. Arends, *Adv. Synth. Catal.*, 2004, **346**, 1051.
21. R. A. Sheldon, I. W. C. E. Arends, G.-J. ten Brink and A. Dijksman, *Acc. Chem. Res.*, 2002, **35**, 774.
22. M. F. Semmelhack, C. R. Schmid, D. A. Cortes and C. S. Chou, *J. Am. Chem. Soc.*, 1984, **106**, 3374.
23. A. Cecchetto, F. Fontana, F. Minisci and F. Recupero, *Tetrahedron Lett.*, 2001, **42**, 6651.
24. F. Minisci, F. Recupero, M. Rodinò, M. Sala and A. Schneider, *Org. Process Res. Dev.*, 2003, **7**, 794.
25. D. H. Hunter, D. H. R. Barton and W. B. Motherwell, *Tetrahedron Lett.*, 1984, **25**, 603.
26. A. E. J. de Nooy, A. C. Besemer and H. van Bekkum, *Synthesis*, 1996, 1153.
27. T. Miyazawa and T. Endo, *J. Mol. Catal.*, 1988, **49**, L31.
28. M. F. Semmelhack, *Tetrahedron Lett.*, 1986, **27**, 119.
29. A. Dijksman, I. W. C. E. Arends and R. A. Sheldon, *Org. Biomol. Chem.*, 2003, **1**, 3232.
30. J.-M. Vatèle, *Synlett*, 2009, 2143.
31. J.-M. Vatèle, *Tetrahedron*, 2010, **66**, 904.
32. G. Yang, L. Wang, J. Li, Y. Zhang, X. Dong, Y. Lv and S. Gao, *Res. Chem. Intermed.*, 2012, **38**, 775.
33. S.-S. Shen, V. Kartika, Y. S. Tan, R. D. Webster and K. Narasaka, *Tetrahedron Lett.*, 2012, **53**, 986.
34. P. Gamez, I. W. C. E. Arends, J. Reedijk and R. A. Sheldon, *Chem. Commun.*, 2003, 2414.
35. J. Laugier, J. M. Latour, A. Caneschi and P. Rey, *Inorg. Chem.*, 1991, **30**, 4474.
36. A. Caneschi, A. Grand, J. Laugier, P. Rey and R. Subra, *J. Am. Chem. Soc.*, 1988, **110**, 2307.
37. P. Gamez, I. W. C. E. Arends, R. A. Sheldon and J. Reedijk, *Adv. Synth. Catal.*, 2004, **346**, 805.
38. D. Geißlmeir, W. G. Jary and H. Falk, *Monatsh. Chem.*, 2005, **136**, 1591.
39. P. J. Figiel, M. Leskelä and T. Repo, *Adv. Synth. Catal.*, 2007, **349**, 1173.
40. C. J. Gartshore and D. W. Lupton, *Adv. Synth. Catal.*, 2010, **352**, 3321.
41. E. T. T. Kumpulainen and A. M. P. Koskinen, *Chem. Eur. J.*, 2009, **15**, 10901.

42. N. Mase, T. Mizumori and Y. Tatemoto, *Chem. Commun.*, 2011, **47**, 2086.
43. J. M. Hoover and S. S. Stahl, *J. Am. Chem. Soc.*, 2011, **133**, 16901.
44. J. M. Hoover, J. E. Steves and S. S. Stahl, *Nat. Protoc.*, 2012, 7, 1161.
45. J. M. Hoover, B. L. Ryland and S. S. Stahl, *J. Am. Chem. Soc.*, 2013, **135**, 2357.
46. N. J. Hill, J. M. Hoover and S. S. Stahl, *J. Chem. Educ.*, 2013, **90**, 102.
47. D. Könning, W. Hiller and M. Christmann, *Org. Lett.*, 2012, **14**, 5258.
48. J. C. A. Flanagan, L. M. Dornan, M. G. McLaughlin, N. G. McCreanor, M. J. Cook and M. J. Muldoon, *Green Chem.*, 2012, **14**, 1281.
49. H. Tian, X. Yu, Q. Li, J. Wang and Q. Xu, *Adv. Synth. Catal.*, 2012, **354**, 2671.
50. N. Jiang and A. J. Ragauskas, *J. Org. Chem.*, 2006, **71**, 7087.
51. S. Mannam, S. K. Alamsetti and G. Sekar, *Adv. Synth. Catal.*, 2007, **349**, 2253.
52. B. Han, X.-L. Yang, C. Wang, Y.-W. Bai, T.-C. Pan, X. Chen and W. Yu, *J. Org. Chem.*, 2012, **77**, 1136.
53. S. Mannam and G. Sekar, *Tetrahedron: Asymmetry*, 2009, **20**, 497.
54. S. K. Alamsetti, S. Mannam, P. Mutupandi and G. Sekar, *Chem. Eur. J.*, 2009, **15**, 1086.
55. W. Brackman and C. J. Gaasbeek, *Recl. Trav. Chim. Pays-Bas*, 1966, **85**, 257.
56. S. Velusamy, A. Srinivasan and T. Punniyamurthy, *Tetrahedron Lett.*, 2006, **47**, 923.
57. S. Striegler, *Tetrahedron*, 2006, **62**, 9109.
58. S. Striegler, N. A. Dunaway, M. G. Gichinga and L. K. Milton, *Tetrahedron*, 2010, **66**, 7927.
59. J. S. Uber, Y. Vogels, D. van den Helder, I. Mutikainen, U. Turpeinen, W. T. Fu, O. Roubeau, P. Gamez and J. Reedijk, *Eur. J. Inorg. Chem.*, 2007, 4197.
60. Z. Lu, J. S. Costa, O. Roubeau, I. Mutikainen, U. Turpeinen, S. J. Teat, P. Gamez and J. Reedijk, *Dalton Trans.*, 2008, 3567.
61. Z. Lu, T. Ladrak, O. Roubeau, J. van der Toorn, S. J. Teat, C. Massera, P. Gamez and J. Reedijk, *Dalton Trans.*, 2009, 3559.
62. P. J. Figiel, A. Sibaouih, J. U. Ahmad, M. Nieger, M. T. Räisänen, M. Leskelä and T. Repo, *Adv. Synth. Catal.*, 2009, **351**, 2625.
63. Q. Wang, Y. Zhang, G. Zheng, Z. Tian and G. Yang, *Catal. Commun.*, 2011, **14**, 92.
64. J. U. Ahmad, P. J. Figiel, M. T. Räisänen, M. Leskelä and T. Repo, *Appl. Catal. A*, 2009, **371**, 17.
65. J. U. Ahmad, M. T. Räisänen, M. Kemell, M. J. Heikkilä, M. Leskelä and T. Repo, *Appl. Catal. A*, 2012, **449**, 153.
66. M. Zhu, B. Li, P. He, X. Wei and Y. Yuan, *Tetrahedron*, 2008, **64**, 9239.
67. P. J. Figiel, A. M. Kirillov, Y. Y. Karabach, M. N. Kopylovich and A. J. L. Pombeiro, *J. Mol. Catal. A: Chem.*, 2009, **305**, 178.

68. K. T. Mahmudov, M. N. Kopylovich, M. F. C. Guedes da Silva, P. J. Figiel, Y. Y. Karabach and A. J. L. Pombeiro, *J. Mol. Catal. A: Chem.*, 2010, **318**, 44.
69. M. M. Hossain and S.-G. Shyu, *Adv. Synth. Catal.*, 2010, **352**, 3061.
70. M. N. Kopylovich, K. T. Mahmudov, M. Haukka, P. J. Figiel, A. Mizar, J. A. L. da Silva and A. J. L. Pombeiro, *Eur. J. Inorg. Chem.*, 2011, **2011**, 4175.
71. M. N. Kopylovich, Y. Y. Karabach, M. F. C. Guedes da Silva, P. J. Figiel, J. Lasri and A. J. L. Pombeiro, *Chem. Eur. J.*, 2012, **18**, 899.
72. I. Gamba, I. Mutikainen, E. Bouwman, J. Reedijk and S. Bonnet, *Eur. J. Inorg. Chem.*, 2013, **2013**, 115.
73. A. Soroceanu, M. Cazacu, S. Shova, C. Turta, J. Kožišek, M. Gall, M. Breza, P. Rapta, T. C. O. Mac Leod, A. J. L. Pombeiro, J. Telser, A. A. Dobrov and V. B. Arion, *Eur. J. Inorg. Chem.*, 2013, **2013**, 1458.
74. Z. Hu and F. M. Kerton, *Appl. Catal. A*, 2012, **413–414**, 332.
75. I. A. Ansari and R. Gree, *Org. Lett.*, 2002, **4**, 1507.
76. B. Betzemeier, M. Cavazzini, S. Quici and P. Knochel, *Tetrahedron Lett.*, 2000, **41**, 4343.
77. G. Ragagnin, B. Betzemeier, S. Quici and P. Knochel, *Tetrahedron*, 2002, **58**, 3985.
78. X.-E. Wu, L. Ma, M.-X. Ding and L.-X. Gao, *Chem. Lett.*, 2005, **34**, 312.
79. N. Jiang and A. J. Ragauskas, *Org. Lett.*, 2005, **7**, 3689.
80. L. Lin, J. Liuyan and W. Yunyang, *Catal. Commun.*, 2008, **9**, 1379.
81. L. Lin, M. Juanjuan, J. Liuyan and W. Yunyang, *J. Mol. Catal. A: Chem.*, 2008, **291**, 1.
82. L. Lin, J. Liuyan and W. Yunyang, *Monatsh. Chem.*, 2008, **139**, 901.
83. N. Jiang and A. J. Ragauskas, *ChemSusChem*, 2008, **1**, 823.
84. K. Soni, A. Kumar and A. K. Sah, *Catal. Commun.*, 2012, **17**, 95.
85. A. K. Sah and K. Soni, *Catal. Commun.*, 2012, **28**, 120.
86. M. Contel, C. Izuel, M. Laguna, P. R. Villuendas, P. J. Alonso and R. H. Fish, *Chem. Eur. J.*, 2003, **9**, 4168.
87. M. Contel, P. R. Villuendas, J. Fernández-Gallardo, P. J. Alonso, J.-M. Vincent and R. H. Fish, *Inorg. Chem.*, 2005, **44**, 9771.
88. N. Lu and Y.-C. Lin, *Tetrahedron Lett.*, 2007, **48**, 8823.
89. C.-X. Miao, L.-N. He, J.-Q. Wang and J. Gao, *Synlett*, 2009, 3291.
90. G. Yang, W. Zhu, P. Zhang, H. Xue, W. Wang, J. Tian and M. Song, *Adv. Synth. Catal.*, 2008, **350**, 542.
91. C. W. Y. Chung and P. H. Toy, *J. Comb. Chem.*, 2007, **9**, 115.
92. K. Akagawa, S. Takigawa, E. Mano and K. Kudo, *Tetrahedron Lett.*, 2011, **52**, 770.
93. M. Herbert, F. Montilla and A. Galindo, *Dalton Trans.*, 2010, **39**, 900.
94. A. Chrobok, S. Baj, W. Pudło and A. Jarzębski, *Appl. Catal. A*, 2010, **389**, 179.
95. A. K. Tucker-Schwartz and R. L. Garrell, *Chem. Eur. J.*, 2010, **16**, 12718.
96. A. Dhakshinamoorthy, M. Alvaro and H. Garcia, *ACS Catal.*, 2011, **1**, 48.

97. L. Liu, J. Ma, Z. Sun, J. Zhang, J. Huang, S. Li and Z. Tong, *Can. J. Chem.*, 2011, **89**, 68.
98. L. Liu, J. Ma, J. Xia, L. Li, C. Li, X. Zhang, J. Gong and Z. Tong, *Catal. Commun.*, 2011, **12**, 323.
99. C. Jallabert and H. Riviere, *Tetrahedron Lett.*, 1977, **14**, 1215.
100. A. M. Sakharov and I. P. Skibida, *J. Mol. Catal.*, 1988, **48**, 157.
101. I. E. Markó, P. R. Giles, M. Tsukazaki, S. M. Brown and C. J. Urch, *Science*, 1996, **274**, 2044.
102. I. E. Markó, M. Tsukazaki, P. R. Giles, S. M. Brown and C. J. Urch, *Angew. Chem. Int. Ed.*, 1997, **36**, 2208.
103. I. E. Markó, A. Gautier, I. Chellé-Regnaut, P. R. Giles, M. Tsukazaki, C. J. Urch and S. M. Brown, *J. Org. Chem.*, 1998, **63**, 7576.
104. I. E. Markó, P. R. Giles, M. Tsukazaki, I. Chellé-Regnaut, A. Gautier, S. M. Brown and C. J. Urch, *J. Org. Chem.*, 1999, **64**, 2433.
105. I. E. Markó, A. Gautier, R. Dumeunier, K. Doda, F. Philippart, S. M. Brown and C. J. Urch, *Angew. Chem. Int. Ed.*, 2004, **43**, 1588.
106. I. E. Markó, A. Gautier, J.-L. Mutonkole, R. Dumeunier, A. Ates, C. J. Urch and S. M. Brown, *J. Organomet. Chem.*, 2001, **624**, 344.
107. M. Davi and H. Lebel, *Org. Lett.*, 2009, **11**, 41.
108. Y. Liu and S. Ma, *Chin. J. Chem.*, 2012, **30**, 29.
109. S. Gao, Y. Liu and S. Ma, *Beilstein J. Org. Chem.*, 2011, **7**, 396.
110. H. Korpi, P. J. Figiel, E. Lankinen, P. Ryan, M. Leskelä and T. Repo, *Eur. J. Inorg. Chem.*, 2007, 2465.
111. O. Das and T. K. Paine, *Dalton Trans.*, 2012, **41**, 11476.
112. M. Munakata, S. Nishibayashi and H. Sakamoto, *J. Chem. Soc. Chem. Commun.*, 1980, 219.
113. C. Jallabert and H. Riviere, *Tetrahedron*, 1980, **36**, 1191.
114. X. Liu, A. Qiu and D. T. Sawyer, *J. Am. Chem. Soc.*, 1993, **115**, 3239.
115. R. Rahimi, E. Gholamrezapor and M. R. Naimi-jamal, *Inorg. Chem. Commun.*, 2011, **14**, 1561.
116. P. Lahtinen, H. Korpi, E. Haavisto, M. Leskelä and T. Repo, *J. Comb. Chem.*, 2004, **6**, 967.
117. H. Korpi, M. Polamo, M. Leskelä and T. Repo, *Inorg. Chem. Commun.*, 2005, **8**, 1181.
118. P. Lahtinen, E. Lankinen, M. Leskelä and T. Repo, *Appl. Catal. A*, 2005, **295**, 177.
119. H.-Y. Shen, L.-Y. Ying, H.-L. Jiang and Z. M. A. Judeh, *Int. J. Mol. Sci.*, 2007, **8**, 505.
120. S. Itoh, M. Taki and S. Fukuzumi, *Coord. Chem. Rev.*, 2000, **198**, 3.
121. S. Bhaduri and N. Y. Sapre, *J. Chem. Soc., Dalton Trans.*, 1981, 2585.
122. N. Kitajima, K. Whang, Y. Moro-oka, A. Uchida and Y. Sasada, *J. Chem. Soc., Chem. Commun.*, 1986, 1504.
123. Y. Wang, J. L. DuBois, B. Hedman, K. O. Hodgson and T. D. P. Stack, *Science*, 1998, **279**, 537.
124. P. Chaudhuri, M. Hess, U. Flörke and K. Wieghardt, *Angew. Chem. Int. Ed.*, 1998, **37**, 2217.

125. P. Chaudhuri, M. Hess, J. Müller, K. Hildenbrand, E. Bill, T. Weyhermüller and K. Wieghardt, *J. Am. Chem. Soc.*, 1999, **121**, 9599.
126. P. Chaudhuri, M. Hess, T. Weyhermüller and K. Wieghardt, *Angew. Chem. Int. Ed.*, 1999, **38**, 1095.
127. T. K. Paine, T. Weyhermüller, K. Wieghardt and P. Chaudhuri, *Dalton Trans.*, 2004, 2092.
128. C. Mukherjee, U. Pieper, E. Bothe, V. Bachler, E. Bill, T. Weyhermüller and P. Chaudhuri, *Inorg. Chem.*, 2008, **47**, 8943.
129. F. Thomas, G. Gellon, I. Gautier-Luneau, E. Saint-Aman and J.-L. Pierre, *Angew. Chem. Int. Ed.*, 2002, **41**, 3047.
130. A. John, M. M. Shaikh and P. Ghosh, *Dalton Trans.*, 2008, 2815.
131. A. K. Nairn, S. J. Archibald, R. Bhalla, B. C. Gilbert, E. J. MacLean, S. J. Teat and P. H. Walton, *Dalton Trans.*, 2006, 172.
132. R. Naik, P. Joshi and R. K. Deshpande, *Catal. Commun.*, 2004, **5**, 195.
133. S. Pande, A. Saha, S. Jana, S. Sarkar, M. Basu, M. Pradhan, A. K. Sinha, S. Saha, A. Pal and T. Pal, *Org. Lett.*, 2008, **10**, 5179.
134. M. Lakshmi Kantam, R. Arundhathi, P. R. Likhar and D. Damodara, *Adv. Synth. Catal.*, 2009, **351**, 2633.
135. Q. Tang, X. Gong, P. Zhao, Y. Chen and Y. Yang, *Appl. Catal. A*, 2010, **389**, 101.
136. C. Han, M. Yu, W. Sun and X. Yao, *Synlett*, 2011, 2363.
137. V. R. Choudhary and D. K. Dumbre, *Top. Catal.*, 2009, **52**, 1677.
138. H. Wang, W. Fan, Y. He, J. Wang, J. N. Kondo and T. Tatsumi, *J. Catal.*, 2013, **299**, 10.
139. K. Premalatha, P. S. Raghavan and B. Viswanathan, *Appl. Catal. A*, 2012, **419–420**, 203.
140. W. Li, A. Wang, X. Liu and T. Zhang, *Appl. Catal. A*, 2012, **433–434**, 146.
141. R. J. Kalbasi, A. A. Nourbakhsh and M. Zia, *J. Inorg. Organomet. Polym. Mater.*, 2012, **22**, 536.

CHAPTER 3

Ruthenium-Based Catalysts for Aerobic Oxidation of Alcohols

TERUYUKI KONDO,*[a] YU KIMURA,[b] HISATSUGU YAMADA[a] AND AKIO TOSHIMITSU[c]

[a] Advanced Biomedical Engineering Research Unit, Center for the Promotion of Interdisciplinary Education and Research, Kyoto University, Katsura, Nishikyo-ku, Kyoto, 615-8510, Japan; [b] Research and Educational Unit of Leaders for Integrated Medical System, Center for the Promotion of Interdisciplinary Education and Research, Kyoto University, Katsura, Nishikyo-ku, Kyoto, 615-8510, Japan; [c] Division of Multidisciplinary Chemistry, Institute for Chemical Research, Kyoto University, Gokanosho, Uji, Kyoto, 611-0011, Japan
*Email: teruyuki@scl.kyoto-u.ac.jp

3.1 Homogeneous Ruthenium-Based Catalysts

3.1.1 Introduction

The selective oxidation of alcohols to carbonyl compounds, such as aldehydes and ketones, is a ubiquitous and pivotal reaction in organic synthesis both in the laboratory and in the chemical industry,[1,2] since carbonyl compounds are important building blocks for medicines, agricultural chemicals, fragrances and many others.[3] Traditionally, oxidation of alcohols has been performed with non-catalytic procedures with stoichiometric inorganic oxidants, represented by CrO_3, $KMnO_4$, MnO_2 and SeO_2.[4] However, these oxidation methods have serious drawbacks, such as their cost and the production of environmentally hazardous/toxic metal salts as waste. In order to minimize waste

RSC Green Chemistry No. 28
Transition Metal Catalysis in Aerobic Alcohol Oxidation
Edited by Francesca Cardona and Camilla Parmeggiani

Published by the Royal Society of Chemistry, www.rsc.org

production, catalytic oxidation using transition metal complexes is the most reliable approach.[5] Among the various transition metal catalysts, much attention has been focused on the ability of ruthenium catalysts, since they have often been used for hydrogenation and hydrogen-transfer reactions[6] and in dehydrogenation and oxidation reactions.[7,8] In fact, some success has been achieved using ruthenium catalysts in the oxidation of alcohols with $NaIO_4$,[9] *tert*-butyl hydroperoxide,[10,11] chloramine-T,[12] amine-*N*-oxides,[12] iodosylbenzene,[13] $NaBrO_3$[14,15] and H_2O_2[16,17] as oxidants. From both economic and environmental viewpoints, there is an urgent demand for greener, more atom-efficient methods that use molecular O_2 and air as ideal oxidants and produce only H_2O as a by-product.[18,19] In combination with transition metal catalysts, the traditional and environmentally unfriendly oxidation processes should be replaced with those using the greener oxidants molecular O_2 and air.[20,21] This section reviews the homogeneous ruthenium-catalyzed aerobic oxidation of alcohols to the corresponding aldehydes and ketones, which should stimulate further development in this important field.

3.1.2 Homogeneous Ruthenium-Catalyzed Aerobic Oxidation of Alcohols

Ruthenium complexes are widely used as catalysts in organic synthesis[22] and have been studied extensively as catalysts for the aerobic oxidation of alcohols.[23] In 1978, Mares and co-workers[24] reported that $RuCl_3 \cdot nH_2O$ catalyzes the aerobic oxidation of secondary alcohols to the corresponding ketones, albeit in modest yields (Equation 3.1). In 1981, Matsumoto and Ito showed that $RuCl_3 \cdot nH_2O$ and $RuCl_2(PPh_3)_3$ catalyze the aerobic oxidation of activated allylic and benzylic alcohols under mild conditions,[25] *e.g.*, the oxidation of retinol was performed at room temperature for 48 h to give retinal in 57% yield (Equation 3.2). Lower aliphatic primary and secondary alcohols were more efficiently oxidized using trinuclear ruthenium carboxylates, $Ru_3O(O_2CR)_6L_n$ ($R = CH_3$, C_2H_5; $L = H_2O$, PPh_3; $n = 0, 1$), as catalyst.[26] Their catalytic activity was about 10 times higher than those of $RuCl_3 \cdot nH_2O$ and $RuCl_2(PPh_3)_3$.

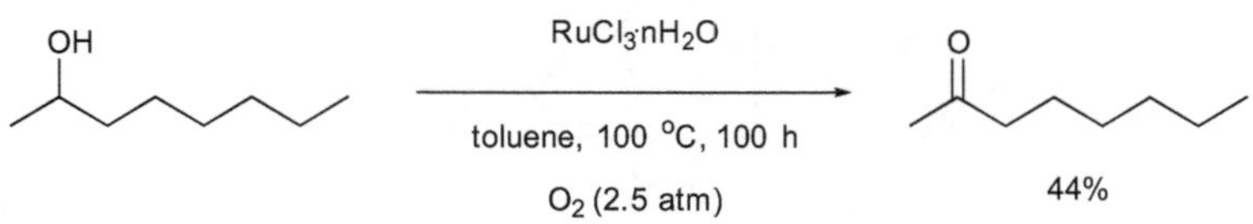

(3.1)

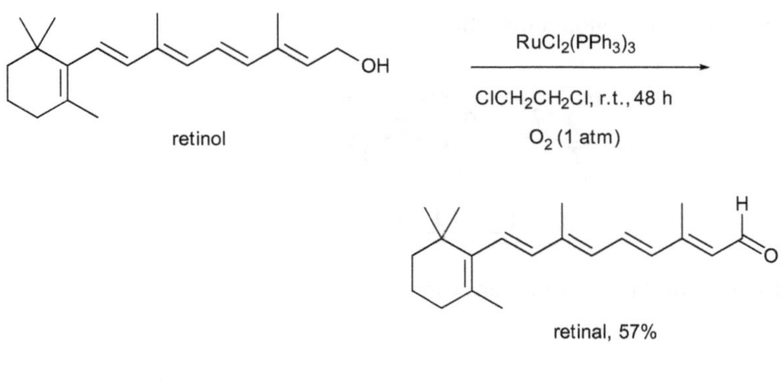

(3.2)

Recently, higher activities were reached using $RuCl_2(PPh_3)_3$ as a catalyst with ionic liquids as solvents. These compounds have been tested as environmentally friendly solvents for a wide variety of catalytic organic reactions.[27] In this particular case, tetramethylammonium hydroxide and Aliquat 336 (tricaprylmethylammonium chloride) were used as solvents and rapid conversion of 1-phenylethanol and benzyl alcohol was observed.[28] Moreover, the tetramethylammonium hydroxide/$RuCl_2(PPh_3)_3$ could be reused after extraction of the product.

Many ruthenium complexes have been widely used as catalysts for hydrogen-transfer reactions. These systems can be readily adapted to the aerobic oxidation of alcohols by employing molecular O_2, in combination with a hydrogen acceptor as a co-catalyst, in a multistep process. For example, Bäckvall and co-workers used low-valent ruthenium complexes, such as $RuCl(OAc)(PPh_3)_3$, in combination with hydroquinone and a cobalt–Schiff's base complex, cobalt(salophen)(PPh_3) [H_2salophen = *N,N*-bis(salicylidene)-*o*-phenylenediamine] (Figure 3.1).[29] The low-valent ruthenium complex reacts with alcohols to give aldehyde and ketone products together with the formation of ruthenium hydride species. The latter undergoes hydrogen-transfer to benzoquinone to give hydroquinone with concomitant regeneration of the low-valent active ruthenium species. The cobalt–Schiff's base complex catalyzes the subsequent aerobic oxidation of hydroquinone to benzoquinone to complete the catalytic cycle. Optimization of the electron-rich quinone, combined with so-called 'Shvo' ruthenium catalyst (Figure 3.2), led to one of the fastest catalytic systems reported for the oxidation of secondary alcohols.[30,31]

The regeneration of benzoquinone can also be achieved with molecular O_2 in the absence of the cobalt co-catalyst. Ishii and co-workers[32] showed that a combination of a catalytic amount of $RuCl_2(PPh_3)_3$, hydroquinone and

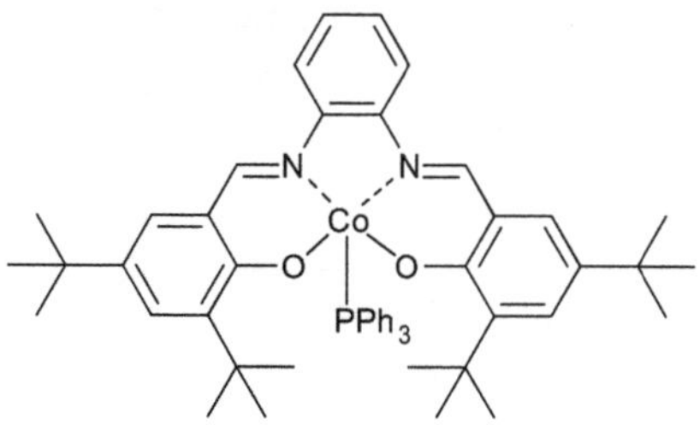

Figure 3.1 Structure of Co(salophen)(PPh_3).

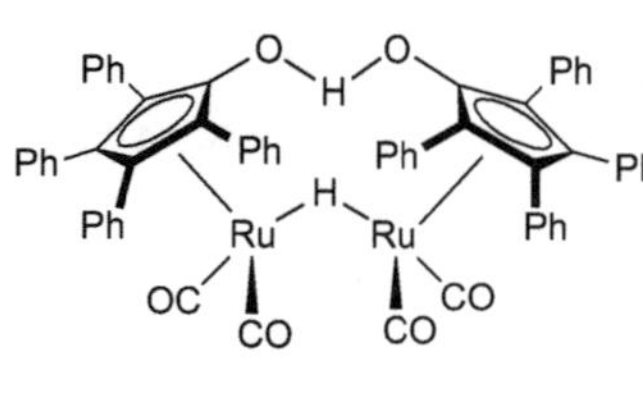

Figure 3.2 Structure of 'Shvo' ruthenium catalyst.

K_2CO_3 under O_2 (1 atm) in $PhCF_3$ as a solvent oxidized primary aliphatic, allylic and benzylic alcohols to the corresponding aldehydes in quantitative yields. Primary alcohols even in the presence of secondary alcohols were selectively oxidized to the corresponding aldehydes in high yield.

Sheldon and co-workers found that a combination of $RuCl_2(PPh_3)_3$ and the stable nitroxyl radical 2,2,6,6-tetramethylpiperidine-1-oxyl (TEMPO) is a remarkably effective catalyst for the aerobic oxidation of a variety of primary and secondary alcohols to give the corresponding aldehydes and ketones, respectively, with >99% selectivity (Equation 3.3).[33,34]

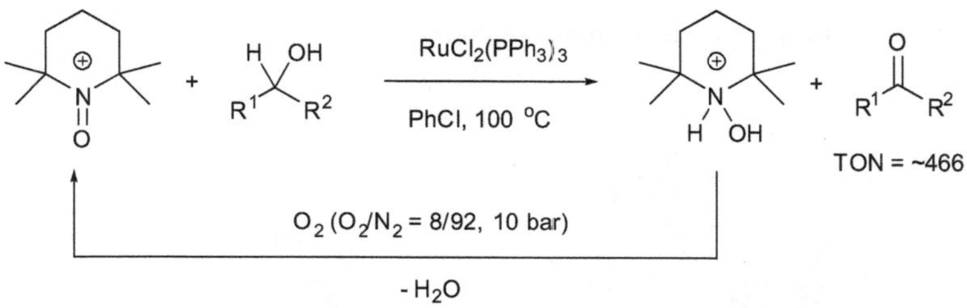

(3.3)

The Ru/TEMPO system displayed a preference for primary *versus* secondary alcohols. Results from Hammett correlation studies ($\rho = -0.58$) and the primary kinetic isotope effect ($k_H/k_D = 5.1$) for the catalytic aerobic oxidation of benzyl alcohol are inconsistent with either an oxoruthenium (O=Ru)- or an oxoammonium-based mechanism. Accordingly, a (hydrido)-metal mechanism, involving a '$RuH_2(PPh_3)_3$' species as the active catalyst, is postulated, in which TEMPO acts as a hydrogen-transfer mediator and is either regenerated by molecular O_2, under catalytic aerobic conditions, or converted to TEMPH under stoichiometric anaerobic conditions.[35,36]

Griffith and co-workers[37] reported the synthesis of the organic soluble tetra-*n*-butylammonium perruthenate (TBAP), $nBu_4N^+RuO_4^-$, in 1985. They later found that tetra-*n*-propylammonium perruthenate (TPAP), $nPr_4{}^+RuO_4^-$, is even easier to prepare, from RuO_4 and nPr_4NOH in water.[38,39] TBAP and TPAP are air stable, non-volatile and soluble in a wide range of organic solvents. The groups of Ley[40] and Marko[41] independently showed that TPAP-catalyzed the oxidation of alcohols using molecular O_2 as an oxidant.

As a more convenient catalyst system of high activity, operational simplicity and low air and moisture sensitivity, a combination of [$RuCl_2$-(*p*-cymene)]$_2$ and Cs_2CO_3 was developed.[42] A related water-soluble diruthenium complex, [$Ru_2(\mu\text{-}OAc)_3(\mu\text{-}CO_3)$], is also effective for the aerobic oxidation of alcohols in water.[43] Using a catalyst system of $RuCl_3/Et_3N$, the oxidation of benzyl alcohol with molecular O_2 could be achieved with a turnover frequency (TOF) of 332 h^{-1} in the absence of solvent.[44] A catalyst system of $RuCl_3 \cdot nH_2O$/dicyclohexylamine displayed a preference for the primary *versus* secondary benzylic alcohols in both intermolecular and intramolecular competition experiments.[45] A (μ-oxo)tetraruthenium cluster (Figure 3.3) showed high catalytic activity for a simple and practical aerobic

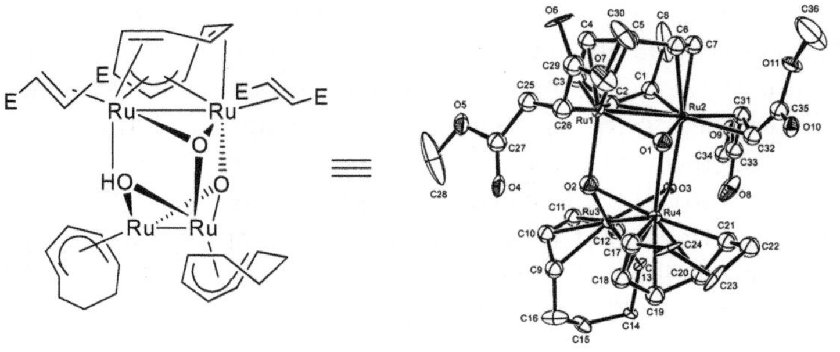

Figure 3.3 Structure of (μ-oxo)tetraruthenium cluster.

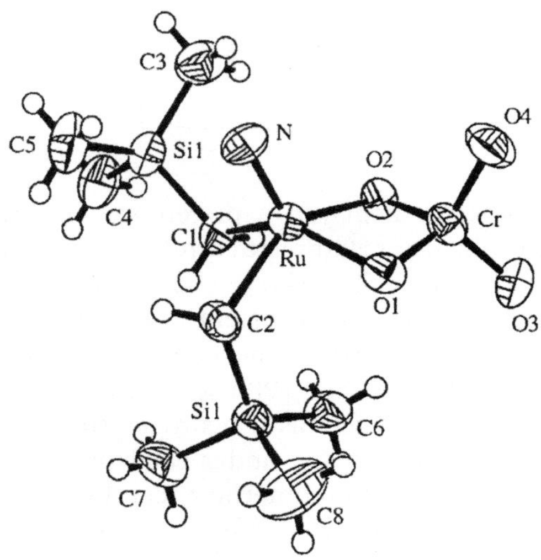

Figure 3.4 ORTEP diagram of $[Ru(N)(CH_2SiMe_3)_2(\mu\text{-}O)_2CrO_2]^-$.

oxidation of alcohols to the corresponding aldehydes and ketones under 1 atm of O_2 or air.[46] The heterobimetallic $[(Bu_4N)(M(N)(CH_2SiMe_3)_2(\mu\text{-}O)_2CrO_2)]$ (M = Ru or Os) catalyzes the selective oxidation of alcohols with molecular O_2 (Figure 3.4).[47] A similar heterobimetallic, $[PPh_4][Ru(N)Me_2(\mu_2\text{-}O)_2Pd((-)\text{-sparteine})]$, prepared by the reaction of $[PPh_4][Ru(N)(OH)_2Me_2]$ and $Pd(OSiMe_3)_2((-)\text{-sparteine})$ (Scheme 3.1), catalyzed the aerobic oxidation of aryl and allylic alcohols to the corresponding carbonyl compounds and also the rearrangement of allylic alcohols to unsaturated aldehydes.[48]

A three-dimensional microporous ruthenium(II,III) polymer complex, $[Ru^{II,III}{}_2(H_2TCPP)]BF_4$, where $H_2TCPP = 4,4',4'',4'''\text{-}(21H,23H)$-porphine-5,10,15,20-tetranyl)tetrakisbenzoic acid (Figure 3.5), exhibited an efficient catalytic performance for the aerobic oxidation of a variety of alcohols.[49] A safe, practical and selective process for the aerobic oxidation of benzylic and allylic alcohols and also *N*- and *S*-heteroatom-containing alcohols to the

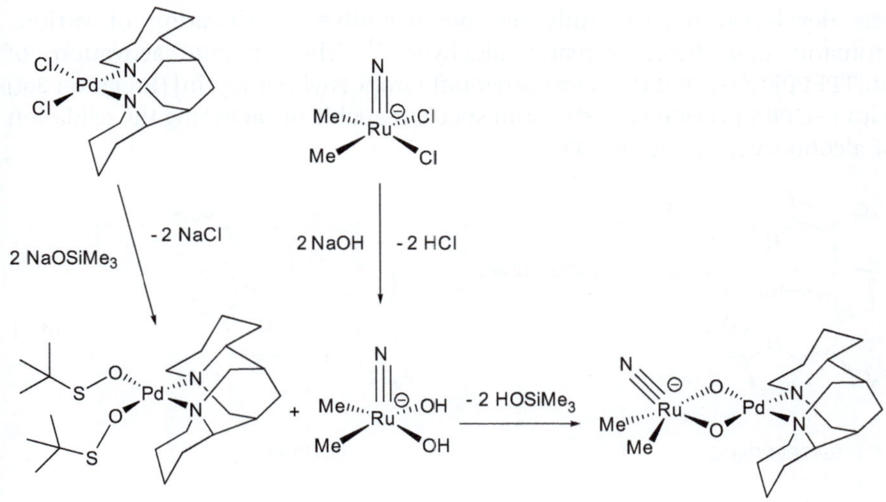

Scheme 3.1 Preparation of $[Ru(N)Me_2(\mu_2\text{-}O)_2Pd((-)\text{-sparteine})]^-$.

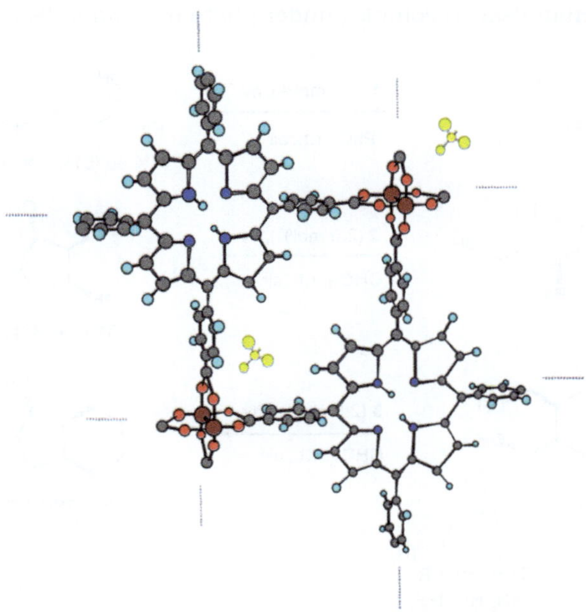

Figure 3.5 Proposed 3D body-centered, interpenetrating network structure for $[Ru^{II,III}{}_2(H_2TCPPT)]BF_4$: carbon (gray), hydrogen (light blue), nitrogen (blue), oxygen (red), ruthenium (reddish brown), boron (yellow) and fluorine (green).

corresponding aldehydes and ketones was developed using a ruthenium catalyst in a continuous-flow reactor.[50] A cooperative NHC (*N*-heterocyclic carbene) and ruthenium-based, $Ru(2,2'\text{-bipyrazine})_3(PF_6)_2$, redox catalyst was developed for the mild aerobic oxidative esterification of various aromatic and heteroaromatic aldehydes.[51] The *in situ* activation of Ru(TPFPP)(CO) [TPFPP = tetrakis(pentafluorophenyl)porphyrin] (Equation 3.4) with *m*-CPBA provided a ruthenium species capable of catalyzing the oxidation of alcohols with molecular O_2.[52]

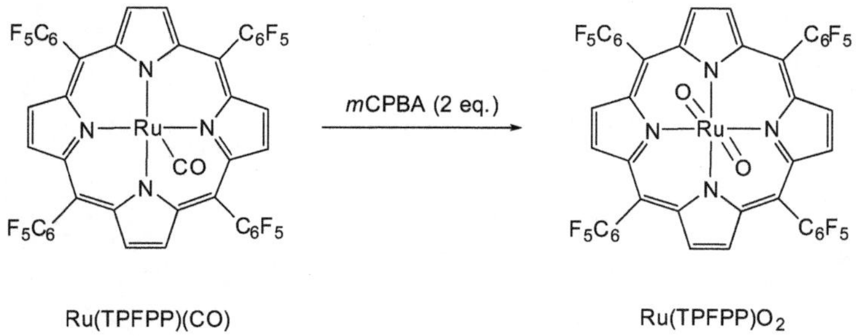

(3.4)

Based on the chemoselective aerobic oxidation of primary alcohols catalyzed by Ru–salen complexes,[53] kinetic resolution of racemic secondary alcohols[54] and desymmetrization of *meso*-diols[55] were realized by using a chiral (nitrosyl)ruthenium(salen) complex under photo-irradiation (Scheme 3.2).[56]

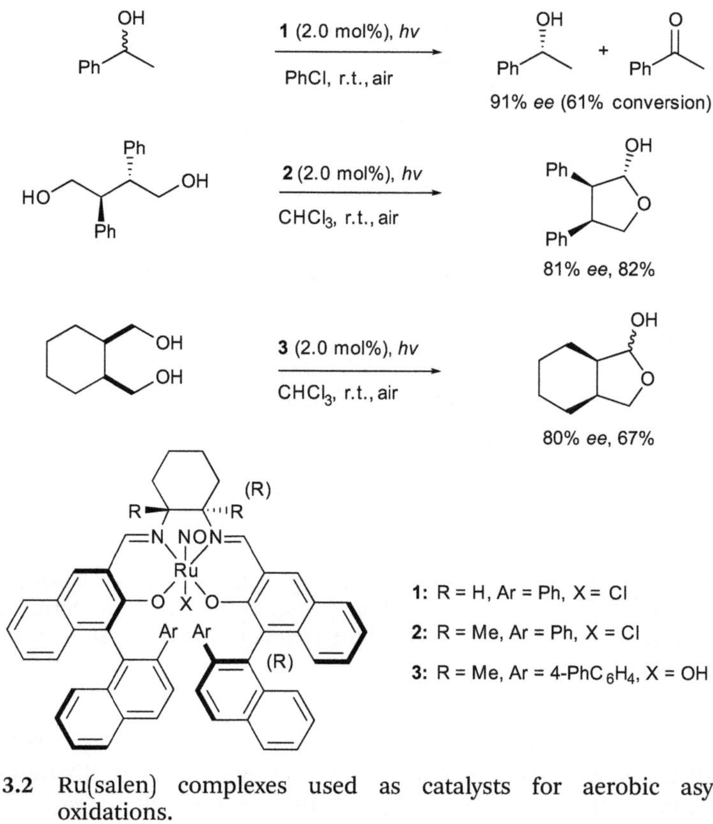

Scheme 3.2 Ru(salen) complexes used as catalysts for aerobic asymmetric oxidations.

3.2 Heterogeneous Ruthenium-Based Catalysts

3.2.1 Introduction

As described in Section 3.1, many homogeneous ruthenium-based catalysts have been developed for the aerobic oxidation of alcohols to the corresponding aldehydes and ketones and they are usually dissolved in reaction solutions making all catalytic sites accessible to the substrates, which realized high catalytic activity and selectivity. Despite these advantages, homogeneous catalysts have a share of only about 20% in industrial processes,[57] since the recovery and reuse of the expensive catalysts are often difficult. In this respect, the development of easily recoverable and recyclable heterogeneous catalysts has attracted considerable research interest and a number of solid-supported catalysts such as ruthenium, palladium, platinum and gold have been investigated and reported.[58,59] The main purpose of this section is to provide an analysis of recent advances in research into the heterogeneous ruthenium-catalyzed oxidation of alcohols with molecular O_2 and the different nature of the catalysts to obtain the desired products in an efficient manner, which are helpful in designing guidelines for next-generation catalysts and developments in this field.

3.2.2 Heterogeneous Ruthenium-Catalyzed Aerobic Oxidation of Alcohols

Among the oxides of ruthenium, RuO_4 is well known as a powerful oxidizing agent of alcohols. It is, however, too strong to be used for the selective oxidation of allylic alcohols to the corresponding unsaturated carbonyl compounds. A lower oxidation state than RuO_4, namely RuO_2 hydrate, acts as an oxidizing agent and a catalyst for the aerobic oxidation of allylic alcohols with higher efficiency than MnO_2.[60,61] A zeolite was used to confine RuO_2 hydrate nanoparticles in its cages, and both saturated and unsaturated alcohols were aerobically oxidized with high activity and selectivity.[62] Zeolite-confined (ZC) nanostructured dinuclear ruthenium clusters (Figure 3.6) were prepared by the following simple three-step procedure: (i) ion exchange of Ru^{3+} ions with the extra-framework Na^+ ions in zeolite-Y, (ii) reduction of

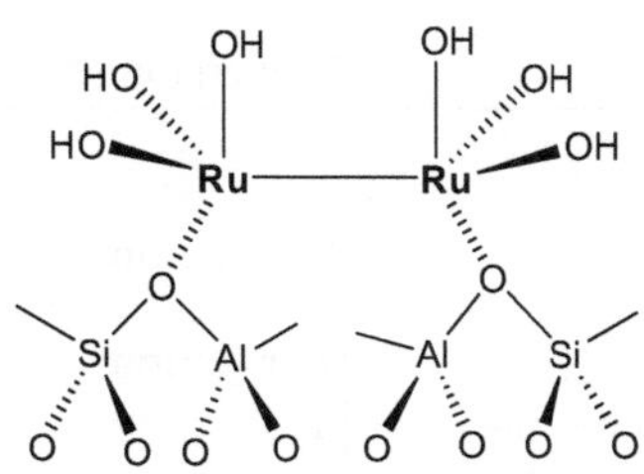

Figure 3.6 Local structure of zeolite-confined (ZC) Ru obtained from Ru K-edge EXAFS analyses.

the Ru^{3+} ions within the cavities of the zeolite with borohydride ions in aqueous solution, all at room temperature, and (iii) drying the isolated samples under aerobic conditions at 100 ± 1.0 °C.[63]

Only a few Ru/C-catalyzed oxidations of alcohols have been developed so far; an example is the oxidation of pantoyllactone to give the corresponding diketone by bubbling O_2 through the lactone at 170 °C (Equation 3.5).[64] More recently, a mild and efficient Ru/C-catalyzed aerobic oxidation of a variety of alcohols was reported.[65] Combination of Ru/C and cyclodextrins was effective for the aerobic oxidation of alcohols in water under neutral conditions.[66] $RuCl_2(PPh_3)_3$ on active carbon catalytically cleaved *vic*-diols to give aldehydes with molecular O_2 in $PhCF_3$ (Table 3.1),[67] and $[RuCl_2(p\text{-cymene})]_2$ on active carbon is a reusable catalyst for the aerobic oxidation of general alcohols to the corresponding aldehydes and ketones.[68]

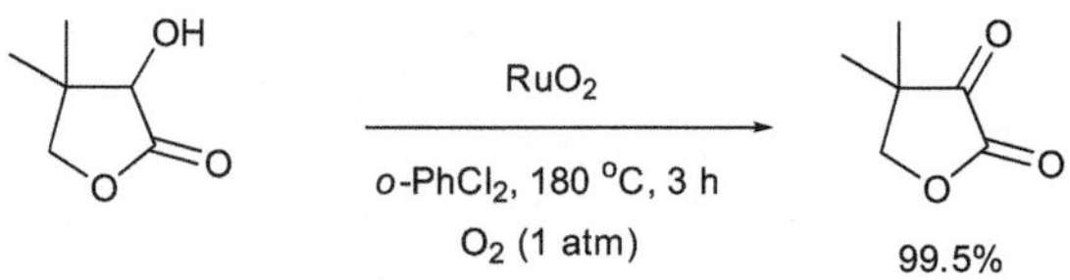

(3.5)

A polymer-incarcerated ruthenium catalyst (PI Ru) (Figure 3.7), prepared from a polystyrene-based copolymer with $RuCl_3 \cdot nH_2O$, showed high catalytic activity for the aerobic oxidation of alcohols to the corresponding aldehydes and ketones in the presence of a catalytic amount of TEMPO.[69] No leaching of ruthenium was observed and the catalyst could be recovered and reused at least 10 times without a decrease in activity.

An efficient polymer-anchored Ru(III) catalyst for the aerobic oxidation of benzyl alcohol was synthesized by sequential attachment of ethylenediamine, salicylaldehyde and $RuCl_3 \cdot nH_2O$ to chloromethylated styrene–divinylbenzene copolymer with 8% cross-linking, which was characterized

Table 3.1 Ruthenium-catalyzed aerobic oxidation of 1,2-octanediol.

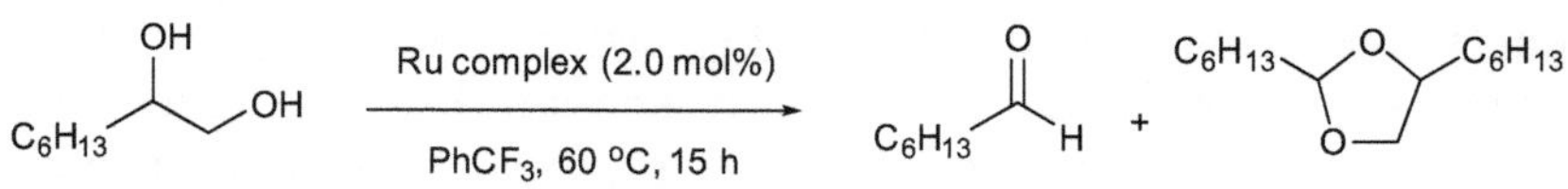

1 2 3

Entry	Ru complex	Conversion (%)	Yield (%) 2	Yield (%) 3
1	TPAP	20	49	Not detected
2	$RuCl_2(PPh_3)_3$	23	Not detected	48
3	$Ru_3(CO)_{12}$	No reaction		
4	$RuCl_3 \cdot nH_2O$	3.9	Not detected	42
5	$RuO_2 \cdot xH_2O$	No reaction		
6	$^{4\,wt\%}RuCl_2(PPh_3)_3/C_{\text{Wako Pure}}$	92	77	9.5
7	$^{4\,wt\%}RuCl_2(PPh_3)_3/C_{\text{Kurare BP-25}}$	87	52	9.0
8	$^{4\,wt\%}RuCl_2(PPh_3)_3/C_{\text{Kurare coal GLC}}$	47	77	3.9
9	$TPAP/C_{\text{Wako Pure}}$	43	74	Not detected

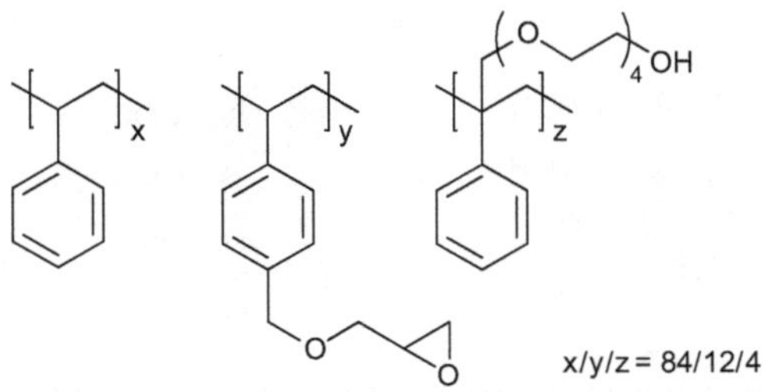

Figure 3.7 Structure of copolymer for encapsulating ruthenium species (PI).

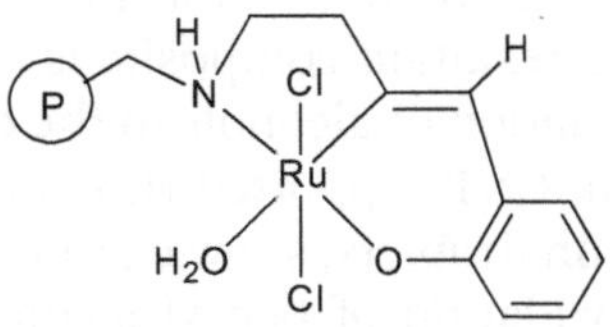

Figure 3.8 Structure of polymer-anchored Ru(III) complex.

by FTIR, reflectance UV–VIS, ESR, ESCA, SEM and TGA techniques (Figure 3.8).[70]

Catalysis by polyoxometallates has attracted much attention because of their strong acidity and redox properties. Another attractive aspect of polyoxometallates in catalysis is their inherent stability towards oxygen donors.[71,72] $RuCl_3 \cdot nH_2O$ successfully reacts with the lacunary silicotungstate $[SiW_{11}O_{39}]^{8-}$ in an organic solvent to give an Ru^{3+}-substituted silicotungstate, which acts as a heterogeneous catalyst for the aerobic oxidation of a wide range of alcohols and also alkanes.[73] An ionic compound, $[RuCl_4(L)_2]_3PW_{12}O_{40}$, containing the Ru(III)-complex cation and α-Keggin-type phosphotungstate anion was synthesized (Equation 3.6) and was found to be an efficient, reusable catalyst for aerobic oxidations of alcohols without bases and/or a nitroxyl radical.[74]

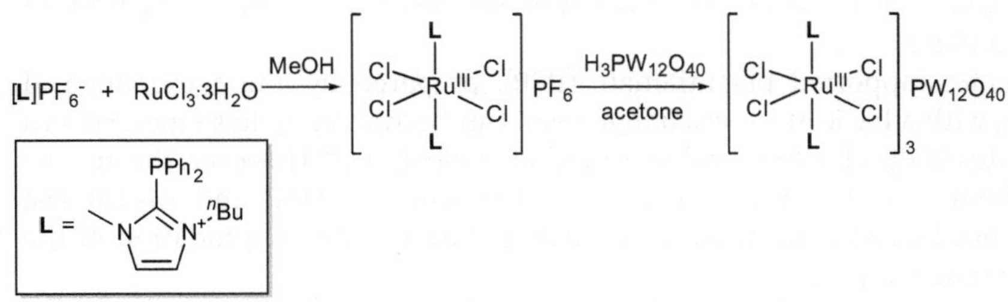

(3.6)

Compared with traditional catalyst supports, carbon nanotubes (CNTs) with high external surface area and aspect ratio display unusual behaviors, such as being able to increase significantly the contact surface between the reactants and the active sites of catalysts and to minimize greatly the diffusion limitations.[75] Herringbone CNT-supported RuO_2 hydrate nanoparticles can be readily prepared by a homogeneous oxidation–precipitation

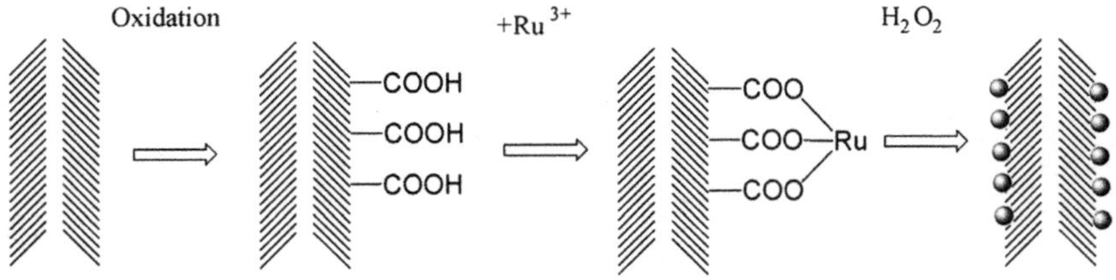

Scheme 3.3 Homogeneous oxidation precipitation of RuO_2 on herringbone CNTs.

(HOP) method using H_2O_2 as both oxidant and precipitant at room temperature (Scheme 3.3). The resultant composite is a very active and selective catalyst for the aerobic oxidation of alcohols to the corresponding aldehydes and ketones.[76] In addition, CNT-supported ruthenium catalysts, assembled at the interfaces of emulsion droplets, show excellent activity, selectivity and stability for the aerobic oxidation of benzyl alcohol to benzaldehyde with molecular O_2 or air in the presence of H_2O.[77] After the reactions, the catalysts can be easily separated and recycled by sedimentation. The activity of RuO_2/CNT for the oxidation of benzyl alcohol can be predicted by the specific capacitance (SC) of RuO_2, suggesting that highly active RuO_2 nanocatalysts are also good supercapacitor materials with SC high up to 1500 $F g_{RuO_2}^{-1}$.[78]

White and co-workers synthesized RuO_2 nanoclusters (1 nm) in the supercages of faujasite (FAU) zeolite (RuO_2/FAU) and they showed high activity and selectivity for the aerobic oxidation of various alcohols.[79] Although bulk RuO_2 showed low activity (16% yield), benzyl alcohol was oxidized quantitatively to benzaldehyde by RuO_2/FAU (7.8 mol%) catalyst under the same reaction conditions. No decrease in yield was observed after the fifth run when using the recovered catalyst. Another interesting feature of the zeolite-based catalysts derived from their uniform pore size is that they behave as a shape-selective catalyst. It was shown that a competitive reaction of benzyl alcohol over 9-hydroxyfluorene resulted in no formation of 9-fluorenone.

Polymer-supported perruthenate (PSP), prepared by anion exchange of $KRuO_4$ with a basic anion-exchange resin (Amberlyst A-26), has emerged as a versatile catalyst for the aerobic oxidation of alcohols.[80] However, the activity was about four times lower than that of homogeneous TPAP and this catalyst could not be recycled, which was attributed to oxidative degradation of the polystyrene support.

A heterogeneous TPAP catalyst that could be recycled successfully and displayed no leaching was developed by tethering the tetraalkylammonium perruthenate to the internal surface of mesoporous silica (MCM-41).[81,82] It was shown to catalyze the selective aerobic oxidation of primary and secondary allylic and benzylic alcohols, but surprisingly, both cyclohexanol and cyclohexenol were unreactive, although these substrates can easily be accommodated in the pores of MCM-41 (Figure 3.9).

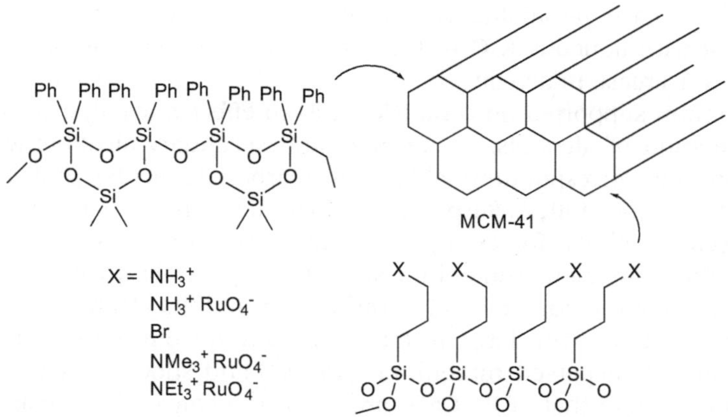

Figure 3.9 Structure of TPAP-tethered MCM-41.

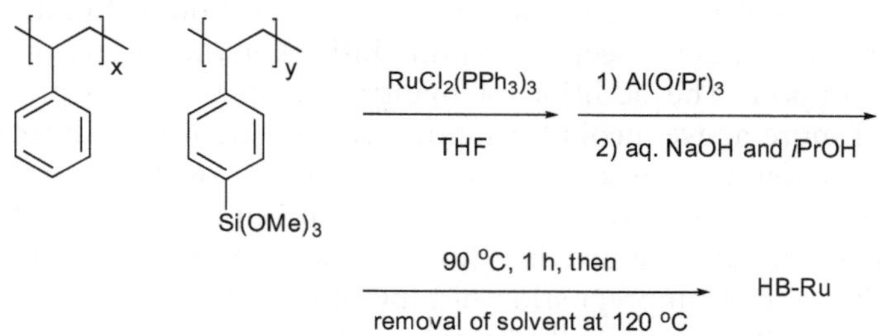

Scheme 3.4 Preparation of HB-Ru.

Another variation on this theme involves straightforward doping of methyl-modified silica, referred to as ormosil, with tetrapropylammonium perruthenate (TPAP) *via* the sol–gel process.[83] A serious disadvantage of this system is the low TOF (1.0 and 1.8 h^{-1}) observed for primary aliphatic alcohols and allylic alcohols, respectively. Sol–gel-encapsulated TPAP (SG-TPAP) was developed and it was found that the surface hydrophobicity and the amount of water and methanol used were key to achieving high catalytic performance. The optimized catalyst showed higher activity than the unsupported TPAP and it could be reused at least six times. To broaden the application range of SG-TPAP, other reaction conditions were investigated, such as conditions of supercritical carbon dioxide ($scCO_2$) and the introduction of ionic liquids or a fluoroalkyl chain to the silica matrix of SG-TPAP.[84] A reusable heterogeneous ruthenium catalyst, organic–inorganic hybrid ruthenium (HB-Ru), was prepared from $RuCl_2(PPh_3)_3$ as the metal source and a copolymer bearing a trimethoxysilyl functionality for the heterogenization of the catalyst (Scheme 3.4). The sol–gel process was employed to initiate hydrolysis by adding aqueous NaOH, followed by polycondensation (cross-linking) to give HB-Ru.[85] The treatment of 4-methylbenzyl alcohol with HB-Ru catalyst (5 mol%) under an O_2

atmosphere gave *p*-tolualdehyde in 97% yield without additives. After treatment with aqueous K_2CO_3, the catalyst was reused at least five times without a decrease in activity.

Ruthenium supported on ceria (CeO_2) is an effective catalyst for the aerobic oxidation of alcohols.[86,87] Primary and secondary alcohols were oxidized to the corresponding aldehydes (carboxylic acids) and ketones, respectively, above 140 °C. Surprisingly, allylic alcohols, such as geraniol and some cyclic alcohols, for example, menthol, were unreactive. The results suggest that low-valent ruthenium species are possibly involved and that coordination of the double bond to ruthenium inhibits alcohol oxidation. A RuO_2/CeO_2 catalyst with large surface area derived from meso- and macropores had well-dispersed ruthenium species[88] and showed high activities, much higher than that of Ru/CeO_2 prepared by a conventional coprecipitation method. A trimetallic $Ru/CeO_2/CoO(OH)$ compound acted as a highly efficient catalyst for the aerobic oxidation of various alcohols to the corresponding carbonyls, including less-reactive primary alcohols.[89,90] Characterization of the catalyst using the K-edge X-ray absorption technique revealed that a monomeric Ru(IV) species was immobilized on the surface of the CeO_2/CoO(OH) support. The addition of manganese species as a promoter resulted in a more active alcohol oxidation catalyst than the comparable cobalt-doped material. The application of Ru/Mn/Ce oxides on redox-active supports, such as TiO_2 or CeO_2, by deposition–precipitation caused a further improvement in catalytic activity. Similarly, a $Ru/MnO_x/CeO_2$ catalyst was prepared by wetness impregnation of MnO_x/CeO_2 precatalyst with an aqueous solution of $RuCl_3 \cdot nH_2O$.[91] The oxidation of alcohols to aldehydes and ketones proceeded smoothly with this catalyst at 300 K under an O_2 atmosphere.

Ruthenium-exchanged hydrotalcites (Ru-HT) were found to be efficient heterogeneous catalysts for the aerobic oxidation of reactive allylic and benzylic alcohols.[92,93] Hydrotalcites are layered anionic clays consisting of a cationic brucite layer with anions (hydroxide or carbonate) situated in the interlayer region. Various cations, such as Ru, Fe, Ni, Mn, V and Cr, can be introduced into the brucite layer by ion exchange. Among them, Ru-HT with the formula $Mg_6Al_2Ru_{0.5}(OH)_{16}CO_3$ showed the highest activity; it was prepared by treating an aqueous solution of $RuCl_3 \cdot nH_2O$, $MgCl_2 \cdot 6H_2O$ and $AlCl_3 \cdot H_2O$ with a solution of NaOH and Na_2CO_3 followed by heating (60 °C, 18 h), filtration, washing with water and drying (110 °C, 12 h). The catalytic activity of Ru-HT was significantly enhanced by further introduction of cobalt(II) into the brucite layer.[94] For example, cinnamyl alcohol underwent complete conversion in 40 min at 60 °C in the presence of Ru/Co-HT, compared with 31% conversion under the same reaction conditions with Ru-HT. The trimetallic Ru/Mn_2-HT was prepared, consisting of Ru^{IV} and Mn^{IV} cations on the HT surface (Figure 3.10).[95] In the absence of Ru, no catalytic activity of Mn_2-HT was observed. Benzyl alcohol was smoothly oxidized to benzaldehyde in the presence of $RuMn_2$-HT (3 mol%) catalyst and the catalyst was reused successfully for a further three runs.

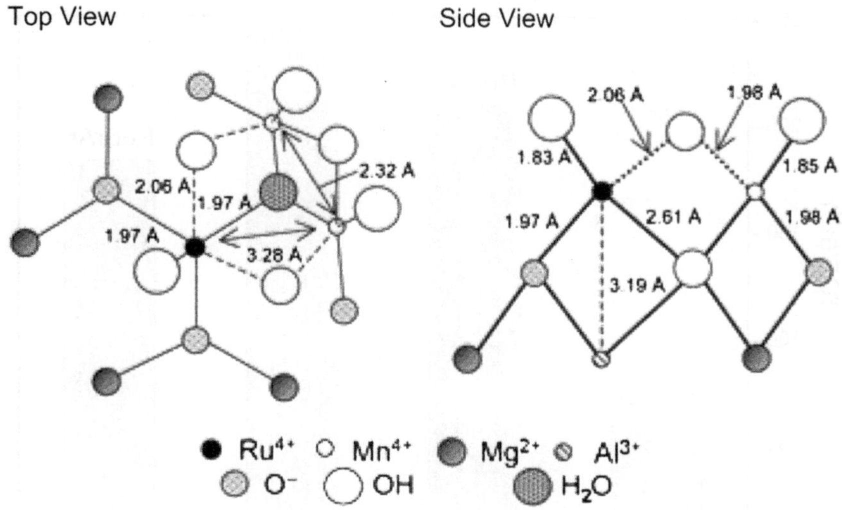

Figure 3.10 Proposed structure of the heterotrimetallic Ru/Mn_2 species on HT.

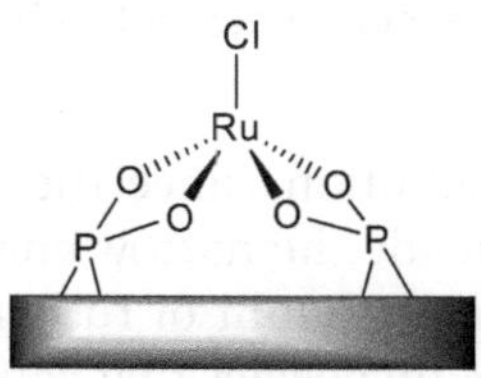

Figure 3.11 Structure of Ru^{III}–hydroxyapatite (Ru-HAP).

Kaneda and co-workers developed a catalyst based on monomeric ruthenium cations on the surface of hydroxyapatite (Ru-HAP) with the formula $(RuCl)_{10}(PO_4)_6(OH)_2$, which could also be recycled and displayed a reasonable substrate scope in aerobic oxidations, although a high loading of ruthenium (17 mol%) was required for the success of the reaction.[96,97] Examinations of the structure by X-ray absorption near-edge structure (XANES), k^3-weighted extended X-ray absorption fine structure (EXAFS) and other analytical techniques were conducted to estimate the environment of Ru^{III} on HAP (Figure 3.11).

The nanorod crystallites of ruthenium-grafted halogenous hydroxyapatite (Ru-XAP, X = F, Cl or Br) were developed as highly efficient catalysts for the aerobic oxidation of alcohols and, compared with Ru-HAP, the existence of F, Cl and Br in the hydroxyapatite dramatically enhanced the catalytic activity, with a high selectivity of more than 99%.[98] In particular, the Ru-ClAP and Ru-FAP catalysts showed excellent catalytic activity with TOFs of ~333 and 210 h^{-1}, respectively, which were nearly three times higher than those with Ru-HAP (Figure 3.12). The incorporation of a metal promoter in the HAP matrix before the introduction of ruthenium dramatically improved the

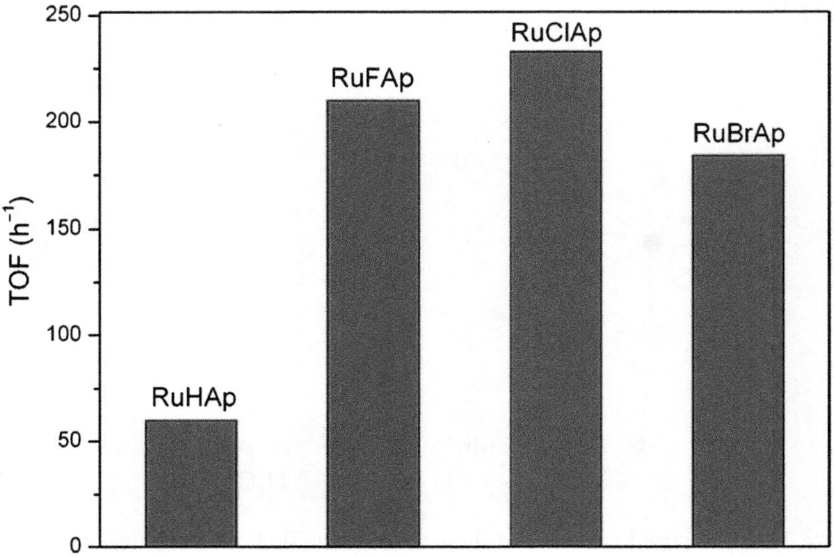

Figure 3.12 Discriminative catalytic performance of the prepared catalysts in the aerobic oxidation of benzyl alcohol to benzaldehyde.

catalytic performance.[99] In addition, since the promoter ion, Co^{II} or Pb^{II}, occupies the 'hidden' sites inside the narrow channels where the substrates are difficult to access, the total amount of ruthenium was decreased by the pretreatment and the ruthenium species that can be involved in the reaction was predominantly located on the surface. Compared with Ru-HAP, a high TOF was observed for RuCo-HAP (1.7 mol%) catalyst.[100]

The experimentally proposed reaction mechanism for the aerobic alcohol oxidation by supported ruthenium hydroxide catalysts [$Ru(OH)_x$/support, support = TiO_2 or Al_2O_3] was investigated theoretically by means of *ab initio* quantum chemistry calculations with model catalysts '$Ru(OH)_3(OH_2)_3$' and '$RuCl_3(OH_2)_3$' for $Ru(OH)_x$/support and $RuCl_x$/support, respectively.[101] The experimentally proposed alcoholate formation and β-hydride elimination steps could be verified. In addition, the structure of ruthenium in Ru/HAP was investigated using infrared spectroscopy in the attenuated total reflection mode (ATR-IR) (Figure 3.13).[102] The spectroscopic data support the existence of a hydrated RuO_x-like phase, which was identified by overtone signals of Ru–O bonds at $\sim$1850 cm^{-1} and is likely organized as a two-dimensional phase on the apatite.

Ruthenium supported on alumina [$Ru(OH)_x/Al_2O_3$] was developed by Yamaguchi and Mizuno as a powerful and recyclable catalyst for the selective aerobic oxidation of both activated and non-activated alcohols.[103–105] Benzyl alcohol was converted into benzaldehyde in over 99% yield and the subsequent six runs with the recovered catalyst gave benzaldehyde quantitatively. The reaction hardly proceeded in the presence of $Ru(OH)_3$ and

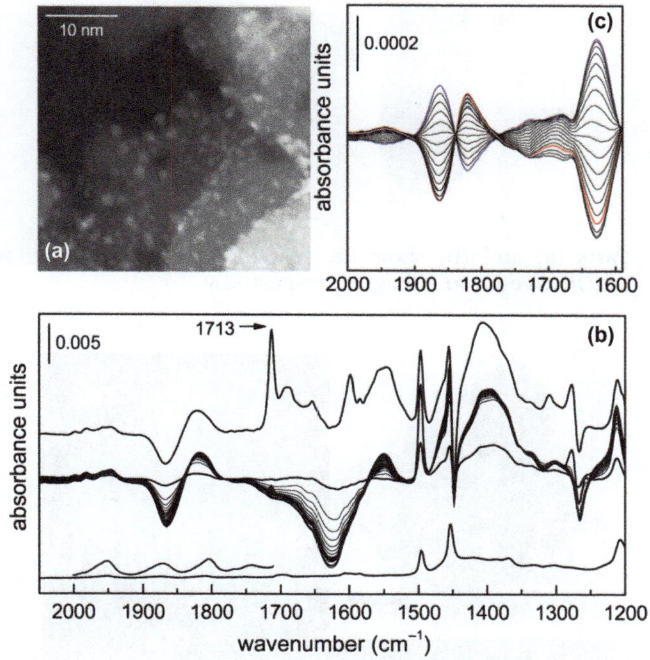

Figure 3.13 (a) HAADF-STEM image of the as-prepared RuHAp; (b) time-dependent ATR-IR spectra of the as prepared RuHAp in contact with an Ar-saturated benzyl alcohol solution (0.02 M) and (offset) the ATR-IR spectrum of RuHAp after contact for 1 h with the reaction mixture under stopped-flow conditions; ATR-IR spectrum of neat benzyl alcohol (bottom); (c) selected phase-resolved spectra with 10° phase angle difference obtained during an MES (Modulation Excitation Spectroscopy) experiment in which the feed was periodically changed between H_2- and O_2-saturated solvent. Conditions: cyclohexane solvent, benzyl alcohol 0.02 M, O_2 (H_2), 333 K, flow rate 0.6 mL min^{-1}.

anhydrous RuO_2. TOFs ranging from 4 h^{-1} (for secondary allylic alcohols) to 18^{-1} (for 2-octanol) were observed in trifluorotoluene, and in the solvent-free oxidation of 2-octanol at 150 °C a TOF of 300 h^{-1} was observed. Since $Ru(OH)_x$ has a one-dimensional chain-like core structure,[106] and dehydrative condensation easily occurs upon heating (>100 °C),[107] dispersion of $Ru(OH)_x$ on metal oxide supports, such as Al_2O_3 and TiO_2,[108,109] is needed to control both the structure and the size of $Ru(OH)_x$. Mizuno and co-workers also synthesized a ruthenium hydroxide species on magnetite [$Ru(OH)_x$/Fe_3O_4].[110] 1-Phenylethanol was quantitatively oxidized to acetophenone and the collection of the catalyst on the wall of the reaction vessel was achieved by using an external permanent magnet (Figure 3.14). Treatment with aqueous NaOH was needed for the second run and a recycling experiment under the same reaction conditions gave a slightly lower yield of acetophenone (93%).

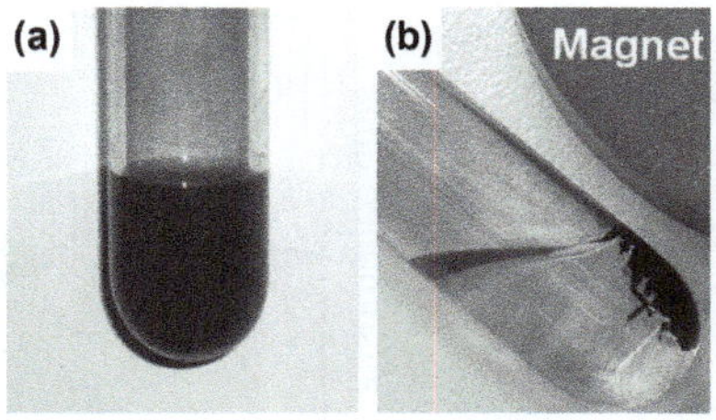

Figure 3.14 Parts (a) and (b) show the reaction mixture and the response of $Ru(OH)_x/Fe_3O_4$ to a magnet, respectively.

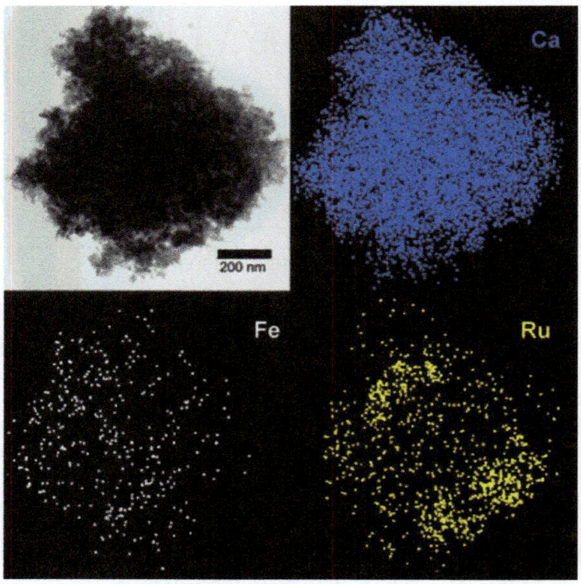

Figure 3.15 TEM image of RuHAP–γ-Fe_2O_3 and STEM elemental mappings.

Kaneda and co-workers developed magnetic γ-Fe_2O_3-containing ruthenium hydroxyapatite (Ru/HAP–γ-Fe_2O_3), which was found to be an effective catalyst for the oxidation of alcohols under atmospheric pressure of molecular O_2.[111] γ-Fe_2O_3 nanocrystallites were dispersed in the HAP matrix and a monomeric ruthenium cation exists uniformly on the outer surface (Figure 3.15). The oxidation state of ruthenium on HAP–γ-Fe_2O_3 was found to be +4, which is different from that in Ru/HAP (+3). The activity of Ru/HAP–γ-Fe_2O_3 was dramatically improved over Ru/HAP and the former was found to be applicable to various alcohols, including allylic, benzylic, aliphatic and heterocyclic alcohols. Compared with Ru/HAP (TOF 2 h^{-1}) and RuCo/HAP (TOF 78 h^{-1}), Ru/HAP–γ-Fe_2O_3 showed a higher catalytic performance (TOF 196 h^{-1}).

The activities of various heterogeneous ruthenium catalysts for the aerobic oxidation of benzyl alcohol to benzaldehyde are summarized in Table 3.2.

Table 3.2 Ruthenium-catalyzed aerobic oxidation of benzyl alcohol to benzaldehyde.

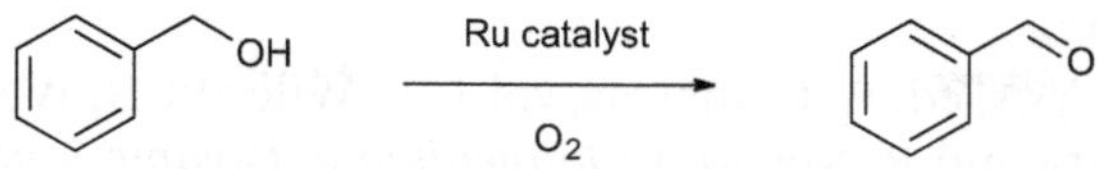

Catalyst	T (K)	Solvent	TOF (h^{-1})	Yield (%)	Selectivity (%)
1.4% Ru/Al_2O_3	356	$PhCF_3$	40	>98	>99
$RuO_2 \cdot xH_2O$	353	$PhCH_3$	1.4	16	100
10% RuO_2/FAU zeolite	353	$PhCH_3$	8.5	>99	>99
$Ru_{0.35}MnFe_{1.5}Cu_{0.15}O_x$	295	$PhCH_3$	3.4	95	100
$Ru_{0.3}Co_2CeO_x$	333	$PhCF_3$	10	>99	>99
Ru–Mg–Al-hydrotalcite	333	$PhCH_3$	1.1	95	97
Ru–Co–Al-hydrotalcite	333	$PhCH_3$	9.3	96	96
Ru^{3+}–hydroxyapatite	353	$PhCH_3$	2	>99	>99
$[RuCl_2(p\text{-cymene})]_2/C$	383	$PhCH_3$	4.7	92	–
RuO_4^-/polymer	348	$PhCH_3$	20	>95	100
$NR_3^+RuO_4^-$/MCM-41	353	$PhCH_3$	170	100	100
$Pr_4N^+RuO_4^-$/ormosil (SiO_2)	348	$PhCH_3$	13	100	100

The selective oxidation of various alcohols to their corresponding aldehydes and ketones was achieved by ruthenium species stabilized on nanocrystalline magnesium oxide (NAP-MgO) by the incorporation of choline hydroxide, a basic ionic liquid, with excellent yields.[112] The catalyst can be used for four cycles with almost constant activity.

3.3 Concluding Remarks

The current trend towards the production of chemicals from feedstocks based on renewable resources rather than fossil fuels such as oil or natural gas has also led to a need for effective methods for the conversion of carbohydrates, from glycerol to polysaccharides, to commercially valuable products. One would expect ruthenium to be a good candidate for catalyzing selective and environmentally benign aerobic oxidation reactions of alcohols and also amines, alkenes and alkanes and, as reviewed in this chapter, considerable progress has been made in the development of synthetically useful ruthenium-catalyzed oxidation reactions. However, there are still many goals to be achieved and it remains a fascinating and challenging area of research.

References

1. I. W. C. E. Arends and R. A. Sheldon, in *Modern Oxidation Methods*, ed. J.-E. Bäckvall, Wiley-VCH, Weinheim, 2004, pp. 83–118.
2. I. W. C. E. Arends and R. A. Sheldon, in *Modern Oxidation Methods*, ed. J.-E. Bäckvall, Wiley-VCH, Weinheim, 2nd edn, 2010, pp. 147–185.
3. *Advances in Oxygenated Processes*, ed. A. L. Baumstark, JAI Press, London, 1995.

4. M. B. Smith and J. March, *March's Advanced Organic Chemistry: Reactions, Mechanisms and Structure*, Wiley, Hoboken, NJ, 6th edn, 2007, pp. 1703–1786.
5. *Green Catalysis*, ed. P. T. Anastas, vol 1–3 , Wiley-VCH, Weinheim, 2009.
6. M. Kitamura and R. Noyori, in *Ruthenium in Organic Synthesis*, ed. S.-I. Murahashi, Wiley-VCH, Weinheim, 2004, pp. 3–52.
7. S.-I. Murahashi and N. Komiya, in *Ruthenium in Organic Synthesis*, ed. S.-I. Murahashi, Wiley-VCH, Weinheim, 2004, pp. 53–93.
8. I. W. C. E. Arends, T. Kodama and R. A. Sheldon, in *Ruthenium Catalysts and Fine Chemistry*, ed. C. Bruneau and P. H. Dixneuf, Springer, Berlin, 2004, pp. 277–320.
9. *Oxidation in Organic Chemistry, Part B*, ed. W. S. Trahanovsky, Academic Press, New York, 1973.
10. Y. Tsuji, T. Ohta, T. Ido, H. Minbu and Y. Watanabe, *J. Organomet. Chem.*, 1984, **270**, 333.
11. S.-I. Murahashi and T. Naota, *Synthesis*, 1993, 433.
12. K. B. Sharpless, K. Akashi and K. Oshima, *Tetrahedron Lett.*, 1976, **17**, 2503.
13. P. Müller and J. Gody, *Tetrahedron Lett.*, 1981, **22**, 2361.
14. Y. Yamamoto, H. Suzuki and Y. Moro-oka, *Tetrahedron Lett.*, 1985, **26**, 2107.
15. S. Kanemoto, H. Tomioka, K. Oshima and H. Nozaki, *Bull. Chem. Soc. Jpn.*, 1986, **59**, 105.
16. F. Shi, M. K. Tse and M. Beller, *Chem. Asian J.*, 2007, **2**, 411.
17. J. Piera and J.-E. Bäckvall, *Angew. Chem. Int. Ed.*, 2008, **47**, 3506.
18. T. Punniyamurthy, S. Velusamy and J. Iqbal, *Chem. Rev.*, 2005, **105**, 2329.
19. T. Matsumoto, M. Ueno, N. Wang and S. Kobayashi, *Chem. Asian J.*, 2008, **3**, 196.
20. P. T. Anastas and J. C. Warner, *Green Chemistry: Theory and Practice*, Oxford University Press, London, 1998.
21. R. A. Sheldon, *Green Chem.*, 2000, **2**, G1.
22. T. Kondo, *Bull. Chem. Soc. Jpn.*, 2011, **84**, 441.
23. R. A. Sheldon, I. W. C. E. Arends and A. Dijksman, *Catal. Today*, 2000, **57**, 157.
24. R. Tang, S. E. Diamond, N. Neary and F. Mares, *J. Chem. Soc., Chem. Commun.*, 1978, 562.
25. M. Matsumoto and S. Ito, *J. Chem. Soc., Chem. Commun.*, 1981, 907.
26. C. Bilgrien, S. Davis and R. S. Drago, *J. Am. Chem. Soc.*, 1987, **109**, 3786.
27. R. Sheldon, *Chem. Commun.*, 2001, 2399.
28. A. Wolfson, S. Wuyts, D. E. De Vos, I. F. J. Vankelecom and P. A. Jacobs, *Tetrahedron Lett.*, 2002, **43**, 8107.
29. J.-E. Bäckvall, R. L. Chowdhury and U. Karlsson, *J. Chem. Soc., Chem. Commun.*, 1991, 473.
30. G.-Z. Wang, U. Andreasson and J.-E. Bäckvall, *J. Chem. Soc., Chem. Commun.*, 1994, 1037.

31. G. Csjernyik, A. H. Ell, L. Fadini, B. Pugin and J.-E. Bäckvall, *J. Org. Chem.*, 2002, **67**, 1657.
32. A. Hanyu, E. Takezawa, S. Sakaguchi and Y. Ishii, *Tetrahedron Lett.*, 1998, **39**, 5557.
33. A. Dijksman, I. W. C. E. Arends and R. A. Sheldon, *J. Chem. Soc., Chem. Commun.*, 1999, 1591.
34. A. Dijksman, A. Marino-Gonzalez, A. Mairata i Payeras, I. W. C. E. Arends and R. A. Sheldon, *J. Am. Chem. Soc.*, 2001, **123**, 6826.
35. R. A. Sheldon, I. W. C. E. Arends, G.-J. T. Brink and A. Dijksman, *Acc. Chem. Res.*, 2002, **35**, 774.
36. A. Dijksman, I. W. C. E. Arends and R. A. Sheldon, *Platinum Met. Rev.*, 2001, **45**, 15.
37. A. C. Dengel, R. A. Hudson and W. P. Griffith, *Transition Met. Chem.*, 1985, **10**, 98.
38. W. P. Griffith, S. V. Ley, G. P. Whitcombe and A. D. White, *J. Chem. Soc., Chem. Commun.*, 1987, 1625.
39. A. C. Dengel, A. M. El-Hendawy and W. P. Griffith, *Transition Met. Chem.*, 1989, **40**, 230.
40. R. Lenz and S. V. Ley, *J. Chem. Soc., Perkin Trans*, 1997, **1**, 3291.
41. I. E. Marko, P. R. Giles, M. Tsukazaki, I. Chelle-Regnaut, C. J. Urch and S. M. Brown, *J. Am. Chem. Soc.*, 1997, **119**, 12661.
42. M. Lee and S. Chang, *Tetrahedron Lett.*, 2000, **41**, 7507.
43. N. Komiya, T. Nakae, H. Sato and T. Naota, *Chem. Commun.*, 2006, 4829.
44. H. Guo, W.-D. Liu and G. Yin, *Appl. Organomet. Chem.*, 2011, **25**, 836.
45. K. Yu, D. Ye, L. Shu, S. Zhang, Q. Hu and L. Liu, *Synth. Commun.*, 2012, **42**, 2318.
46. T. Kondo, Y. Kimura, T. Kanda, D. Takagi, K. Wada and A. Toshimitsu, *Green Sus. Chem.*, 2011, **1**, 149.
47. P. A. Shapley, N. Zhang, J. L. Allen, D. H. Pool and H.-C. Liang, *J. Am. Chem. Soc.*, 2000, **122**, 1079.
48. J. L. Kuiper and P. A. Shapley, *J. Organomet. Chem.*, 2007, **692**, 1653.
49. C. N. Kato, M. Ono, T. Hino, T. Ohmura and W. Mori, *Catal. Commun.*, 2006, 7, 673.
50. N. Zotova, K. Hellgardt, G. H. Kelsall, A. S. Jessiman and K. K. Hill, *Green Chem.*, 2010, **12**, 2157.
51. J. Zhao, C. Mueck-Lichtenfeld and A. Studer, *Adv. Synth. Catal.*, 2013, **355**, 1098.
52. V. N. Korotchenko, K. Severin and M. R. Gagne, *Org. Biomol. Chem.*, 2008, **6**, 1961.
53. H. Mizoguchi, T. Uchida, K. Ishida and T. Katsuki, *Tetrahedron Lett.*, 2009, **50**, 3432.
54. Y. Nakamura, H. Egami, K. Matsumoto, T. Uchida and T. Katsuki, *Tetrahedron*, 2007, **63**, 6383.
55. H. Shimizu, S. Onitsuka, H. Egami and T. Katsuki, *J. Am. Chem. Soc.*, 2005, **127**, 5396.
56. R. Irie and T. Katsuki, *Chem. Rec.*, 2004, **4**, 96.

57. J. Hagen, *Industrial Catalysis: a Practical Approach*, Wiley-VCH, Weinheim, 1999.
58. T. Mallat and A. Baiker, *Chem. Rev.*, 2004, **104**, 3037.
59. *Fine Chemicals Through Heterogeneous Catalysis*, ed. R. A. Sheldon and H. van Bekkum, Wiley-VCH, Weinheim, 2001.
60. M. Matsumoto and S. Ito, *J. Chem. Soc., Chem. Commun.*, 1981, 907.
61. M. Matsumoto and N. Watanabe, *J. Org. Chem.*, 1984, **49**, 3435.
62. B. Z. Zhan, M. A. White, T. K. Sham, J. A. Pincock, R. J. Doucet, K. V. R. Rao, K. N. Robertson and T. S. Cameron, *J. Am. Chem. Soc.*, 2003, **125**, 2195.
63. M. Zahmakiran and S. Ozkar, *J. Mater. Chem.*, 2009, **19**, 7112.
64. M. Matsumoto and S. Ito, *Synth. Commun.*, 1984, **14**, 697.
65. S. Mori, M. Takubo, K. Makida, T. Yanase, S. Aoyagi, T. Maegawa, Y. Monguchi and H. Sajiki, *Chem. Commun.*, 2009, 5159.
66. V. K. Akkilagunta, V. P. Reddy and R. R. Kakulapati, *Synlett*, 2010, 2571.
67. E. Takezawa, S. Sakaguchi and Y. Ishii, *Org. Lett.*, 1999, **1**, 713.
68. E. Choi, C. Lee, Y. Na and S. Chang, *Org. Lett.*, 2002, **4**, 2369.
69. T. Matsumoto, M. Ueno, J. Kobayashi, H. Miyamura, Y. Mori and S. Kobayashi, *Adv. Synth. Catal.*, 2007, **349**, 531.
70. M. K. Dalal, M. J. Upadhyay and R. N. Ram, *J. Mol. Catal. A: Chem.*, 1999, **142**, 325.
71. R. Neumann, *Prog. Inorg. Chem.*, 1998, **47**, 317.
72. C. L. Hilland and C. M. Prosser-McCartha, *Coord. Chem. Rev.*, 1995, **143**, 407.
73. K. Yamaguchi and N. Mizuno, *New J. Chem.*, 2002, **26**, 972.
74. S.-S. Wang, J. Zhang, C.-L. Zhou, G. Vo-Thanh and Y. Liu, *Catal. Commun.*, 2012, **28**, 152.
75. S. Iijima, *Nature*, 1991, **354**, 56.
76. X. Fu, H. Yu, F. Peng, H. Wang and Y. Qian, *Appl. Catal. A: Gen.*, 2007, **321**, 190.
77. X. Yang, X. Wang and J. Qiu, *Appl. Catal. A: Gen.*, 2010, **382**, 131.
78. H. Yu, X. Fu, C. Zhou, F. Peng, H. Wang and J. Yang, *Chem. Commun.*, 2009, 2408.
79. B.-Z. Zhan, M. A. White, T.-K. Sham, J. A. Pincock, R. J. Doucet, K. V. R. Rao, K. N. Robertson and T. S. Cameron, *J. Am. Chem. Soc.*, 2003, **125**, 2195.
80. B. Hinzen, R. Lenz and S. V. Ley, *Synthesis*, 1998, 977.
81. M. Pagliaro and R. Ciriminna, *Tetrahedron Lett.*, 2001, **42**, 4511.
82. A. Bleloch, B. F. G. Johnson, S. V. Ley, A. J. Price, D. S. Shephard and A. W. Thomas, *Chem. Commun.*, 1999, 1907.
83. R. Ciriminna and M. Pagliaro, *Chem. Eur. J*, 2003, **9**, 5067.
84. S. Campestrini, M. Carraro, R. Ciriminna, M. Pagliaro and U. Tonellato, *Adv. Synth. Catal.*, 2005, **347**, 825.
85. T. Matsumoto, M. Ueno, N. Wang and S. Kobayashi, *Chem. Asian J.*, 2008, **3**, 239.
86. F. Vocanson, Y. P. Guo, J. L. Namy and H. B. Kagan, *Synth. Commun.*, 1998, **28**, 2577.

87. K. Wada, S. Hosokawa and M. Inoue, *Catal. Surv. Asia*, 2011, **15**, 1.
88. Y. Hayashi, S. Hosokawa, S. Imamura and M. Inoue, *J. Ceram. Soc. Jpn.*, 2007, **115**, 592.
89. H.-B. Ji, T. Mizugaki, K. Ebitani and K. Kaneda, *Tetrahedron Lett.*, 2002, **43**, 7179.
90. K. Ebitani, H.-B. Ji, T. Mizugaki and K. Kaneda, *J. Mol. Catal. A: Chem.*, 2004, **212**, 161.
91. T. Sato and T. Komanoya, *Catal. Commun.*, 2009, **10**, 1095.
92. K. Kaneda, T. Yamashita, T. Matsushita and K. Ebitani, *J. Org. Chem.*, 1998, **63**, 1750.
93. K. Ebitani, T. Mizugaki, K. Mori and K. Kaneda, *Curr. Org. Chem.*, 2006, **10**, 241.
94. T. Matsushita, K. Ebitani and K. Kaneda, *Chem. Commun.*, 1999, 265.
95. K. Ebitani, K. Motokura, T. Mizugaki and K. Kaneda, *Angew. Chem. Int. Ed.*, 2005, **44**, 3423.
96. K. Yamaguchi, K. Mori, T. Mizugaki, K. Ebitani and K. Kaneda, *J. Am. Chem. Soc.*, 2000, **122**, 7144.
97. K. Kaneda, K. Mori, T. Hara, T. Mizugaki and K. Ebitani, *Catal. Surv. Asia*, 2004, **8**, 231.
98. Y. Zhang, J. Wang, J. Yin, K. Zhao, C. Jin, Y. Huang, Z. Jiang and T. Zhang, *J. Phys. Chem. C*, 2010, **114**, 16443.
99. Z. Opre, J.-D. Grunwaldt, M. Maciejewski, D. Ferri, T. Mallat and A. Baiker, *J. Catal.*, 2005, **230**, 406.
100. Z. Opre, J.-D. Grunwaldt, T. Mallat and A. Baiker, *J. Mol. Catal.: Chem.*, 2005, **242**, 224.
101. F. Nikaidou, H. Ushiyama, K. Yamaguchi, K. Yamashita and N. Mizuno, *J. Phys. Chem. C*, 2010, **114**, 10873.
102. C. Mondelli, D. Ferri and A. Baiker, *J. Catal.*, 2008, **258**, 170.
103. K. Yamaguchi and N. Mizuno, *Angew. Chem. Int. Ed.*, 2002, **41**, 4538.
104. K. Yamaguchi and N. Mizuno, *Chem. Eur. J.*, 2003, **9**, 4353.
105. K. Yamaguchi and N. Mizuno, *Synlett*, 2010, 2365.
106. B.-Z. Zhan, M. A. White, T.-K. Sham, J. A. Pincock, R. J. Doucet, K. V. R. Rao, K. N. Robertson and T. S. Cameron, *J. Am. Chem. Soc.*, 2003, **125**, 2195.
107. D. A. McKeown, P. L. Hagans, L. P. L. Carette, A. E. Russell, K. E. Swider and D. R. Rolison, *J. Phys. Chem. B*, 1999, **103**, 4825.
108. K. Yamaguchi, T. Koike, J. W. Kim, Y. Ogasawara and N. Mizuno, *Chem. Eur. J.*, 2008, **14**, 11480.
109. K. Yamaguchi, J. W. Kim, J. He and N. Mizuno, *J. Catal.*, 2009, **268**, 343.
110. M. Kotani, T. Koike, K. Yamaguchi and N. Mizuno, *Green Chem.*, 2006, **8**, 735.
111. K. Mori, S. Kanai, T. Hara, T. Mizugaki, K. Ebitani, K. Jitsukawa and K. Kaneda, *Chem. Mater.*, 2007, **19**, 1249.
112. M. L. Kantam, U. Pal, B. Sreedhar, S. Bhargava, Y. Iwasawa, M. Tada and B. M. Choudary, *Adv. Synth. Catal.*, 2008, **350**, 1225.

CHAPTER 4

Selective Palladium-Catalysed Aerobic Oxidation of Alcohols

LEE J. DURNDELL,[a] ADAM F. LEE,*[a] DAVID S. BAILIE[b] AND MARK J. MULDOON*[b]

[a] European Bioenergy Research Institute, Aston University, Aston Triangle, Birmingham B4 7ET, UK; [b] School of Chemistry and Chemical Engineering, Queen's University Belfast, David Keir Building, Stranmillis Road, Belfast BT9 5AG, UK
*Email: a.f.lee@aston.ac.uk; m.j.muldoon@qub.ac.uk

4.1 General Introduction

Oxidation of alcohols by palladium has been known for many years, with reports dating back to 1828 when Berzelius noted the precipitation of Pd metal when K_2PdCl_4 was refluxed in an ethanol solution.[1] In more recent times, there have been significant efforts aimed at developing efficient and selective palladium catalysts. From both an economic and environmental point of view, molecular oxygen is the ideal terminal oxidant for catalytic oxidations, but using O_2 for Pd-catalysed selective oxidations is not without its challenges. There is now a substantial body of work on Pd-catalysed aerobic oxidations and this chapter gives an overview of the two major approaches: heterogeneous Pd catalysts and homogeneous, ligand-modulated, Pd(II) systems.

RSC Green Chemistry No. 28
Transition Metal Catalysis in Aerobic Alcohol Oxidation
Edited by Francesca Cardona and Camilla Parmeggiani

Published by the Royal Society of Chemistry, www.rsc.org

4.2 Aspects of Heterogeneous Palladium Catalysts in the Selective Oxidation of Alcohols

4.2.1 Introduction

The heterogeneously catalysed, aerobic selective oxidation (selox) of alcohols, carbohydrates and related α,β-unsaturated substrates is particularly demanding owing to the requirement to activate molecular oxygen and C–O bonds in close proximity at a surface in a solid–liquid–gas environment.[2,3] The past decade has seen significant progress in understanding the fundamental mode of action of palladium-derived heterogeneous catalysts for such transformations and the associated reaction pathways and deactivation processes.[4] This insight has been aided by advances in analytical methodologies, notably the development of *in situ* or *operando* (under working conditions) spectroscopic[5–7]/microscopic[8–11] tools able to provide quantitative, spatio-temporal information on structure–function relations of solid catalysts in the liquid and vapour phase. Parallel improvements in inorganic synthetic protocols offer finer control over preparative methods to direct the nanostructure (composition, morphology, size, valence and support architecture) of palladium catalysts[12–14] and thereby enhance activity, selectivity and lifetime in an informed manner.

4.2.2 Practical Considerations for Heterogeneous Selox Catalysts

The scientific, technological and commercial importance of green chemistry presents a significant challenge to traditional selox methods, which employ hazardous and toxic stoichiometric oxidants, including permanganates, chromates and peroxides, with concomitant poor atom efficiencies, and require energy-intensive separation steps to obtain the desired carbonyl or acid product. Alternative heterogeneous catalysts utilizing oxygen must not only offer high activity and selectivity in alcohol selox, but also be scalable in terms of both catalyst synthesis and implementation. For example, continuous flow microreactors have been implemented in both homogeneous and heterogeneous oxidations (Figure 4.1), and offer facile catalyst recovery from feed streams for the latter,[15,16] but their scale-up/out requires complex manifolding and to ensure adequate oxygen dissolution and uniform reactant mixing and delivery.[17,18] Efforts to overcome mass-transport and solubility issues inherent to three-phase palladium-catalysed oxidations have centred around the use of supercritical carbon dioxide ($scCO_2$) to facilitate rapid diffusion of substrates to and products from the active catalyst site at modest temperatures[19] affording enhanced turnover frequencies (TOFs), selectivity and on-stream performance *versus* conventional batch operation in liquid organic solvents.[20–24]

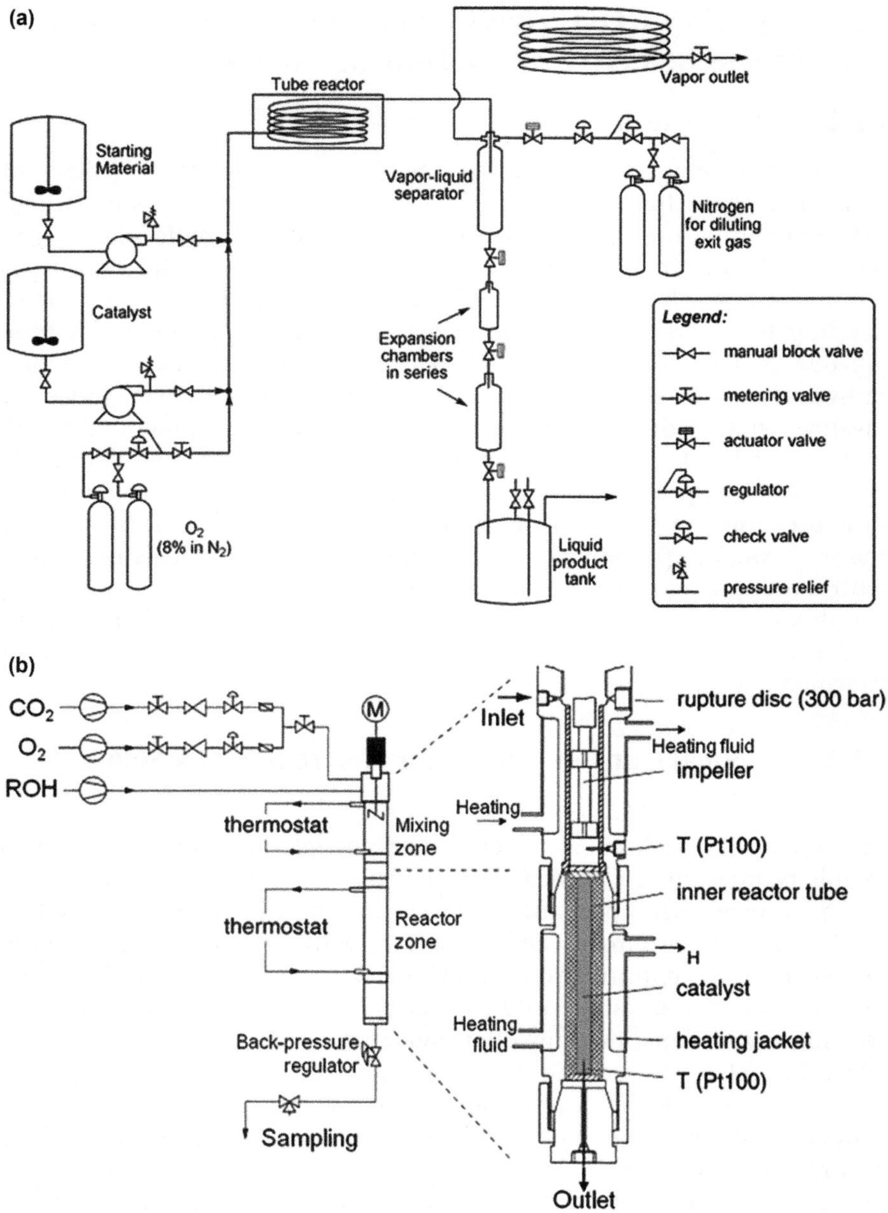

Figure 4.1 Schematic of continuous flow reactors for Pd-catalysed aerobic alcohol oxidations. (a) Employing $Pd(OAc)_2$/pyridine in toluene. Reproduced from Ref. 19 with permission from Elsevier. (b) Employing a Pd/Al_2O_3 packed bed under $scCO_2$.
Adapted with permission from Ref. 21. Copyright 2006 American Chemical Society.

Ultimately, heterogeneous palladium catalysts may offer significant advantages over homogeneous analogues in respect of initial catalyst cost, product separation and metal recovery and recyclability.[25] Catalyst development can therefore no longer be considered simply a matter of reaction kinetics, but as a clean technology where all aspects of process design, such as solvent selection, batch/flow operation, catalyst recovery and waste production and disposal are taken into account.[26] The efficacy of platinum-group metal (PGM) surfaces towards the liquid-phase oxidation of alcohols has been known for over 50 years,[27] and the development of heterogeneous platinum selox catalysts (and more recently coinage metals such as gold)[28,29] are the subject of several reviews.[2,30–32] Hence here we focus on palladium selox catalysis.

4.2.3 Particle Size Effects

In the field of nanocatalysis, it is well known that particle size is an important parameter influencing both activity and selectivity. This reflects the combination of quantum and geometric effects associated with the respective evolution of electronic properties from atomistic to condensed matter band structure and transition from low to high coordination surface atoms, with increasing nanoparticle size and dimensionality. Kaneda and co-workers hypothesized that the unique reactivity of Pd_{2060} clusters supported on titania towards aromatic alcohol selox arose from a distribution of Pd^0, Pd^+ and Pd^{2+} surfaces sites, with π-bonding interactions between the phenyl group and Pd^{2+} species facilitating subsequent oxidative addition of the O–H bond by neighbouring Pd^0 and eventual β-hydride elimination.[33] Surface palladium hydride was then hypothesized to react with oxygen from a neighbouring Pd_2O centre forming H_2O and regenerating the metal site (Figure 4.2). Optimal activity for cinnamyl alcohol selox to cinnamaldehyde coincided with clusters possessing the maximum fraction of Pd^+ character.

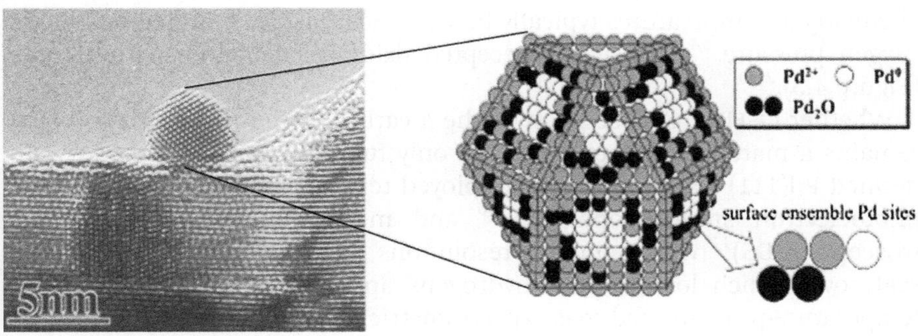

Figure 4.2 HRTEM image of Pd/TiO_2 selox catalyst and the suggested Pd nanoparticle model predicting the presence of different oxidation states and surface environments.
Adapted from Ref. 33 with permission from The Royal Society of Chemistry.

A strict particle size dependence was also reported for the catalytic transformation of benzyl alcohol over Pd nanoparticles dispersed on alumina, SiO_2 and NaX zeolite supports.[34,35] For Pd/NaX and Pd/SiO_2–Al_2O_3, benzyl alcohol selox was maximized for particle sizes between 3 and 5 nm, whereas geraniol and 2-octanol were insensitive to structure. Systematic studies of particle size with cinnamyl and crotyl alcohols over amorphous and mesostructured alumina and silica supports have likewise uncovered pronounced size effects in both initial selox rates and TOFs,[36–39] which increase monotonically with shrinking nanoparticle diameters even down to single atoms.[40] High-angle annular dark-field scanning transmission electron microscopy (HAADF-STEM) measurements show that atomically dispersed palladium exhibits maximum rates towards benzyl, cinnamyl and crotyl alcohols, with selectivities to their corresponding aldehydes of >70%. We revisit the origin of such size effects later in this chapter. The application of colloidal Pd nanoclusters for selox in aqueous media is rather limited[41–43] owing to the high affinity for Pd aggregation and subsequent formation of Pd black, which hinders catalytic performance. However, Lu and co-workers successfully stabilized 3.6 nm Pd nanoclusters using Pluronic P123, an amphiphilic, non-ionic triblock copolymer. In the selective oxidation of benzyl alcohol, 100% aldehyde selectivity and high conversion rates were obtained, even after 13 recycling reactions, indicating no decrease in activity due to sintering and a route to overcome the stability problems which normally plague colloidal Pd nanoparticles.[44]

4.2.4 Surface Reaction Mechanism

The rational design and optimization of active and selective palladium selox catalysts require a microscopic understanding of both the active catalytic species responsible for alcohol and oxygen activation and the associated reaction pathway to the aldehyde product and any competing processes. A key characteristic of palladium is its ability to perform the desired selox chemistry at temperatures typically between 60 and 160 °C and atmospheric oxygen pressure,[2,45] *via* a long-accepted oxidative dehydrogenation route (Figure 4.3).[2,30]

Whether O–H or C–H scission of the α-carbon is the first oxidation step remains a matter of debate, since the only fundamental studies over well-defined Pd(111) surfaces to date employed temperature-programmed X-ray photoelectron spectroscopy (XPS)[46] and metastable de-excitation spectroscopy (MDS)[47] with temporal resolutions on the second–minute timescale over which loss of both hydrogens appeared coincident. However, temperature-programmed mass spectrometric[48] and vibrational[49] studies of unsaturated C_1–C_3 alcohols implicated O–H cleavage and attendant alkoxy formation over Pd single-crystal surfaces.[45,50] It is generally held that the resultant hydrogen adatoms react with dissociatively absorbed oxygen to form water, which immediately desorbs at ambient temperature, thereby shifting the equilibrium to carbonyl formation.[2,30] In any event,

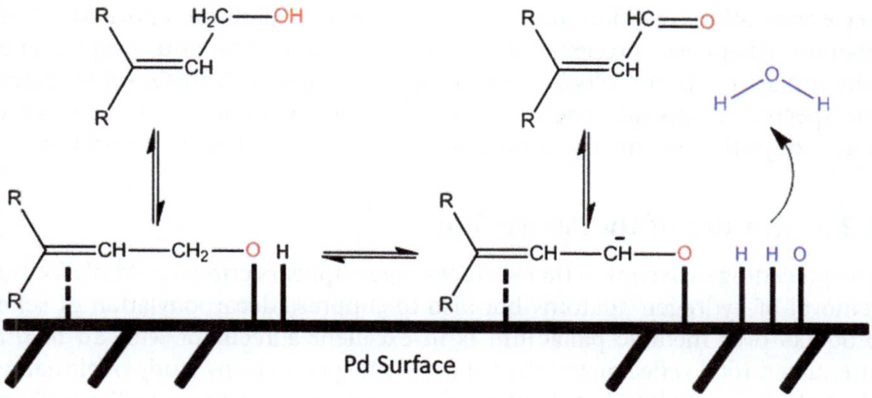

Figure 4.3 Oxidative dehydrogenation of allylic alcohols over palladium surfaces.

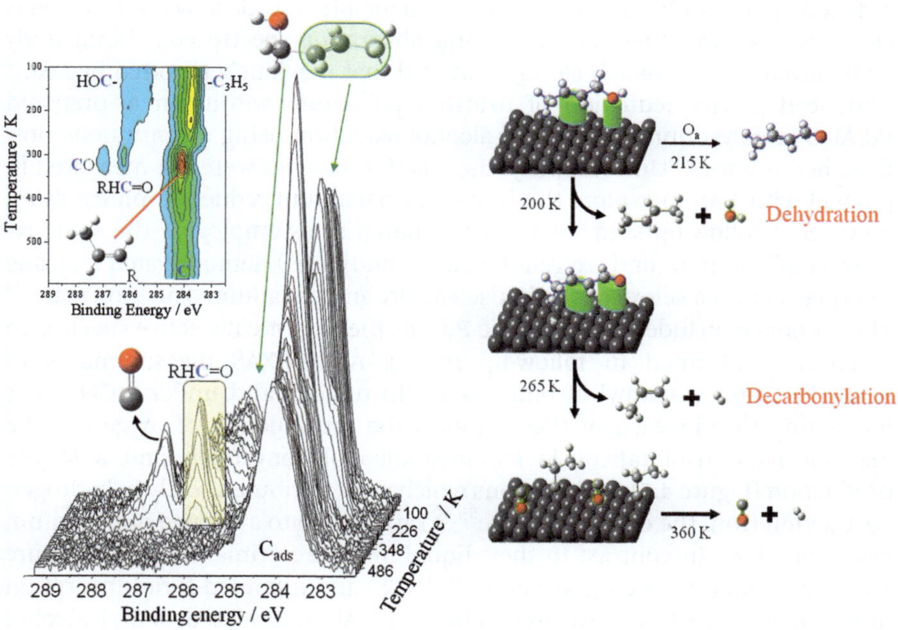

Figure 4.4 Temperature-programmed C 1s XP spectra of a saturated crotyl alcohol adlayer over Pd(111) highlighting primary dehydrogenation pathway and competing decarbonylation.
Reprinted with permission from Ref. 46. Copyright 2007 American Chemical Society.

temperature-programmed XPS measurements of crotyl alcohol over clean Pd(111)[46] surfaces proved that oxidative dehydrogenation to crotonaldehyde occurs at temperatures as low as −60 °C (Figure 4.4), with alcohol dehydration only a minor pathway. These ultra-high-vacuum measurements also revealed that reactively formed crotonaldehyde undergoes a competing decarbonylation reaction over clean, metallic palladium above 0 °C to liberate gas-phase propene and strongly bound CO and propylidene, which may act as site blockers poisoning subsequent catalytic selox cycles. Unexpectedly, pre-adsorbed atomic oxygen switched off undesired decarbonylation chemistry, promoting facile crotonaldehyde desorption.

4.2.5 Nature of the Active Site

The preceding observation that surface oxygen appears critical not only for the removal of hydrogen adatoms but also to suppress decarbonylation of selox products over metallic palladium is in excellent agreement with an *in situ* attenuated total reflectance infrared (ATR-IR) spectroscopy study of cinnamyl alcohol selox over Pd/Al_2O_3 by Baiker and co-workers.[51] In an earlier body of work employing aqueous electrochemical analysis, the same group postulated that oxidative dehydrogenation of alcohols requires PGM catalysts in a reduced state, with 'over-oxidation' responsible for deactivation of selox catalysts.[32] A subsequent *operando* X-ray absorption spectroscopy (XAS) study by Grunwaldt *et al.*, which closely mirrored that previously reported by Lee,[52] evidenced *in situ* reduction of oxidized palladium within an as-prepared Pd/Al_2O_3 catalyst during cinnamyl alcohol oxidation using a continuous flow fixed-bed reactor.[53] Unfortunately, the reaction kinetics were not measured in parallel with XAS to explore the impact of palladium reduction on catalysis; however, a follow-up study of 1-phenylethanol selox employing the same reactor configuration (and oxygen-deficient conditions) demonstrated a strong interplay between selox conversion/selectivity and palladium oxidation state.[54] The authors concluded that metallic Pd was the catalytically active species, an assertion re-affirmed in follow-up *in situ* ATR-IR/XAS measurements of benzyl[55–57] and cinnamyl alcohol[58] selox in toluene and under $scCO_2$. It is interesting that in each of these studies the introduction of oxygen to the reactant feed instantaneously *improved* alcohol conversion and aldehyde production (Figure 4.5), observations which were attributed solely to hydrogen abstraction from the catalyst surface[59,60] rather than to a change in palladium oxidation state. In contrast to their liquid-phase experiments, high-pressure X-ray absorption near edge structure (XANES) and extended X-ray absorption fine structure (EXAFS) measurements on Pd/Al_2O_3-catalysed benzyl alcohol selox in $scCO_2$ led Grunwaldt *et al.* to conclude that maximum activity arose from particles mainly oxidized in the surface/shelf-edge.[61]

In a parallel programme of research, Lee and co-workers set out to characterize systematically the physicochemical properties of palladium nanoparticles as a function of size over non-reducible supports in order to quantify structure–function relations in allylic alcohol selox.[36–40,62,63] The combination of XPS and XAS measurements revealed that freshly prepared alumina-[37,40] and silica-supported[38,62] nanoparticles are prone to oxidation as their diameter falls below $\sim$4 nm, with the fraction of PdO proportional to the support surface area and interconnectivity. Complementary kinetic analyses uncovered a direct correlation between the surface PdO content and

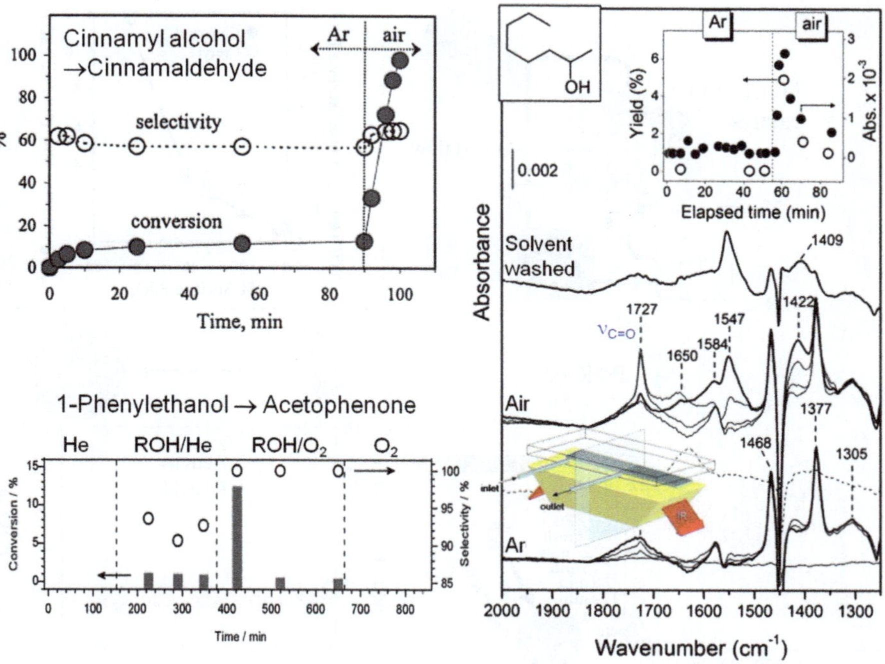

Figure 4.5 Impact of oxygen on the selective oxidation of (top left) cinnamyl alcohol, (bottom left) 1-phenylethanol and (right) 2-octanol.
Adapted from Refs 51, 54 and 57 with permission from Elsevier.

TOFs towards cinnamyl and crotyl alcohol (Figure 4.6).[37,40] In accordance with the observations of Grunwaldt *et al.*, *operando* liquid-phase XAS of Pd/C and Pd/Al_2O_3-SBA-15 catalysts during cinnamyl alcohol selox evidenced *in situ* reduction of PdO (Figure 4.6); however, by virtue of simultaneously measuring the rate of alcohol selox, Lee and co-workers were able to prove that this oxide → metal structural transition was accompanied by coincident deactivation. Together, these findings strongly implicate a (surface) PdO active phase, consistent with surface science predictions that metallic palladium favours aldehyde decarbonylation and consequent self-poisoning by CO and organic residues,[46,64] akin to that reported during fatty acid decarboxylation over Pd/MCF.[65]

In order to establish conclusively whether oxide or metal is responsible for alcohol selox catalysed by dispersed palladium nanoparticles, a multi-dimensional spectroscopic investigation of vapour-phase crotyl alcohol selox was undertaken (since XAS is an averaging technique, a complete understanding of catalyst operation requires multiple analytical techniques).[66–68] Synchronous, time-resolved diffuse reflectance infrared Fourier transform spectroscopy (DRIFTS)/mass spectrometry (MS)/XAS measurements of supported and colloidal palladium were performed in a bespoke environmental cell[69] to interrogate simultaneously adsorbates on the catalyst surface, Pd oxidation state and reactivity under transient conditions in the absence of

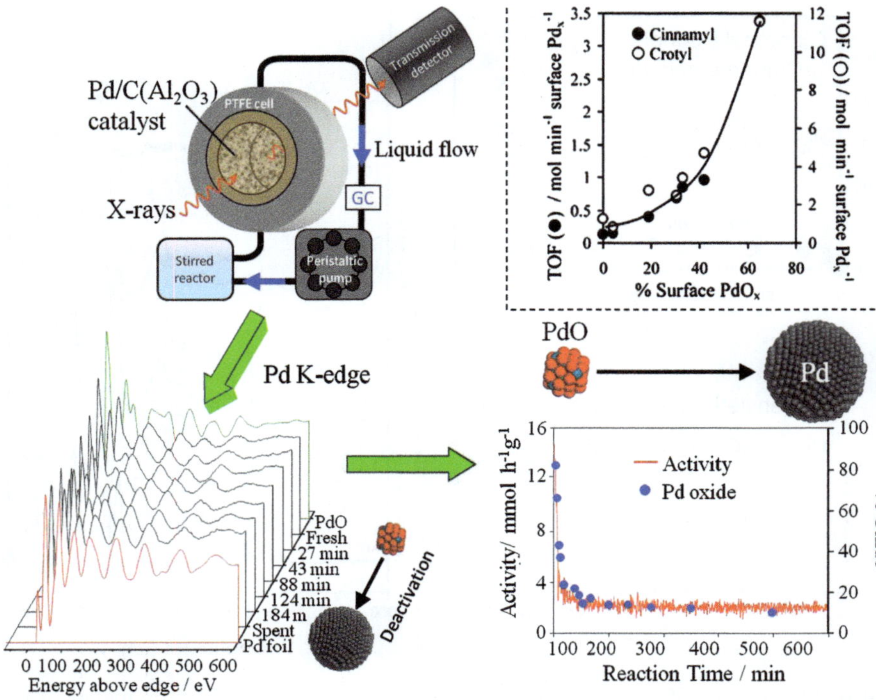

Figure 4.6 (Top right) dependence of allylic alcohol selox rate upon surface PdO; (top left) schematic of *operando* liquid-phase reactor; (bottom left) evolution of Pd K-edge XAS of Pd/Al_2O_3 catalyst during cinnamyl alcohol aerobic selox; (bottom right) temporal correspondence between Pd oxidation state and selox activity in cinnamyl alcohol selox.
Adapted from Refs 36 and 37 with permission from The Royal Society of Chemistry.

competitive solvent effects.[70,71] Under mild reaction temperatures, palladium nanoparticles were partially oxidized and unperturbed by exposure to sequential alcohol or oxygen pulses (Figure 4.7). Crotonaldehyde formed immediately upon contact of crotyl alcohol with the oxide surface, but only desorbed upon oxygen co-adsorption. Higher reaction temperatures induced PdO reduction in response to crotyl alcohol exposure, mirroring that observed during liquid-phase selox; however, this reduction could be fully reversed by subsequent oxygen exposure. Such reactant-induced restructuring was exhibited by all palladium nanoparticles, but the magnitude was inversely proportional to particle size.[72] These dynamic measurements decoupled the relative reactivity of palladium oxide from metal, revealing that PdO favoured crotyl alcohol selox to crotonaldehyde and crotonic acid, whereas metallic palladium drove secondary decarbonylation to propene and CO in accordance with surface science predictions.[46]

Recent ambient-pressure XPS investigations of crotyl alcohol–O_2 gas mixtures over metallic and oxidized Pd(111) single-crystal surfaces

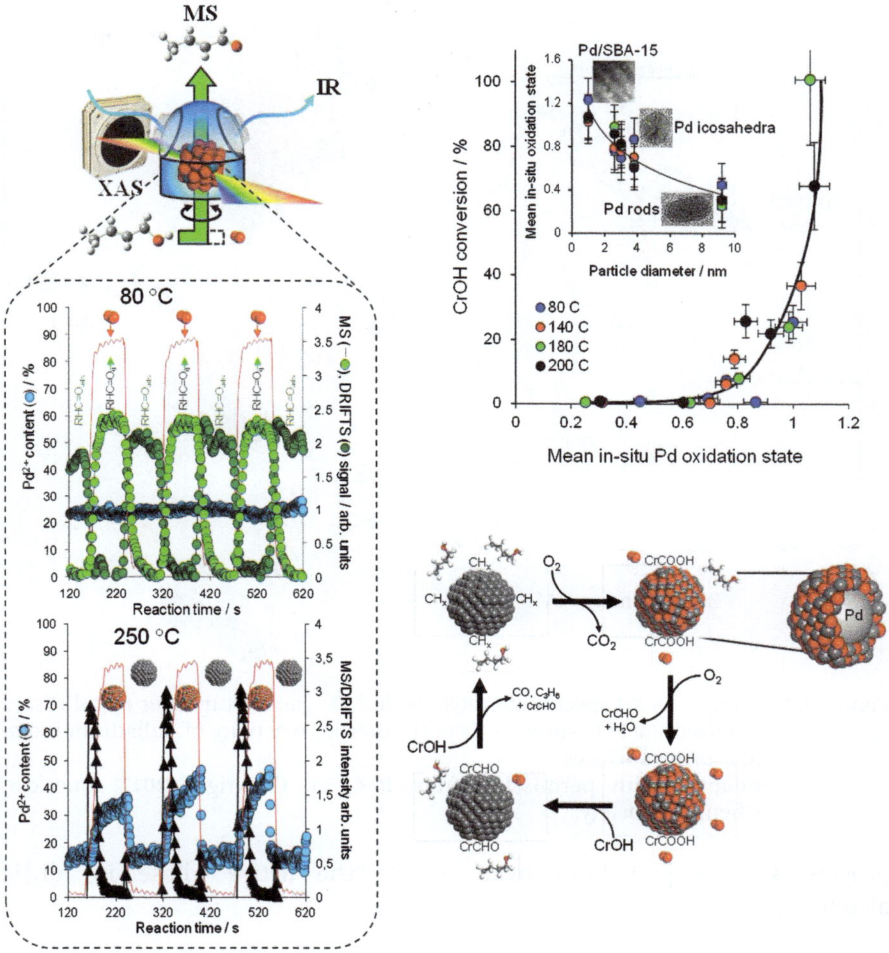

Figure 4.7 (Left) cartoon of *operando* DRIFTS/MS/XAS reaction cell and resulting temperature-dependent behaviour of Pd oxidation state and associated reactivity towards crotyl alcohol oxidation over a Pd/meso-Al_2O_3 catalyst. Only selective oxidation over surface PdO occurs at 80 °C, whereas crotonaldehyde decarbonylation and combustion dominate over Pd metal at 250 °C. (Top right) relationship between Pd oxidation derived *in situ* and crotyl alcohol conversion. (Bottom right) summary of reaction-induced redox processes in Pd-catalysed crotyl alcohol selox. Adapted with permission from Refs 70 and 72. Copyright 2011 and 2012 American Chemical Society.

confirmed that only two-dimensional Pd_5O_4 and three-dimensional PdO_x surfaces were capable of crotonaldehyde production (Figure 4.8).[73] However, even under oxygen-rich conditions, on-stream reduction of the Pd_5O_4 monolayer oxide occurred at >70 °C accompanied by surface poisoning by hydrocarbon residues. In contrast, PdO_x multilayers were capable of sustained catalytic turnover of crotyl alcohol to crotonaldehyde, conclusively

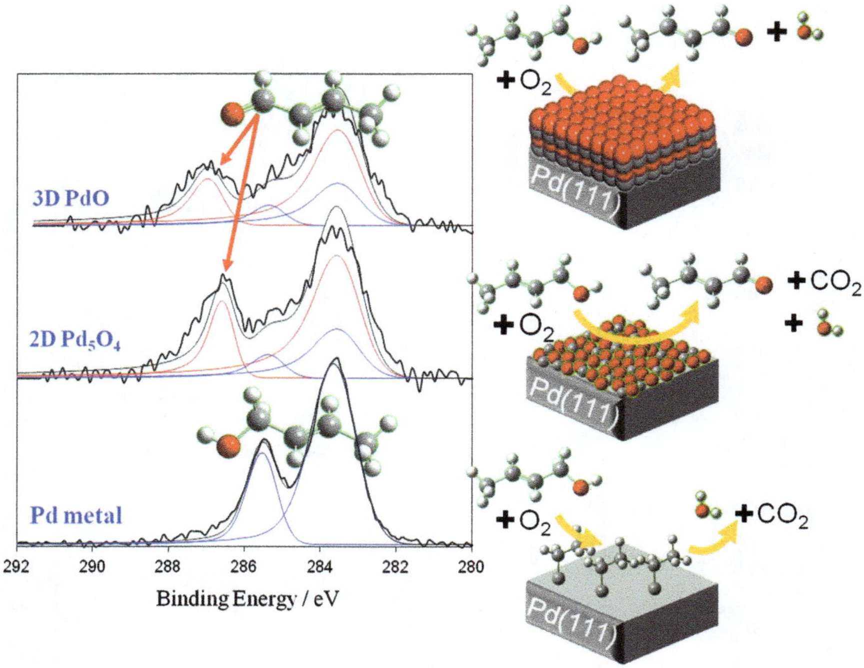

Figure 4.8 (Left) C 1s XP spectra of crotyl alcohol–O_2 gas mixture over metallic and oxidized Pd(111) surfaces; (right) differing reactivity of palladium metal and oxide surfaces.
Adapted with permission from Ref. 73. Copyright 2012 American Chemical Society.

proving surface palladium oxide to be the active phase in allylic alcohol selox.

4.2.6 Establishing Support Effects

Anchoring Pd nanoparticles to support structures offers an effective means to tune their physicochemical characteristics and prevent on-stream deactivation, *e.g.* by sintering. Supports employing porous architectures, acid–base character and/or surface redox chemistry, *e.g.* strong metal–support interaction (SMSI), afford further opportunities to influence catalyst reactivity.[74–77]

Mesoporous silicas are widely used to disperse metal nanoparticles.[38,39,75,78,79] The transition from low surface area, amorphous silica (200 m^2 g^{-1}) to two-dimensional non-interconnected pore channels (SBA-15)[80] and three-dimensional interconnected porous frameworks (SBA-16, KIT-6)[80–82] improved the dispersion of Pd nanoparticles and hence the degree of surface oxidation and thus activity in allylic alcohol selox, but had little impact on the mass transport of small alcohols to/from the active site.[38,39] The high thermal and chemical stability of such mesoporous silica[83,84] makes such supports well suited to commercialization. Pd nanoparticles confined within such mesoporous silicas demonstrate good selectivity in crotyl and cinnamyl alcohol selox to their respective aldehydes (>70%) and excellent TOFs of 7000 and 5000 h^{-1} for the respective alcohols. Similar activities have been reported for secondary and tertiary allylic alcohols, highlighting the versatility of silica-supported Pd nanoparticles.[24,38,39,85,86]

The benefits of mesostructured supports are not limited to silica. Hackett *et al.* showed that ultra-low loadings of palladium impregnated on a surfactant template mesoporous alumina (350 $m^2\ g^{-1}$) resulted in atomically dispersed Pd^{2+} species.[40] Such single-site catalysts were 10 times more active in crotonaldehyde and cinnamaldehyde production than comparable materials employing conventional (100 $m^2 \cdot g^{-1}$) γ-alumina, owing to the preferential stabilization of higher concentrations of electron-deficient palladium[37,40] due to either pinning at cation vacancies or metal$\rightarrow$support charge transfer.[87] These Pd/meso-Al_2O_3 catalysts exhibited similar TOFs to their silica counterparts (7080 and 4400 h^{-1} for crotyl and cinnamyl alcohol selox, respectively),[40] consistent with a common active site and reaction mechanism (Figure 4.9).

Mesoporous titania and ceria supports have attracted recent interest as novel catalyst supports. The oxygen storage capacity of ceria-derived materials is of particular interest due to their facile $Ce^{3+} \leftrightarrow Ce^{4+}$ redox chemistry.[77,88–92] Sacrificial reduction of the ceria supports by reactively formed hydrogen liberated during the oxidative dehydrogenation of alcohols could mitigate *in situ* reduction of oxidized palladium and hence maintain selox activity and catalyst lifetime, with Ce^{4+} sites regenerated by dissociatively adsorbed gas-phase oxygen.[91,93,94] Owing to its high density, conventional nanocrystalline cerias possess meagre surface areas

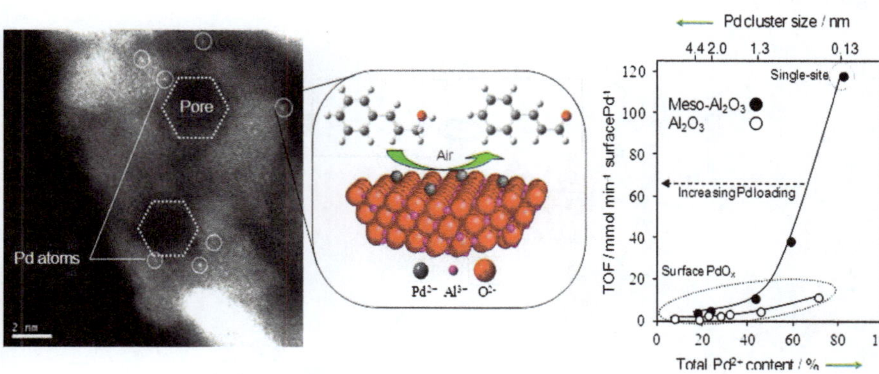

Figure 4.9 (Left) HAADF-STEM image of atomically dispersed Pd atoms on a mesoporous Al_2O_3 support; and (right) associated relationship between Pd^{2+} content/dispersion and activity in crotyl alcohol selox over Pd/alumina catalysts.
Adapted with permission from Ref. 40. Copyright 2007 Wiley-VCH Verlag GmbH & Co. KGaA.

(typically $\sim 5\ m^2\ g^{-1}$), hence Pd/CeO_2 typically exhibits poor selox behaviour owing to the resultant low nanoparticle dispersions that favour (self-poisoning) metallic Pd.[93,95,96]

Fabrication of hierarchical, porous ceria supports is possible through the evaporation-induced self-assembly protocol (EISA),[97] permitting highly tuneable surface areas between 100 and 250 $m^2\ g^{-1}$.[88,90,98,99] Increasing the ceria surface area and internal porosity confers similar benefits to those observed for silica-supported palladium catalysts due to an enhanced PdO surface density and electronic perturbation *via* Pd–Ce bonding,[96] which is known to improve activity and selectivity during CO oxidation.[100–103] Although still in its infancy, ceria remains a promising support for palladium-catalysed allylic alcohol selox.[77]

The popularity of titania as a reducible metal oxide support for Pd-based oxidation and photocatalysis has led to several elegant porous analogues. Careful choice of the titania precursor, surfactant and synthesis conditions permits extensive control over the support architecture and palladium–support interactions.[104–109] PdO nanoparticles impregnated on high surface area, porous titania by Mitsudo and co-workers demonstrated excellent activity and selectivity (87%) in benzyl alcohol selox.[110]

Mori and co-workers also explored the use of hydroxyapatite $[Ca_{10}(PO_4)_6(OH)_2]$ as a catalyst support,[111,112] comprising Ca^{2+} ions surrounded by PO_4^{3-} tetrahedra. This structure facilitates isomorphic substitution of calcium for transition metal cations such as Pd, thereby generating stable monomeric phosphate complexes with an inherent proclivity for aerobic selox.[111] Recent computational and *in situ* XAFS studies showed that palladium crystallites exposing high-energy facets (110) supported on hydroxyapatite promote O_2 dissociation in the solventless selox of DL-*sec*-phenethyl alcohol, benzyl alcohol and cinnamyl alcohol (Figure 4.10).[113]

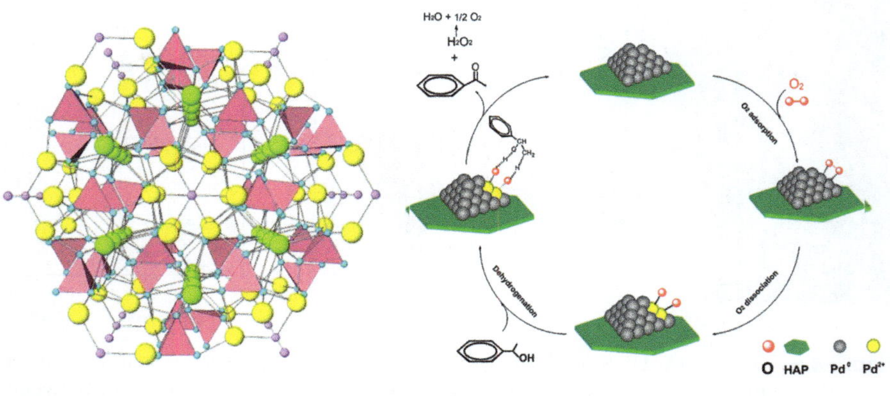

Figure 4.10 (Left) crystal structure of hydroxyapatite and (right) proposed reaction mechanism for the oxidation of alcohols over Pd-HAP systems. Adapted from Ref. 113 with permission from The Royal Society of Chemistry.

4.2.7 Bimetallic Palladium-Catalysed Selective Oxidation

Incorporation of a second metal in palladium catalysts can improve both alcohol selox stability and selectivity. Typical promoters such as Ag, Bi, Pb and Sn[60,114–117] enhance oxidation performance towards challenging substrates such as propylene glycol[118] and allylic and benzylic alcohols. Wenkin *et al.* reported that the rate of oxidation of glucose to gluconates was increased by a factor of 20 over Pd–Bi/C catalysts ($Bi{:}Pd_s = 0.1$) *versus* Pd/C counterparts.[115] *In situ* XAS and ATR-IR suggested that Bi residing at the catalyst surface protects palladium from deactivation by either over-oxidation (a hypothesis since disproved)[70,71,73] or site blocking by aromatic solvents.[56] Prati and co-workers first reported significant rate enhancements and resistance to deactivation phenomena in the liquid phase selox of D-sorbitol to gluconic/gulonic acids upon addition of Au to Pd/C and Pt/C materials,[119] subsequently extended to polyols and long-chain aliphatic alcohols.[120] A strong synergy between Pd and Au centres was also demonstrated by Hutchings and co-workers, who found that Au–Pd alloy nanoparticles supported on titania exhibited increased reactivity towards a diverse range of primary, allylic and benzylic alkyl alcohols compared with monometallic palladium analogues.[121]

The versatility of Au–Pd catalysts has also been shown in selox of saturated hydrocarbons,[122] ethylene glycol,[123] glycerol[124] and methanol,[125] where high selectivity and resistance to on-stream deactivation were noted.

The effect of Au–Pd composition has been extensively studied for bimetallic nanoparticles stabilized by PVP surfactants.[126] An optimal Au:Pd composition of 1:3 was identified for 3 nm particles towards the aqueous-phase aerobic selox of benzyl alcohol, 1-butanol, 2-butanol, 2-buten-1-ol and 1,4-butanediol; in each case the bimetallic catalysts were superior to palladium alone. Mertens *et al.* examined similar systems utilizing 1.9 nm nanoparticles and an optimal Au content of around 80% was determined for benzyl alcohol selox.[127] The synergic interaction between Au and Pd therefore appears to be dependent on nanoparticle size. It is well-known that the catalytic activity of Au nanoparticles increases dramatically with a particle size <2 nm,[128] hence it is interesting to compare phase-separated and alloyed catalysts systematically. Lee *et al.* prepared titania-supported Au shell (five-layer)–Pd core (20 nm) bimetallic nanoparticles for the liquid-phase selox of crotyl alcohol and systematically studied the evolution of their bulk and surface properties as a function of thermal processing by *in situ* XPS, DRIFTS, EXAFS and XRD and *ex situ* HRTEM.[129] Limited Au–Pd alloying occurred below 300 °C in the absence of particle sintering. Higher temperatures induced bulk and surface alloying, with concomitant sintering and surface roughening. Migration of Pd atoms from the core to the surface dramatically enhanced activity and selectivity, with the most active and selective surface alloy containing 40 at.% Au (Figure 4.11), in excellent agreement with the surface chemistry of crotyl alcohol,[130] crotonaldehyde and propene over Au–Pd(111) single-crystal alloys.[64]

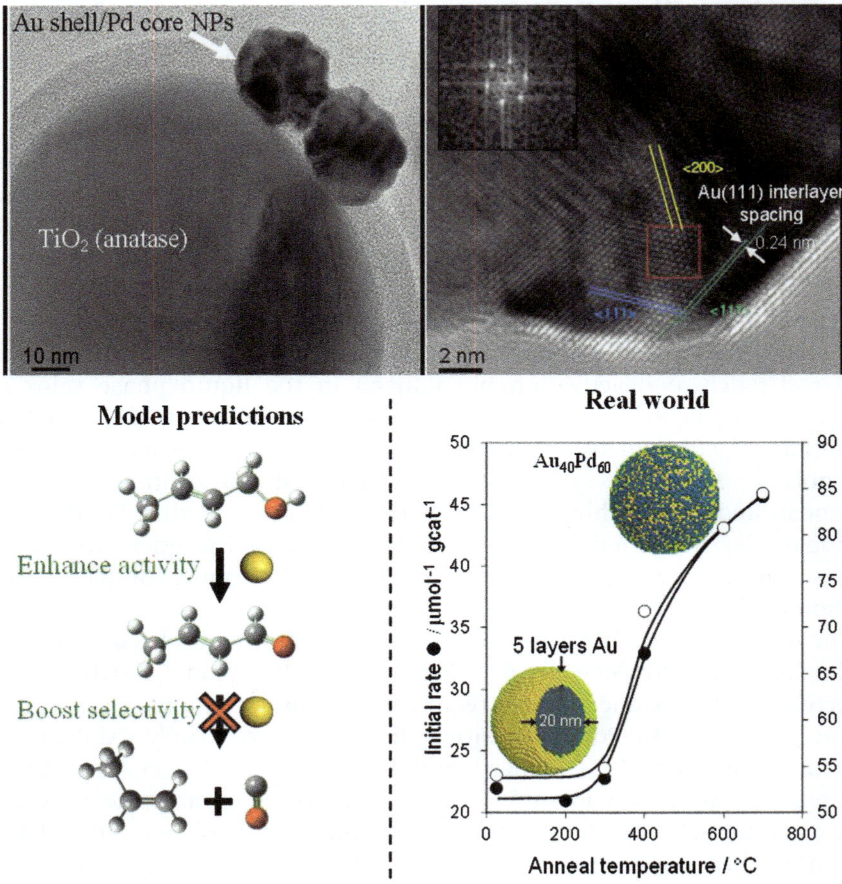

Figure 4.11 (Top) HRTEM images of titania-supported Au shell–Pd core nanoparticles; and (bottom) impact of annealing and Au/Pd surface alloying upon catalytic activity and selectivity towards cinnamyl alcohol aerobic selox.
Adapted from Ref. 129 with permission from Elsevier.

Scott and co-workers prepared Au core–Pd shell nanoparticles and explored the catalytic cycle for alcohol selox in order to assess their associated stability.[131–135] *In situ* Pd K and Pd L_{III} edge XAS of a Au nanoparticle–Pd(II) salt solution was undertaken in order to discriminate two possible reaction mechanisms (Figure 4.12). No evidence was found that crotyl alcohol oxidation was accompanied by Pd^{2+} reduction on to Au nanoparticles, resulting in the formation of a metallic Pd shell (with oxygen subsequently regenerating electron-deficient palladium), therefore β-H elimination was proposed as the favoured pathway. Scott and co-workers proposed that the Au core prevents the re-oxidation of surface Pd^0 atoms; no Pd–O and Pd–Cl contributions were observed by EXAFS.

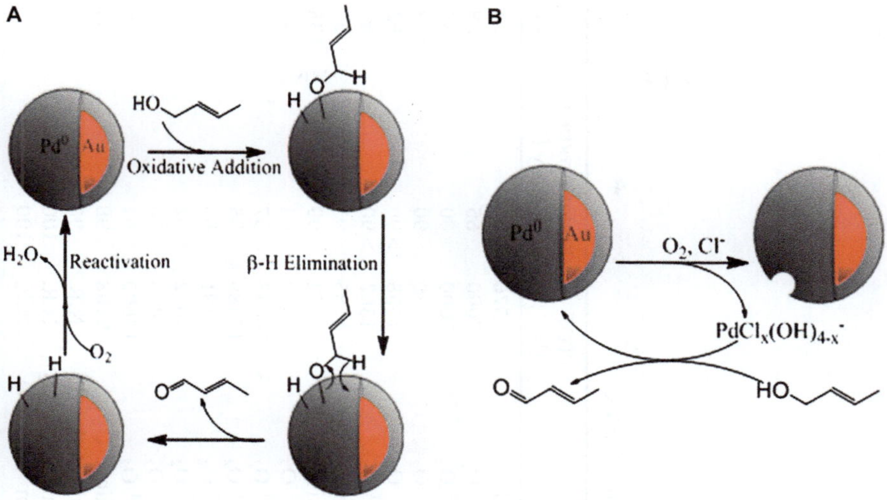

Figure 4.12 Proposed mechanisms for crotyl alcohol aerobic selox over Au core–Pd shell nanoparticles: (left) β-H elimination mechanism; (right) redox mechanism.
Adapted with permission from Ref. 135. Copyright 2013 American Chemical Society.

4.2.8 Substrate Scope

The preceding sections have outlined the mechanisms underpinning the aerobic selox of alcohols over heterogeneous Pd catalysts and associated structure–function relations. An important consideration for the commercialization of such catalysts is the range of alcohols for which they afford an economic and speedy synthesis of aldehydes in high yields.

Pd/mesoporous silicas show excellent activity towards primary and secondary allylic alcohols.[34,38,39] TOFs generally decrease with increasing molecular mass, likely reflecting slower substrate diffusion into such mesoporous frameworks rather than intrinsically slower selox kinetics (Table 4.1). The introduction of 300 nm macropores into SBA-15 led to a >400% increase in TOF for the long-chain allylic phytol and farnesol (1258 and 1502 h^{-1}, respectively, with macropores *versus* 252 and 480 h^{-1} without) and still delivering >90% selectivity to the respective aldehydes.[62]

Alumina's ability to support sub-2 nm PdO_x clusters leads to superior selox performance compared with silica-supported palladium created *via* identical protocols. Similar trends are apparent with regard to the influence of mesoporosity and increased surface area in conferring faster TOFs for benzyl, crotyl and cinnamyl alcohols compared with amorphous low surface area γ-alumina analogues.[40,63,137] Despite their higher activities, alumina-supported palladium nanoparticles exhibit slightly lower selectivity (by around 10%) towards crotyl and cinnamyl alcohol than over mesoporous silicas, due to enhanced C=C hydrogenation to saturated products.

Table 4.1 Comparative performance of heterogeneous palladium catalysts for aerobic alcohol selox.

Catalyst	Substrate: metal ratio	Substrate	Reaction parameters	TOF (h^{-1})	Major product selectivity (%)	Ref.
Pd/meso-SiO_2	4061	Benzyl alcohol	Solvent free, 353 K, 3 $cm^3\ min^{-1}$ O_2	576	98	34
	17872	Crotyl alcohol	Toluene, 363 K, 3 $cm^3\ min^{-1}$ O_2	7000	65	38
	17872	Cinnamyl alcohol	Toluene, 363 K, 3 $cm^3\ min^{-1}$ O_2	5000	80	38
	3885	Prenol	Toluene, 363 K, 3 $cm^3\ min^{-1}$ O_2	920	90	39
Pd/mesomacro-SiO_2	260	Phytol	Toluene, 363 K, 3 $cm^3\ min^{-1}$ O_2	1258	>90	62
	260	Farnesol	Toluene, 363 K, 3 $cm^3\ min^{-1}$ O_2	1502	>90	62
Pd/γ-Al_2O_3	N.R.	Benzyl alcohol	$scCO_2$, 373 K, 15 MPa O_2, flow setup	1585	95	20
	16167	Benzyl alcohol	Solvent free, 361 K, 3 $cm^3\ min^{-1}$ O_2	1952	99	135
	14893	Crotyl alcohol	Toluene, 363 K, 3 $cm^3\ min^{-1}$ O_2	7080	64	40
Pd/meso-Al_2O_3	17872	Benzyl alcohol	Toluene, 363 K, 3 $cm^3\ min^{-1}$ O_2	16643	97	63
	17872	Crotyl alcohol	Toluene, 363 K, 3 $cm^3\ min^{-1}$ O_2	17389	55	
	17872	Cinnamyl alcohol	Toluene, 363 K, 3 $cm^3\ min^{-1}$ O_2	13391	66	
	604	Geraniol	Toluene, 363 K, 3 $cm^3\ min^{-1}$ O_2	1052	62	
	604	Farnesol	Toluene, 363 K, 3 $cm^3\ min^{-1}$ O_2	425	70	
Pd/Al–SBA-15	1161	Cinnamyl alcohol	Toluene, 363 K, 3 $cm^3\ min^{-1}$ O_2	13600	64	136
Pd/SiO_2–Al_2O_3	7446	Benzyl alcohol	Solvent free, 353 K, 3 $cm^3\ min^{-1}$ O_2	9108	98	34
	7446	1-Phenylethanol	Solvent free, 353 K, 3 $cm^3\ min^{-1}$ O_2	N.R.	90	
	307	Geraniol	Trifluorotoluene, 373 K, 3 $cm^3\ min^{-1}$ O_2	N.R.	100	
	307	2-Thiophenemethanol	Trifluorotoluene, 353 K, 3 $cm^3\ min^{-1}$ O_2	N.R.	100	

Pd/TiO_2	1000	Benzyl alcohol	Toluene, 353 K, air		87	110
Pd/HAP	1054	1-Phenylethanol	Solvent free, 433 K, 0.1 MPa O_2	11600	>90	121
	500	Benzyl alcohol	Trifluorotoluene, 363 K, 0.1 MPa O_2	N.R.	99	111
	500	Cinnamyl alcohol	Toluene, 363 K, 0.1 MPa O_2	N.R.	87	
	500	Oct-3-en-2-ol	Toluene, 363 K, 0.1 MPa O_2	N.R.	80	
PEG-functionalized Pd	1000	Benzyl alcohol	H_2O, 353 K, 0.1 MPa O_2, 1.45 mmol K_2CO_3	56	99	137
Colloidal Pd	500	Benzyl alcohol	H_2O, 308 K, 0.1 M Pa O_2, 0.375 mmol K_2CO_3, pH 3.5	250	100	138
AuPd/C	1000	D-Sorbitol	H_2O, 323 K, 0.3 M Pa O_2, 0.3 M NaOH, pH 11	333	57	119
	25000	Glycerol	H_2O, 333 K, 1.0 M Pa O_2, 0.6 M NaOH	25200	80	139
$AuPd/TiO_2$	192354[a]	Benzyl alcohol	Solvent free, 433 K, 0.1 MPa O_2	86500	92	121
	184000[a]	Crotyl alcohol	Solvent free, 433K, 0.5 MPa O_2	12600	>90	
	15463[a]	Cinnamyl alcohol	H_2O, 363 K, 0.1 MPa O_2	97	100	
	184000[a]	1,2-Butanediol	Solvent free, 433 K, 0.1 MPa O_2	1520	N.R.	
	184000[a]	1,4-Butanediol	Solvent free, 433 K, 0.1 MPa O_2	104200	>90	
	192354[a]	1-Phenylethanol	Solvent free, 433 K, 0.1 MPa O_2	269000	>90	
Au_{Shell}–Pd_{Core}/TiO_2	500[a]	Benzyl alcohol	H_2O, 358 K, 0.1 MPa O_2, 0.375 mmol K_2CO_3, pH 11.5	57	100	133
		1,4-Butanediol	H_2O, 335 K, 0.1 MPa O_2, 0.375 mmol K_2CO_3, pH 11.5	234	100	
Pd–Bi/C	7876[b]	Glucose	H_2O, 323 K, 0.4 l cm^3 min^{-1} O_2, pH 9.25	N.R.	38	140

N.R. = not reported
[a]Calculated from total Au and Pd content.
[b]Ratio calculated from total Bi and Pd content.

Although highly ordered, mesoporous alumina facilitates geraniol and farnesol selox, the lack of macroporosity results in slower TOFs and lower selectivities (<70%) than achieved with hierarchical macroporous-mesoporous silica, further evidencing substrate mass transport as the rate-limiting factor for the oxidation of bulky alcohols and indicating a competing hydrogenation pathway. Pd-impregnated hydroxyapatites are also versatile catalysts for the selox of primary, secondary and long-chain alcohols with >80% selectivity to aldehydes and ketones, although they have only been trialled at low substrate : catalyst ratios, hence their viability for large-scale production remains to be proved.[111,121]

A number of groups have attempted solvent-free aerobic selox, particularly of benzyl alcohol, wherein the substrate necessarily is employed at molar ratios far in excess of the catalyst. Palladium on silica, γ-alumina and mixed SiO_2–Al_2O_3 have been applied in this respect, with TOFs increasing across this sequence of oxide supports. Pd/SiO_2–Al_2O_3 was also reported to oxidize secondary alcohols, including phenylethanol, non-activated alcohols such as cyclooctanol, 2-octanol and 1-octanol, and even the heterocyclic 2-thiophenemethanol, to their various aldehydes and ketones.[34] Support functionalization, whether *via* modified pore architecture or surface acid-base properties, is thus an effective means for optimizing catalyst activity without sacrificing selectivity.

In summary, supported monometallic Pd catalysts can selectively oxidize primary and secondary allylic and aromatic alcohols, limited heterocyclic alcohols and (upon incorporation of macroporous networks) longer chain alcohols including phytol, geraniol and farnesol. However, significant selox activities have yet to be reported for linalool, a tertiary alcohol, and non-allylic alcohols, leading to the postulate that the allyl bond is a prerequisite for activation of the –OH bond, possibly by ensuring the correct molecular orientation to bring the alcohol function close to the active site or strengthening adsorption of the molecularly bound parent alcohols or a reactive intermediate to facilitate thermal activation.

Palladium promotion by gold is advantageous in terms of both enhancing activity and suppressing unwanted side reactions (hydrogenolysis and decarbonylation).[47,64,130] Hutchings and co-workers demonstrated this efficacy under solvent-free conditions.[121] Palladium surface enrichment of Au catalysts increased the selectivity and TOFs towards benzyl alcohol, crotyl alcohol, 1,4-butanediol and 1-phenylethanol with respect to those possible over the individual metal components; product selectivities were reported to exceed 90%, a marked increase on those reported for palladium. In some instances, these impressive solvent-free performances were attained for extremely high substrate : catalyst ratios, exceeding 180 000, improving the green credentials of heterogeneously catalysed aldehyde/ketone production. Unfortunately, these bimetallics were inert towards the selox of secondary alcohols such as 2-octanol and 1,2-butanediol, attributed to a combination of steric factors, strongly bound ketone adsorption and self-poisoning. Au–Pd systems prepared *via* simple wetness impregnation were also surprisingly unreactive towards cinnamyl alcohol (a prototypical allylic alcohol for aerobic selox) and vanillyl alcohol, with TOFs of 10 and 97, respectively, although selectivity to the desired aldehydes was 100% for both. Polyols represent the ultimate challenge for selective oxidation catalysis, with the coexistence of multiple –OH functions in close proximity providing numerous competing reaction pathways to navigate in order to yield a single desired product. In this context, Villa and co-workers[119] highlighted the potential of Au–Pd bimetallic systems for D-sorbitol (an important sugar alcohol) and Davis and co-workers[140] similarly for glycerol, with 57 and 80% selectivity to gluconate and glyceric acid, respectively.

Mono- and bimetallic Pd catalysts are effective for the selective aerobic oxidation of a broad range of primary and secondary allylic and aromatic alcohols. Developments in catalyst design and the use of structured support architectures/stabilizers are enabling the selox of more complex polyols and sterically demanding alcohols to be achieved, but further selectivity improvements and energy-efficient routes to the activation of tertiary and saturated alcohols are still sought in order to enhance the profile and commercial implementation of palladium in selox.

4.2.9 Concluding Remarks on Heterogeneous Pd Systems

The selective oxidation of complex alcohol substrates can be accomplished through Pd-mediated heterogeneous catalysis with high turnover and product selectivity. Application of *in situ* and *operando* techniques, such as *in situ* X-ray and IR spectroscopy, has helped to elucidate the mechanism of the oxidative dehydrogenation of alcohols and competing aldehyde decarbonylation. Surface PdO was identified as the active catalytic species and deactivation the result of reduction to metallic palladium and concomitant self-poisoning by strongly bound CO and carbonaceous residues. Although this field is still in its relative infancy, breakthroughs in analytical tools and synthetic approaches to engineering nanoporous supports and shape/size-controlled nanoparticles have already delivered significant progress towards improved atom and energy efficiency and catalyst stability.

4.3 Ligand-Modulated Pd Systems

4.3.1 Introduction

Traditionally, the challenge for Pd(II)-catalysed oxidation reactions was that the re-oxidation of Pd(0) species by O_2 back to the active Pd(II) was found to be slower than the aggregation process. A major breakthrough in the field of Pd(II) oxidations was the Wacker process, which was developed for the oxidation of ethylene to acetaldehyde.[141] In this case, Pd(0) is oxidized back to Pd(II) using Cu(II) salts and the resultant Cu(I) species can be re-oxidized directly with O_2. This is an approach that is still exploited today in academic research; however, the presence of the copper co-catalysts is often

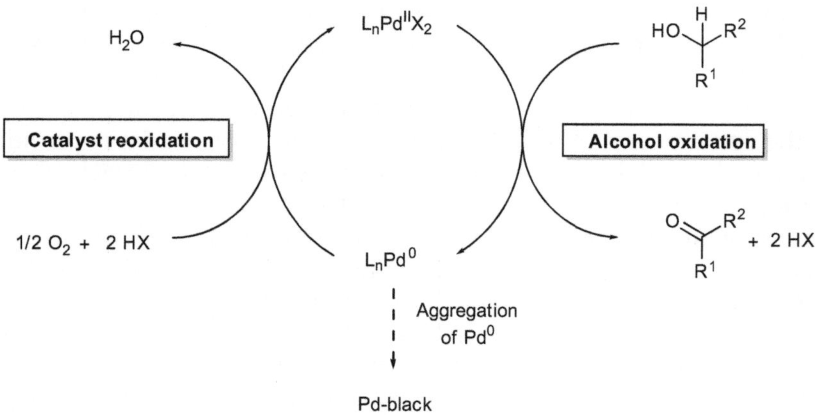

Figure 4.13 Catalytic cycle for aerobic ligand-modulated Pd(II)-catalysed alcohol oxidation.

undesirable; for example, it can lead to reduced selectivity in some cases. To address such issues, methods have been developed that utilize redox co-catalysts other than copper. Bäckvall has been one of the pioneers in this area and his group have developed systems which use a series of 'electron-transfer mediators' to enable O_2 to be used as the terminal oxidant.[142] For example, Pd(0) is oxidized to Pd(II) by 1,4-benzoquinone and the resultant 1,4-hydroquinone is oxidized by a metal complex (*e.g.* cobalt salophen or iron porphyrin complexes) which can be oxidized readily by O_2. Once again, this is an approach that is still utilized by researchers today, as it can deliver very impressive results, with reactions having excellent selectivity with low loadings of Pd. Nonetheless, it would be highly advantageous if it were possible to use a single Pd(II) complex as a catalyst and re-oxidize it directly with O_2, as shown in Figure 4.13.

A single complex should be simpler and arguably more attractive for larger-scale use and, importantly, the use of ligands should permit the tuning of the catalytic activity and selectivity. Ligand modulation of catalyst performance is well established in other areas of homogeneous catalysis such as hydrogenation and hydroformylation; however, this paradigm was slower to take hold in homogeneous palladium oxidations. In this part of the chapter, we focus on ligand-modulated Pd(II) catalysis for alcohol oxidation. It is not possible to discuss all the catalysts that have been reported to date; therefore, we will highlight examples of leading catalyst systems and outline the current mechanistic understanding for these systems.

4.3.2 Examples of Ligand-Modulated Pd(II) Catalysts

Perhaps the first report of co-catalyst-free Pd-catalysed aerobic oxidation of alcohols was in 1977 by Blackburn and Schwartz.[143] In this case, $PdCl_2$ was used in conjunction with 5 mol% NaOAc and the reaction rates and turnover numbers (TONs) were modest. Despite this report, developments in this area really only picked up pace in the late 1990s.[144,145] In 1998, two different systems were reported that seemed to mark the start of an intense period of research in this area. Peterson and Larock reported a $Pd(OAc)_2$–dimethyl sulfoxide (DMSO) system,[146] and Uemura and co-workers reported the use of pyridines.[147,148] In the DMSO case, the solvent is most likely acting as both a solvent and a ligand, while the pyridine system allows the use of weakly coordinating solvents (in that case toluene). The pyridine system was conceivably the first to explore the use of oxidatively stable ligands in this way, and since that time this is an approach that has been built on by others. Uemura and co-workers screened a number of pyridine-based ligands, but simple pyridine was the most effective and their general system consisted of $Pd(OAc)_2$ (5 mol%), pyridine (20 mol%) and MS3A (3 Å molecular sieves) in toluene at 80 °C under 1 atm of O_2. It was shown to be effective for oxidizing a wide range of benzylic and aliphatic alcohols, including diols. However, it is desirable to reduce the loading of Pd, as a loading of 5 mol% (the loading used in both the DMSO and pyridine systems) is too high to be practical for many applications. Therefore, although there have been a number of studies exploiting pyridine-based ligands, we highlight here other ligand systems that enable lower catalyst loadings to be employed.

The use of bidentate ligands will lead to a more stable catalyst complex and this means that catalysts do not require an excess of ligand in the same way that the pyridine system does. Some of the most active systems reported are based on the use of phenanthroline-type ligands, with initial reports in this area being published in 2000.[149,150] Sheldon, Arends and co-workers studied phenanthroline-based systems in detail, examining the influence of factors such as ligands, pH, solvents, gas pressure/composition and substrates.[151–154] Ligands with methyl groups in the 2- and 9- positions gave the best performance and the potential of 2,9-dimethyl-1,10-phenanthroline (neocuproine) was examined in detail. The optimized system is notable because it can oxidize even unactivated aliphatic substrates with comparatively low catalyst loadings (down to 0.1 mol%), as shown in Table 4.2.

In the case of aliphatic primary alcohols, the aldehyde is not stable under the reaction conditions and is common with such substrates it produces the over-oxidized product, the carboxylic acid. It was demonstrated that the aldehyde could be produced selectively if the stable free radical TEMPO (2,2,6,6-tetramethylpiperidine-1-oxyl) was added, as this prevents the autoxidation of the aldehyde to the acid. It was shown that (neocuproine)-$Pd(OAc)_2$ could oxidize smoothly a range of substrates (including unactivated aliphatic substrates). The catalyst could tolerate the presence of functional groups such as alkenes, alkynes, halides, ethers, thioethers, sulfoxides, sulfones, sulfonates, amines, cyanides, amides, carbonates and esters, although in many of these cases their presence resulted in slower reaction rates.

More recently, Waymouth and co-workers studied neocuproine-based catalysts, and prepared $[(neocuproine)Pd(\mu\text{-}OAc)]_2[OTf]_2$ (Figure 4.14), which

Table 4.2 Examples from the optimized (neocuproine)Pd(OAc)$_2$ system from Ref. 154.

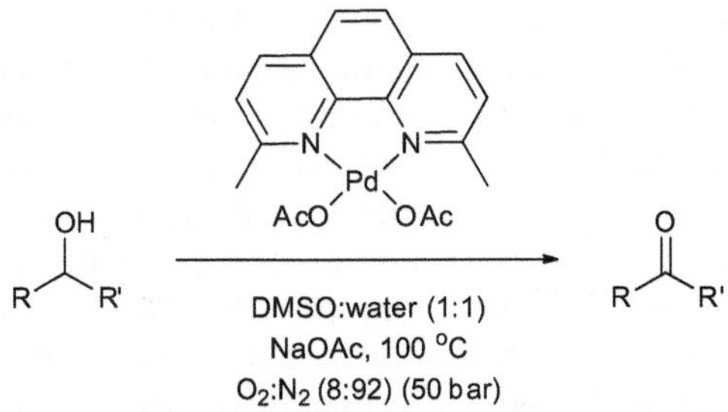

Substrate	Time (h)	TOF_0 (h^{-1})	Conversion (selectivity) (%)
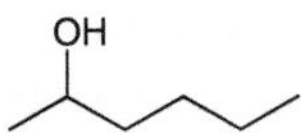	2	>500	100 (100)
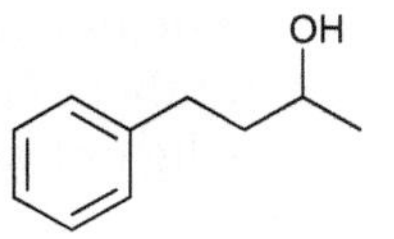	2	>500	99 (99)[b]
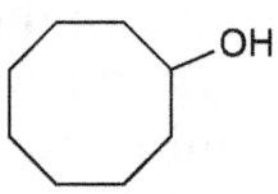	3	400	93 (>99)[c]
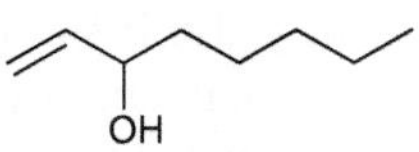	3	400	95 (96)[d]
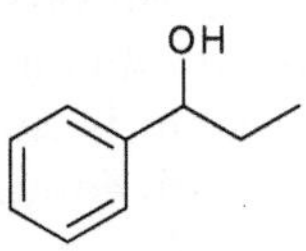	3	300	80 (>99)
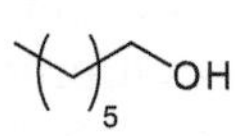	2.5	200	40 (75)[e]
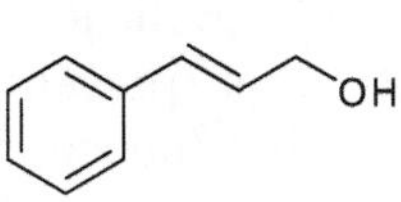	10	135	88 (>99)
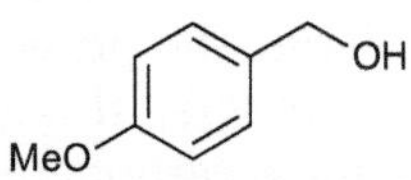	3	>330	100 (100)

[a]Reaction conditions: 0.1 mol% (neocuproine)Pd(OAc)$_2$, 20 mmol alcohol, 50 mL water–DMSO 1:1 (v/v), 5 mol% NaOAc, 100 °C, 50 bar 8% O_2/N_2, 750 rpm stirring speed. [b]Side-product 1-phenyl-1-buten-3-one. [c]Side-product cyclooctenone. [d]Side-products 1-hydroxy-3-octanone (2%) and 3-octanone (1%). [e]Selectivity to heptanoic acid, side products heptanal (20%), 2-heptenal (5%), 60 bar 8% O_2/N_2, Pd^0 formed.

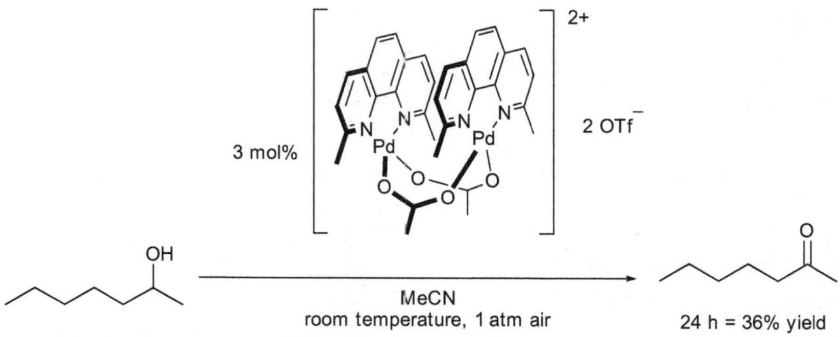

Figure 4.14 [(neocuproine)Pd(μ-OAc)]$_2$[OTf]$_2$ catalyst developed by Waymouth and co-workers.[156]

forms monomeric cationic complexes. This catalyst was designed to permit more facile alcohol oxidation owing to the open coordination site provided by the weakly coordinating triflate ion.[155] It was found that the catalyst could deliver very fast initial rates at room temperature for the oxidation of 2-heptanol (Figure 4.14), but it produced a low TON as the catalyst became inactive owing to oxidation of one of the methyl groups on the neocuproine ligand. Oxidation of the ligand was due to a hydroperoxy intermediate that is formed in the catalytic cycle.

This hydroperoxy species is perhaps longer lived under these conditions due to a lack of protons in the system [see details about the general Pd(II) mechanism in Section 4.3.3]. Such ligand oxidation was not an issue under the previously described Sheldon conditions, but under those conditions there is an excess of both NaOAc and water and the reaction is conducted at higher temperatures (80–100 °C).[154] In an attempt to address the issue of ligand oxidation, Waymouth's group examined fluorinated phenanthroline derivatives, which should be more oxidatively stable.[156] Although these fluorinated ligands did lead to an improvement compared with neocupruoine, the overall performance (in terms of TON) was not as good as under the original Sheldon conditions. More recently, Waymouth and co-workers found that their neocuproine cationic complex system shows potential for the oxidation of polyols.[157,158] The catalyst performed the regioselective and chemoselective oxidations of unprotected polyols with α-hydroxy ketones being formed in high yield. In particular, the tetralols erythritol and threitol were converted to erythrulose with minimal racemization (Figure 4.15). The catalyst also performed the oxidative lactonization of 1,5-diols to cyclic lactones, and in the case of cyclic alcohols the catalyst exhibited a preference for hydroxyl groups in axial positions over those in equatorial positions.

As part of their studies, they carried out mechanistic investigations to gain greater insight. Although the catalyst selectively oxidized secondary alcohols in polyols, it was found to oxidize 1-propanol faster than 2-propanol. They concluded that in the case of vicinal diols, β-hydride elimination from the secondary Pd alkoxide is favoured to give hydroxy ketones. In the case of

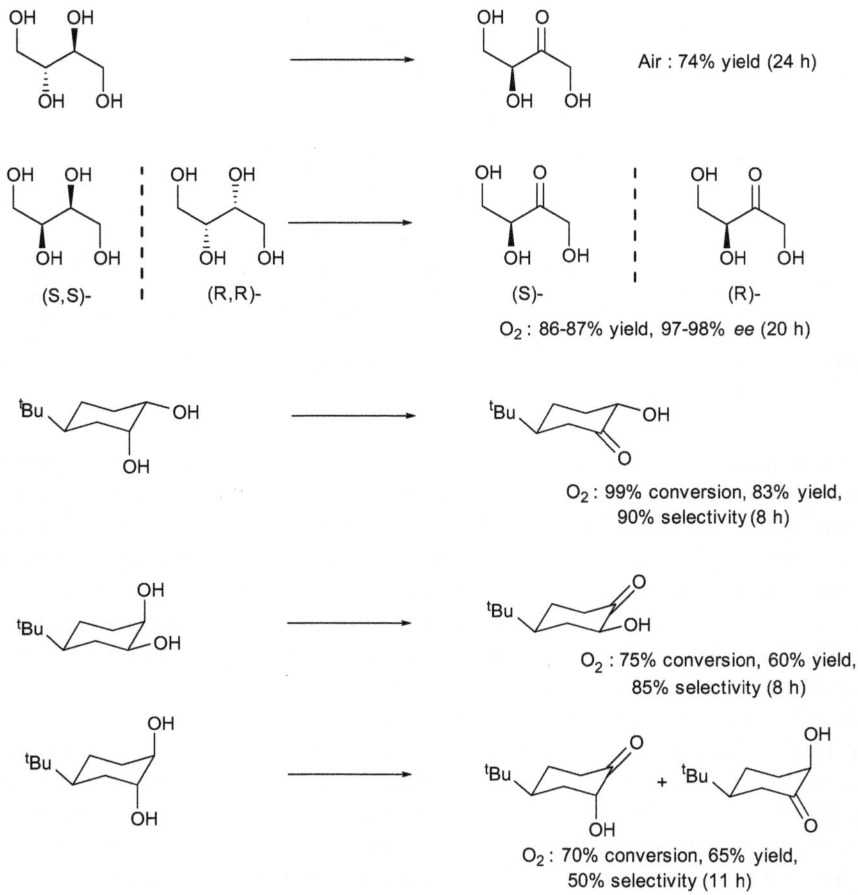

Figure 4.15 Selective oxidation of polyols by Waymouth and co-workers.[159] General conditions: 10 mol% $[(\text{neocuproine})\text{Pd}(\mu\text{-OAc})]_2(\text{OTf})_2$, 9 : 1 (v/v) MeCN–$H_2O$, 25 °C.

mixtures of primary and secondary alcohols, the observed preference for the oxidation of the primary alcohols was believed to be due to pre-equilibria favouring primary alkoxides. At the present time, the catalyst loadings are relatively high when it is used aerobically, although it was found that this could be dramatically reduced when benzoquinone was used as the terminal oxidant (as low as 0.12 mol% on a 10 g scale). Nonetheless, such studies are extremely valuable in the design of the next generation of chemoselective aerobic catalysts. Indeed, the selective oxidation of polyols is very useful and it is an area that will surely continue to be explored. In this area, it is also worth highlighting the work of Oberhauser and co-workers, who reported the chemoselective oxidation of unprotected diols using neutral and cationic Pd(II) complexes in the presence pyridine and pyridine derivatives.[159] This utilized a readily accessible catalyst system that could oxidize diols to the

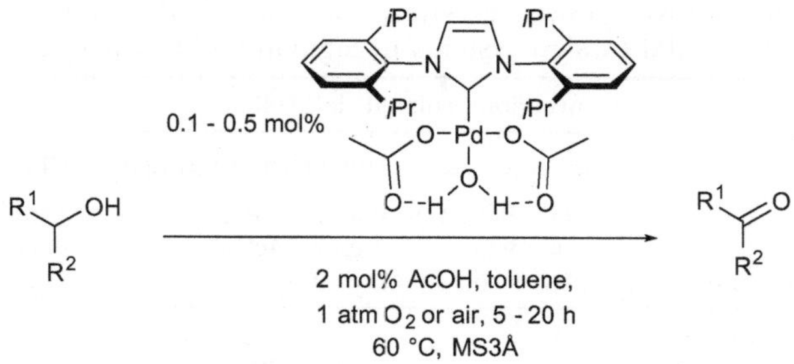

Figure 4.16 Sigman's (IPr-NHC)Pd$(OAc)_2(H_2O)$ catalyst.

corresponding hydroxy ketones with good selectivity with catalyst loadings as low as 1 mol%.

Sigman's group have made significant contributions to the area of homogeneous Pd(II) alcohol oxidation. They developed the first Pd/NHC (NHC = *N*-heterocyclic carbene) catalyst system for aerobic alcohol oxidations using the 1,3-bis(2,6-diisopropylphenyl)imidazol-2-ylidene (IPr) ligand (Figure 4.16).[160] Other NHC systems have since been reported,[161] but the original system exhibits some of the best performances reported to-date.

With this catalyst, acetate acts as an internal base, conducting an intramolecular deprotonation of the alcohol and simultaneously leaving a vacant coordination site for β-hydride elimination. The crystal structure of the catalyst involves a water molecule coordinated to the Pd in the vacant site and hydrogen bonding to the neighbouring acetate ligands. The (IPr-NHC)Pd$(OAc)_2(H_2O)$ catalyst is capable of performing at a catalyst loading of 0.5 mol% for a variety of alcohols, with 4-methoxybenzyl alcohol being oxidized in 20 h at loadings as low as 0.1 mol%. Catalytic quantities of acetic acid were found to be beneficial for the system and they examined the influence of acid–base chemistry in detail.[162] The use of acid enables ambient air to be used as the terminal oxidant source as opposed to pure O_2, although needing slightly extended reaction times in order to achieve comparable yields. The use of open air makes this an accessible method to synthetic chemists working on a small laboratory scale.

Sigman and co-workers also developed a second-generation NHC system, the pivalate derivative (IPr-NHC)Pd$(OPiv)_2$, which was found to function well under the mild conditions of room temperature and atmospheric air, albeit with a higher loading of 1 mol%. The group had also developed a system that used triethylamine as a ligand and operated at room temperature,[163] and they compared the performance of these three catalyst systems.[164] Table 4.3 gives some examples of the catalyst performance for these three systems for a range of substrates. It can be seen that the catalysts are generally suitable for a broad range of alcohols. It is worth noting that the NHC systems with the larger pivalate counterions (as opposed to acetate) struggled with more sterically demanding substrates. Also worth mentioning is that for some substrates, when the oxidation with (IPr-NHC)Pd$(OAc)_2(H_2O)$ was

Table 4.3 Comparative substrate study of the triethylamine/$Pd(OAc)_2$ and (IPr-NHC)Pd catalyst systems by Sigman and co-workers.[165]

		Conversion (isolated yield) (%)		
		$NEt_3/Pd(OAc)_2$	**(IPr-NHC)$Pd(OAc)_2(H_2O)$**	**(IPr-NHC)$Pd(OPiv)_2$**
	Substrate	3 mol% $Pd(OAc)_2$, 6 mol% NEt_3, O_2, RT, 12 h.	0.5 mol% Pd, 2 mol% AcOH, O_2, 60 °C	1 mol% Pd, 0.5 mol% PivOH, air, RT, 14 h
Benzylic	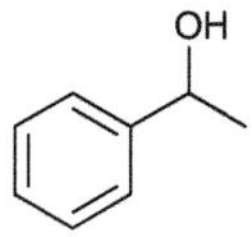	(93)	5 h: >99 14 h: >99 (97)[a]	>99
	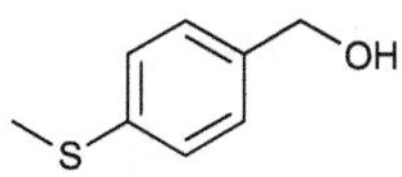	35	13 h: 90 14 h: >95 (90)[b,c]	>99
	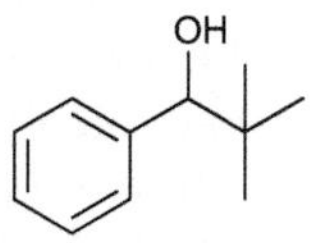	(50)	14 h: 91[b,c]	23
	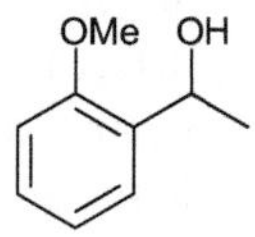	<5	14 h: <5	N/D[j]
Aliphatic	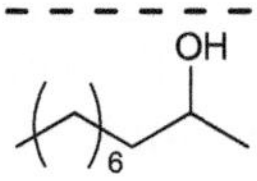	(97)	13 h: 99 (93)[d]	97
	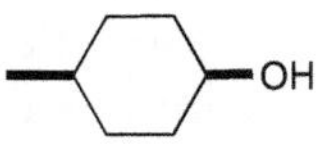	N/D	13 h: <99[d] 14 h: 96[e]	63
	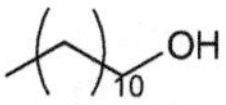	(85)[e]	10 h: 85 (76)[c]	N/D
Allylic	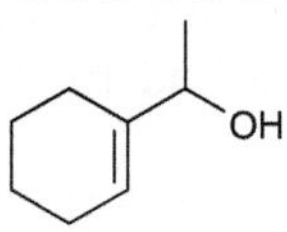	N/D[j]	12 h: 91 (84)	62
	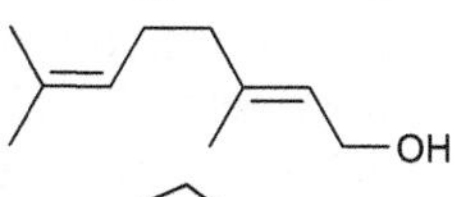	33	18 h: (72)[c,f]	68
	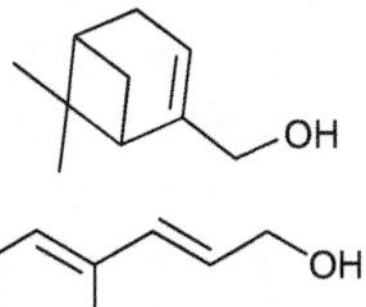	82[g]	20 h: 97[c]	99
	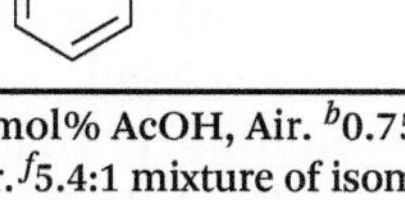	(30)[h] (39)[i]	N/D[j]	N/D[j]

[a] 5 mol% AcOH, Air. [b] 0.75 mol% Pd. [c] 5 mol% $[N_{4444}][OAc]$ in place of AcOH. [d] 1 mol% AcOH. [e] Air. [f] 5.4:1 mixture of isomers measured by ^1H NMR. [g] 18 mol% NEt_3. [h] 50 mol% NEt_3. [i] 300 mol% NEt_3. [j] N/D: no data.

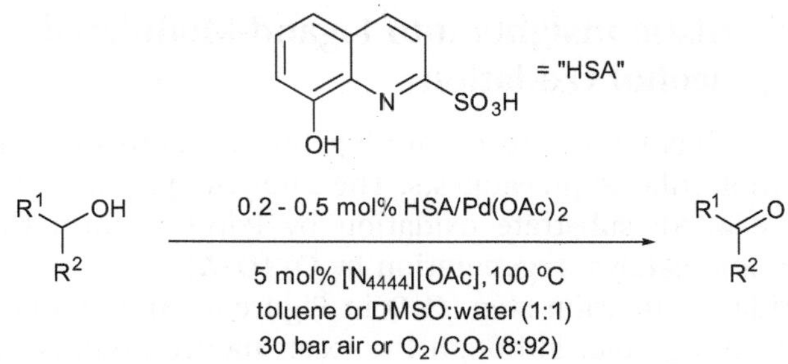

Figure 4.17 HSA/Pd(OAc)$_2$-catalysed oxidation of secondary alcohols.

comparatively slow, the amount of acetic acid was reduced or even replaced with tetrabutylammonium acetate ([N$_{4444}$][OAc]). This resulted in improved yields of products, but the use of the acetate salt, especially in high concentrations, increased the extent of catalyst decomposition observed.

Compared with other areas of homogeneous catalysis, Pd(II) alcohol oxidations have utilized a relatively narrow selection of ligand structures. It is therefore worth highlighting the use of *N,O*-ligands. Muldoon and co-workers screened a selection of commercially available *N,O*-ligands and found that 8-hydroxy-2-quinolinesulfonic acid (HSA) gave good performances.[165] The system (Figure 4.17) was most suitable for secondary alcohols, but impressively it could oxidize even unactivated aliphatic alcohols with loadings as low 0.2 mol%.[165b] Similar to the work of Sheldon and co-workers (see Table 4.2), they attempted to switch from using pressurized air to safer, more dilute O_2 mixtures (*e.g.* 8% O_2).

When 0.5 mol% of catalyst was employed, it was found that better performance could be achieved when O_2–CO_2 (8:92) was used rather than O_2–N_2 (8:92). It was thought that the CO_2 mixture resulted in superior O_2 solubility and mass transfer, resulting in a performance comparable to that when the equivalent pressure of air was used. O_2–N_2 (8:92) resulted in slower reaction rates and greater catalyst decomposition owing to the system being mass transfer limited in O_2.

Cámpora and co-workers also investigated the use of *N,O*-ligands and prepared a range of well-defined Pd complexes of the type [Pd(CH$_2$CMe$_2$Ph)(N–O)(L)], where L = pyridine or a pyridine derivative and N–O is a bidentate, anionic ligand (*e.g.* 2-pyridinecarboxylate).[166] Oxidation reactions generally required a catalyst loading of 1 mol%, 100 °C and 12 h; however, they also carried out mechanistic investigations as part of their study and these gave further insight into the influence of such *N,O*-ligands. It appears that having ligands with an acidic group means that the ligand can become involved in key steps in the catalytic cycle, 'facilitating the proton transfer from the substrate (alcohol) to the final electron acceptor (oxygen)'.[166] This conclusion also fits well the studies of Muldoon and co-workers, as they found that the best performing ligands of those screened had acidic groups in positions adjacent to where the Pd would bind.[165]

4.3.3 Mechanistic Insights into Ligand-Modulated Pd(II) Alcohol Oxidation

Significant work has gone into examining the mechanism of alcohol oxidation using molecular Pd(II) catalysts. The catalytic cycle involves two main phases (Figure 4.18): substrate oxidation by Pd(II) to form the carbonyl product (**A–D**) and catalyst regeneration by O_2 (**D–A**).

The β-hydride elimination step (**C–D** in Figure 4.18) is understood to be the rate-determining step, although the exact nature process of β-hydride elimination may depend on the catalyst system. For example, Sigman and co-workers suggested that with their monodentate NHC catalyst (IPr-NHC)Pd(OAc)$_2$(H$_2$O) system, β-hydride elimination occurs *via* a dissociative pathway, whereby a ligand/counterion dissociates prior to β-hydride elimination.[162] The kinetic isotope effect values for this catalyst were substantially larger than those for other catalyst systems (for example, pyridine- and phenanthroline-type systems). It was suggested that with these

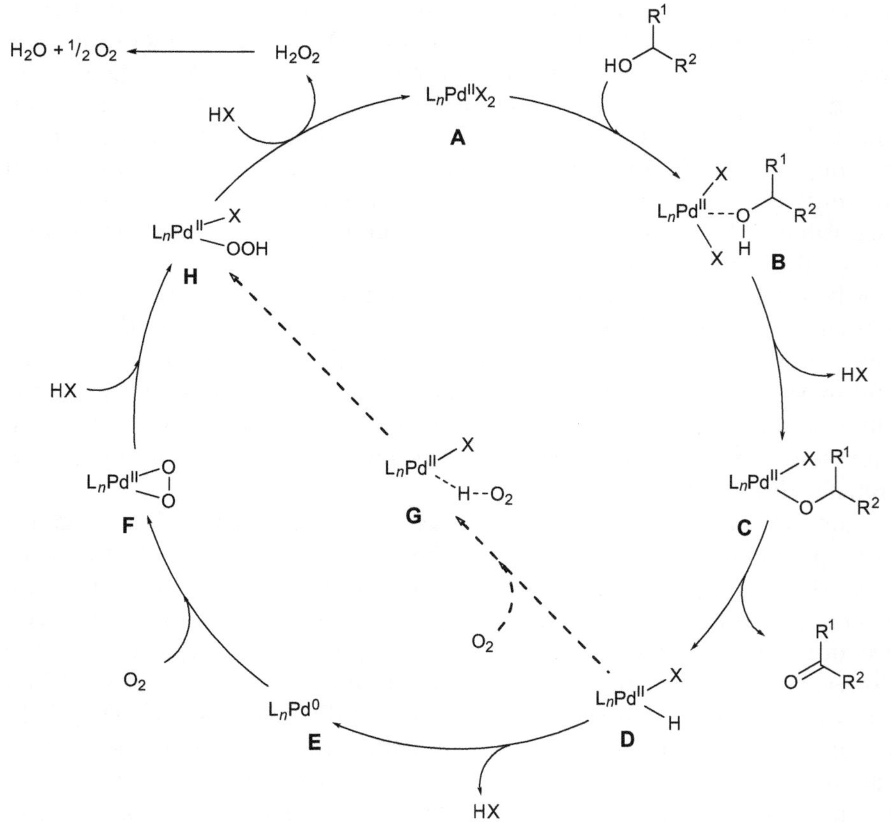

Figure 4.18 Catalytic cycle for palladium-catalysed aerobic oxidation reactions with ligands, highlighting potential pathways for the reaction of molecular oxygen to bring about catalyst regeneration.

other systems, β-hydride elimination proceeds *via* an associative pathway where the ligand/counterion is partially coordinated to Pd(II), in other words, a five-coordinate Pd species transition state.

A number of studies have investigated the mechanism by which the catalyst is regenerated (**D–A** in Figure 4.18). Under the reaction conditions, re-oxidation is of Pd(0) is generally not found to be rate determining. First, this demonstrates the significant influence of the ligands, because as we mentioned at the beginning, re-oxidation of Pd(0) species in the absence of ligands is sluggish and aggregation of Pd(0) to Pd black dominates. In the case of ligand-modulated systems, it has been found that unless the system is mass transfer limited in O_2, then the rate of alcohol oxidation is independent of O_2 pressure.[154,167] Due to facile re-oxidation under the reaction conditions, determining the exact mechanism for this part of the catalytic cycle is not straightforward. There are a number of possibilities, but arguably the two most likely pathways proposed have been reductive elimination to form Pd(0) (**E**) followed by oxidation with O_2 or direct hydrogen atom abstraction by molecular oxygen which bypasses the Pd(0) state (*i.e.* the pathway involving **G**). There have been a number of studies which show that for some systems a hydrogen atom abstraction pathway is possible,[168] but there is now considerable evidence (both experimental and computational) to support the theory that the mechanism proceeds *via* a reductive elimination pathway and a Pd(0) state.[169] The Pd(0) intermediate reacts with O_2 to form an η^2-peroxopalladium(II) complex (i.e. such as shown in **F** in Figure 4.18). Stahl *et al.* managed to isolate such a complex from the direct reaction of a Pd(0) complex, Pd(bathocuproine)(dba), with O_2.[170] They also demonstrated such peroxo species when NHC–Pd(0) complexes react with O_2.[171] In this study they further demonstrated that the addition of acetic acid resulted in the formation of the hydroperoxopalladium(II) complex (*i.e.* such as shown in **H** in Figure 4.18). Indeed, the role of acids and bases is crucial in this catalytic cycle. It can be seen that at different stages of the catalytic cycle, both Brønsted acid and Brønsted base are involved in key steps of the catalytic cycle. As is probably evident from the examples we have given, the use of a carboxylate salt such as acetate is common practice. Even in the pyridine system, it is thought that acetate from $Pd(OAc)_2$ is responsible for the generation of the alkoxide.[172] Although acetate is not strong enough to deprotonate free alcohols, once the alcohol is coordinated to the metal centre (**B**) it is activated and can be deprotonated by acetate. It can be seen in Figure 4.18 that a Brønsted acid is required to complete the re-oxidation of the catalyst. DFT studies on the oxygenation of an NHC–Pd^0 complex revealed that the formation of the peroxopalladium species is reversible, which would result in the re-formation of the Pd(0) species and further contribute to catalyst deactivation by aggregation.[173] Consequently, it is important that acid is present in order to protonate the peroxo species and lead to regeneration of the catalyst. This was touched on earlier when we discussed Waymouth and co-workers' studies with cationic complexes,[155] and it appears that the *N*,*O*-ligands with acidic groups are beneficial.[165,166] The influence of acid and base chemistry was examined in detail by Sigman and co-workers for their (IPr-NHC)Pd$(OAc)_2(H_2O)$ catalyst system.[162] In their system, they added acetic acid for optimal performance. Without the addition of acetic acid, protonation of the peroxopalladium species can become rate limiting and reversible formation of Pd(0) could lead to increased catalyst decomposition. With increasing addition of acetic acid, the β-hydride elimination step becomes more influential until it is rate limiting. They also suggested that acetic acid leads to more stable catalytic turnover by slowing the alcohol oxidation step of the reaction, thereby preventing a build-up of Pd(0), the species thought to be responsible for catalyst decomposition in most cases. This study highlights the delicate balance in terms of acid–base chemistry and the challenges in preventing catalyst decomposition, as although Pd(0) is generally not the rate-determining factor in the cycle, it does have a lifetime and this means that over time it can and does aggregate.[167]

4.3.4 Nanoparticles and Ligands

As we have just highlighted, aggregation of Pd(0) species is thought to be a major pathway of catalyst decomposition in most cases. As we discussed, the Sigman NHC system utilizes acetic acid to help reduce this pathway, whereas others have suggested that the use of bulky ligands is a possible method for preventing aggregation.[174] Mechanistic studies by Stahl and co-workers found that molecular sieves that were used in the original Uemura/pyridine system are beneficial as they provide a surface that interacts with the catalyst and reduces the rate of Pd(0) aggregation.[175] In the first half of this chapter we focused on heterogeneous catalysts, which are predominantly solid-supported Pd nanoparticles. The use of molecular sieves in the pyridine system demonstrates that there is an overlap between 'homogeneous' and 'heterogeneous' systems, and in homogeneous systems the active species may well not be a defined molecular species. Indeed, Sheldon and co-workers recently discussed that stabilized nanoparticles may be the active catalyst in the case of their neocuproine aqueous–DMSO system.[176] As they pointed out, such catalysts are similar to the ligand-stabilized 'giant palladium clusters' (before the term 'nanoparticle' was as prevalent) previously described by others such as Moiseev and co-workers.[177]

There is also a link between homogeneous and heterogeneous systems when we look at the use of ligand-stabilized nanoparticles on solid supports. For example, in 1999, Kaneda and co-workers reported the use of 1,10-phenanthroline ligand-stabilized 'giant palladium clusters' uniformly immobilized on a neutral TiO_2.[178] A number of groups have explored the use of silica-tethered amine ligands for the stabilization of Pd nanoparticles (Figure 4.19).[179]

The work to-date demonstrates (*via* TEM) that such systems result in nanoparticles rather than ligated discrete molecular complexes. From these studies, there are certainly similar trends to those observed with the types of heterogeneous systems that we discussed in the first half the chapter.

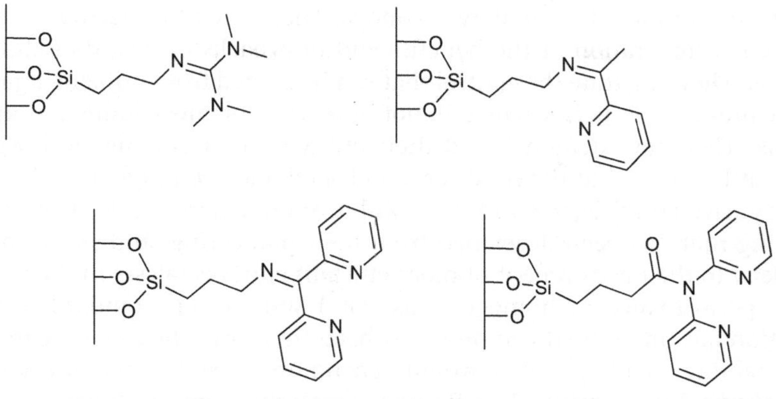

Figure 4.19 Examples of tethered ligand systems for stabilizing Pd nanoparticles: immobilized guanidine[180f] and bipyridyl and pyridyl variants.[180d]

For example, studies show that better/more stable catalysts are produced when mesoporous silicas such as SBA-15 and SBA-16 are used as the support. Additionally, although such catalysts can be recycled and reused, they are really only feasible for activated alcohols; these ligand-stabilized nanoparticles systems have relatively poor performance for unactivated aliphatic alcohols.

4.4 Concluding Remarks

In the last 15 years, there has been significant progress in the performance and understanding of aerobic palladium catalyst systems for alcohol oxidation. This applies to heterogeneous and homogeneous/ligand-modulated systems. At the present time there are still a number of challenges and both heterogeneous and ligand-modulated systems have their own advantages and disadvantages. In terms of utilizing Pd-based catalysts on a larger scale, the use of a heterogeneous catalyst is undoubtedly preferred. They enable the catalyst to be readily separated from the product and continually re-used or indeed employed in a continuous flow system. As we have highlighted, at present, heterogeneous Pd catalysts need improved performance for challenging substrates such as polyols and unactivated alcohols. In the case of homogeneous/ligand-modulated systems, the performance of these systems can have a number of advantages. A number of catalysts have been shown to oxidize challenging unactivated alcohols effectively and can often operate under mild conditions. Furthermore, the use of ligands offers the chance to develop systems that demonstrate excellent selectivity, and we have given some examples where the catalyst exhibits excellent chemoselectivity. Although we have not discussed this and we have focused on standard alcohol oxidation, it is worth highlighting that ligands also offer opportunities for developing chiral catalysts. For example, the groups of Sigman[180] and Stoltz[181] have (primarily) exploited the use of (–)-sparteine to carry out the kinetic resolution of secondary alcohols. These enantioselective systems are a good illustration of the benefits and drawbacks of ligand-modulated systems. They illustrate the control that can be gained from utilizing ligands, but at present the performance is not adequate for most industrial applications. They have demonstrated their utility at an academic level (*e.g.* in total synthesis)[181g] but the need for 5 mol% Pd and even higher loadings of an expensive chiral ligand will be prohibitive on a larger scale. The understanding that has been developed from their pioneering studies will hopefully lead to the development of more efficient chiral catalysts in the future.

The performance of homogeneous Pd(II) systems for standard alcohol oxidation has undoubtedly improved substantially over the last 15 years and their viability on a larger scale would depend on the substrate, for example, inexpensive fine chemicals (*e.g.* fragrance molecules) may well need catalyst loadings even lower than 0.1 mol%. The TONs that are achieved still lag significantly behind those in other areas of homogeneous catalysis (*e.g.* hydrogenation catalysts), and hopefully this can be improved by further reducing the aggregation of Pd(0) species. An ideal catalyst may be some sort of solid-supported/covalently tethered Pd(II) complex that does not aggregate and exhibits good performance for unactivated alcohols. It is hoped that continued efforts will result in these remaining challenges being overcome.

References

1. J. J. Berzelius, *Ann. Phys. Chim.*, 1828, **B13**, 435.
2. T. Mallat and A. Baiker, *Chem. Rev.*, 2004, **104**, 3037.
3. A. F. Lee, in *Heterogeneous Catalysts for Clean Technology: Spectroscopy, Design and Monitoring*, ed. K. Wilson and A. F. Lee, Wiley-VCH, Weinheim, 2013, pp. 11–33.
4. C. P. Vinod, K. Wilson and A. F. Lee, *J. Chem. Technol. Biotechnol.*, 2011, **86**, 161.
5. A. M. Beale, S. D. M. Jacques and B. M. Weckhuysen, *Chem. Soc. Rev.*, 2010, **39**, 4656.
6. J. D. Grunwaldt and C. G. Schroer, *Chem. Soc. Rev.*, 2010, **39**, 4741.
7. A. F. Lee, *Aust. J. Chem.*, 2012, **65**, 615.
8. P. L. Gai, R. Sharma and F. M. Ross, *MRS Bull.*, 2008, **33**, 107.
9. K. L. Jungjohann, J. E. Evans, J. A. Aguiar, I. Arslan and N. D. Browning, *Microsc. Microanal.*, 2012, **18**, 621.
10. N. D. Browning, M. A. Bonds, G. H. Campbell, J. E. Evans, T. LaGrange, K. L. Jungjohann, D. J. Masiel, J. McKeown, S. Mehraeen, B. W. Reed and M. Santala, *Curr. Opin. Solid State Mater. Sci.*, 2012, **16**, 23.
11. E. D. Boyes, M. R. Ward, L. Lari and P. L. Gai, *Ann. Phys. Berlin*, 2013, **525**, 423.
12. Y. Xiong, H. Cai, B. J. Wiley, J. Wang, M. J. Kim and Y. Xia, *J. Am. Chem. Soc.*, 2007, **129**, 3665.
13. H. Zhang, M. Jin, Y. Xiong, B. Lim and Y. Xia, *Acc. Chem. Res.*, 2012, **46**, 1783.

14. X. Xia, S.-I. Choi, J. A. Herron, N. Lu, J. Scaranto, H.-C. Peng, J. Wang, M. Mavrikakis, M. J. Kim and Y. Xia, *J. Am. Chem. Soc.*, 2013, **135**, 15706.
15. S. Kobayashi, H. Miyamura, R. Akiyama and T. Ishida, *J. Am. Chem. Soc.*, 2005, **127**, 9251.
16. K. Kaizuka, K. Y. Lee, H. Miyamura and S. Kobayashi, *J. Flow Chem.*, 2012, **2**, 1.
17. X. A. Ye, M. D. Johnson, T. N. Diao, M. H. Yates and S. S. Stahl, *Green Chem.*, 2010, **12**, 1180.
18. A. Ayude, J. Cechini, M. Cassanello, O. Martinez and P. Haure, *Chem. Eng. Sci.*, 2008, **63**, 4969.
19. R. Tschan, R. Wandeler, M. S. Schneider, M. M. Schubert and A. Baiker, *J. Catal.*, 2001, **204**, 219.
20. M. Caravati, J. D. Grunwaldt and A. Baiker, *Catal. Today*, 2004, **91–92**, 1.
21. J. D. Grunwaldt, M. Caravati and A. Baiker, *J. Phys. Chem. B*, 2006, **110**, 9916.
22. M. Burgener, T. Tyszewski, D. Ferri, T. Mallat and A. Baiker, *Appl. Catal. A*, 2006, **299**, 66.
23. Z. S. Hou, N. Theyssen and W. Leitner, *Green Chem.*, 2007, **9**, 127.
24. Z. Hou, N. Theyssen, A. Brinkmann, K. V. Klementiev, W. Grünert, M. Bühl, W. Schmidt, B. Spliethoff, B. Tesche and C. Weidenthaler, *J. Catal.*, 2008, **258**, 315.
25. D. Astruc, F. Lu and J. R. Aranzaes, *Angew. Chem. Int. Ed.*, 2005, **44**, 7852.
26. I. Hermans, E. S. Spier, U. Neuenschwander, N. Turrà and A. Baiker, *Top. Catal.*, 2009, **52**, 1162.
27. K. Heyns and H. Paulsen, *Angew. Chem.*, 1957, **69**, 600.
28. X. Y. Liu, R. J. Madix and C. M. Friend, *Chem. Soc. Rev.*, 2008, **37**, 2243.
29. N. Dimitratos, J. A. Lopez-Sanchez and G. J. Hutchings, *Chem. Sci.*, 2012, **3**, 20.
30. M. Besson and P. Gallezot, *Catal. Today*, 2000, **57**, 127.
31. J. Kluytmans, A. Markusse, B. Kuster, G. Marin and J. Schouten, *Catal. Today*, 2000, **57**, 143.
32. T. Mallat and A. Baiker, *Catal. Today*, 1994, **19**, 247.
33. K.-M. Choi, T. Akita, T. Mizugaki, K. Ebitani and K. Kaneda, *New J. Chem.*, 2003, **27**, 324.
34. J. Chen, Q. Zhang, Y. Wang and H. Wan, *Adv. Synth. Catal.*, 2008, **350**, 453.
35. F. Li, Q. Zhang and Y. Wang, *Appl. Catal. A*, 2008, **334**, 217.
36. A. F. Lee and K. Wilson, *Green Chem.*, 2004, **6**, 37.
37. A. F. Lee, S. F. Hackett, J. S. Hargreaves and K. Wilson, *Green Chem.*, 2006, **8**, 549.
38. C. M. A. Parlett, D. W. Bruce, N. S. Hondow, A. F. Lee and K. Wilson, *ACS Catal.*, 2011, **1**, 636.
39. C. M. A. Parlett, D. W. Bruce, N. S. Hondow, M. A. Newton, A. F. Lee and K. Wilson, *ChemCatChem*, 2013, **5**, 939.

40. S. F. Hackett, R. M. Brydson, M. H. Gass, I. Harvey, A. D. Newman, K. Wilson and A. F. Lee, *Angew. Chem.*, 2007, **119**, 8747.
41. J. Baeza, L. Calvo, M. Gilarranz, A. Mohedano, J. Casas and J. Rodriguez, *J. Catal.*, 2012, **293**, 85.
42. X. Wang, H. Yang, B. Feng, Z. Hou, Y. Hu, Y. Qiao, H. Li and X. Zhao, *Catal. Lett.*, 2009, **132**, 34.
43. S. Yudha, S. R. N. Dhital and H. Sakurai, *Tetrahedron Lett.*, 2011, **52**, 2633.
44. R. Dun, X. Wang, M. Tan, Z. Huang, X. Huang, W. Ding and X. Lu, *ACS Catal.*, 2013, 3063.
45. M. K. Weldon and C. M. Friend, *Chem. Rev.*, 1996, **96**, 1391.
46. A. F. Lee, Z. Chang, P. Ellis, S. F. Hackett and K. Wilson, *J. Phys. Chem. C*, 2007, **111**, 18844.
47. J. Naughton, A. Pratt, C. W. Woffinden, C. Eames, S. P. Tear, S. M. Thompson, A. F. Lee and K. Wilson, *J. Phys. Chem. C*, 2011, **115**, 25290.
48. J. L. Davis and M. A. Barteau, *Surf. Sci.*, 1987, **187**, 387.
49. J. L. Davis and M. A. Barteau, *Surf. Sci.*, 1990, **235**, 235.
50. F. Zaera, *Catal. Lett.*, 2003, **91**, 1.
51. C. Keresszegi, T. Burgi, T. Mallat and A. Baiker, *J. Catal.*, 2002, **211**, 244.
52. A. F. Lee, *Abstr. Pap. Am. Chem. Soc.*, 2001, **221**, U335.
53. J.-D. Grunwaldt, C. Keresszegi, T. Mallat and A. Baiker, *J. Catal.*, 2003, **213**, 291.
54. C. Keresszegi, J.-D. Grunwaldt, T. Mallat and A. Baiker, *J. Catal.*, 2004, **222**, 268.
55. J.-D. Grunwaldt, M. Caravati and A. Baiker, *J. Phys. Chem. B*, 2006, **110**, 25586.
56. C. Mondelli, D. Ferri, J.-D. Grunwaldt, F. Krumeich, S. Mangold, R. Psaro and A. Baiker, *J. Catal.*, 2007, **252**, 77.
57. C. Keresszegi, D. Ferri, T. Mallat and A. Baiker, *J. Catal.*, 2005, **234**, 64.
58. M. Caravati, D. M. Meier, J.-D. Grunwaldt and A. Baiker, *J. Catal.*, 2006, **240**, 126.
59. T. Mallat and A. Baiker, *Catal. Today*, 1995, **24**, 143.
60. T. Mallat, Z. Bodnar, P. Hug and A. Baiker, *J. Catal.*, 1995, **153**, 131.
61. J.-D. Grunwaldt, M. Caravati and A. Baiker, *J. Phys. Chem. B*, 2006, **110**, 9916.
62. C. M. A. Parlett, P. Keshwalla, S. G. Wainwright, D. W. Bruce, N. S. Hondow, K. Wilson and A. F. Lee, *ACS Catal.*, 2013, **3**, 2122.
63. C. M. A. Parlett, L. J. Durndell, K. Wilson, D. W. Bruce, N. S. Hondow and A. F. Lee, *Catal. Commun.*, 2014, **44**, 40.
64. J. Naughton, A. F. Lee, S. Thompson, C. P. Vinod and K. Wilson, *Phys. Chem. Chem. Phys*, 2010, **12**, 2670.
65. E. W. Ping, J. Pierson, R. Wallace, J. T. Miller, T. F. Fuller and C. W. Jones, *Appl. Catal. A*, 2011, **396**, 85.
66. M. A. Newton, B. Jyoti, A. J. Dent, S. G. Fiddy and J. Evans, *Chem. Commun.*, 2004, 2382.

67. D. Ferri and A. Baiker, *Top. Catal.*, 2009, **52**, 1323.
68. M. A. Newton, *Top. Catal.*, 2009, **52**, 1410.
69. M. A. Newton, A. J. Dent, S. G. Fiddy, B. Jyoti and J. Evans, *Catal. Today*, 2007, **126**, 64.
70. A. F. Lee, C. V. Ellis, J. N. Naughton, M. A. Newton, C. M. Parlett and K. Wilson, *J. Am. Chem. Soc.*, 2011, **133**, 5724.
71. C. M. A. Parlett, C. V. Gaskell, J. N. Naughton, M. A. Newton, K. Wilson and A. F. Lee, *Catal. Today*, 2013, **205**, 76.
72. C. V. Gaskell, C. M. A. Parlett, M. A. Newton, K. Wilson and A. F. Lee, *ACS Catal.*, 2012, **2**, 2242.
73. A. F. Lee, J. N. Naughton, Z. Liu and K. Wilson, *ACS Catal.*, 2012, **2**, 2235.
74. S. Tauster, *Acc. Chem. Res.*, 1987, **20**, 389.
75. E. Ghedini, F. Menegazzo, M. Signoretto, M. Manzoli, F. Pinna and G. Strukul, *J. Catal.*, 2010, **273**, 266.
76. R. F. Hicks, H. Qi, M. L. Young and R. G. Lee, *J. Catal.*, 1990, **122**, 295.
77. J. Beckers and G. Rothenberg, *Green Chem.*, 2010, **12**, 939.
78. K. Inumaru, K. Nakamura, K. Ooyachi, K. Mizutani, S. Akihara and S. Yamanaka, *J. Porous Mater.*, 2010, **18**, 455.
79. Y.-M. Liu, Y. Cao, N. Yi, W.-L. Feng, W.-L. Dai, S.-R. Yan, H.-Y. He and K.-N. Fan, *J. Catal.*, 2004, **224**, 417.
80. D. Zhao, Q. Huo, J. Feng, B. F. Chmelka and G. D. Stucky, *J. Am. Chem. Soc.*, 1998, **120**, 6024.
81. T.-W. Kim, F. Kleitz, B. Paul and R. Ryoo, *J. Am. Chem. Soc.*, 2005, **127**, 7601.
82. D. Zhao, J. Feng, Q. Huo, N. Melosh, G. H. Fredrickson, B. F. Chmelka and G. D. Stucky, *Science*, 1998, **279**, 548.
83. O. Sneh and S. M. George, *J. Phys. Chem.*, 1995, **99**, 4639.
84. J. M. Kim, J. H. Kwak, S. Jun and R. Ryoo, *J. Phys. Chem.*, 1995, **99**, 16742.
85. K. Kaneda, M. Fujii and K. Morioka, *J. Org. Chem.*, 1996, **61**, 4502.
86. V. Polshettiwar and R. S. Varma, *Org. Biomol. Chem.*, 2009, **7**, 37.
87. F. Behafarid, L. K. Ono, S. Mostafa, J. R. Croy, G. Shafai, S. Hong, T. S. Rahman, S. R. Bare and B. R. Cuenya, *Phys. Chem. Chem. Phys.*, 2012, **14**, 11766.
88. Q. Yuan, H. H. Duan, L. L. Li, L. D. Sun, Y. W. Zhang and C. H. Yan, *J. Colloid. Interface Sci.*, 2009, **335**, 151.
89. Y. Lykhach, T. Staudt, M. P. A. Lorenz, R. Streber, A. Bayer, H.-P. Steinrück and J. Libuda, *ChemPhysChem*, 2010, **11**, 1496.
90. Tana, M. Zhang, J. Li, H. Li, Y. Li and W. Shen, *Catal. Today*, 2009, **148**, 179.
91. A. Bensalem, F. Bozon-Verduraz and V. Perrichon, *J. Chem. Soc., Faraday Trans.*, 1995, **91**, 2185.
92. B. Harrison, A. Diwell and C. Hallett, *Platinum Met. Rev.*, 1988, **32**, 73.
93. Y. Zhu, S. Zhang, J.-j. Shan, L. Nguyen, S. Zhan, X. Gu and F. Tao, *ACS Catal.*, 2013, **3**, 2627.

94. T. Maillet, Y. Madier, R. Taha, J. Barbier Jr. and D. Duprez, *Stud. Surf. Sci. Catal.*, 1997, **112**, 267.
95. S.-H. Oh and G. B. Hoflund, *J. Phys. Chem. A*, 2006, **110**, 7609.
96. A. Badri, C. Binet and J.-C. Lavalley, *J. Chem. Soc., Faraday Trans.*, 1996, **92**, 1603.
97. W.-C. Chu, S.-F. Chiang, J.-G. Li and S.-W. Kuo, *RSC Adv.*, 2014, **4**, 784.
98. H. Li, L. Zhang and H. He, *Inorg. Chem.*, 2009, **48**, 4421.
99. D. M. Lyons, K. M. Ryan and M. A. Morris, *J. Mater. Chem.*, 2002, **12**, 1207.
100. H. Zhu, Z. Qin, W. Shan, W. Shen and J. Wang, *J. Catal.*, 2004, **225**, 267.
101. R. S. Monteiro, L. C. Dieguez and M. Schmal, *Catal. Today*, 2001, **65**, 77.
102. M.-F. Luo, Z.-Y. Hou, X.-X. Yuan and X.-M. Zheng, *Catal. Lett.*, 1998, **50**, 205.
103. F. Mariño, C. Descorme and D. Duprez, *Appl. Catal. B*, 2004, **54**, 59.
104. M. S. Hamdy, R. Amrollahi and G. Mul, *ACS Catal.*, 2012, 2641.
105. R. Zhang, A. A. Elzatahry, S. S. Al-Deyab and D. Zhao, *Nano Today*, 2012, **7**, 344.
106. I. Tamiolakis, I. N. Lykakis, A. P. Katsoulidis and G. S. Armatas, *Chem. Commun.*, 2012, **48**, 6687.
107. A. A. Ismail and D. W. Bahnemann, *J. Mater. Chem.*, 2011, **21**, 11686.
108. G. Soler-Illia, A. Louis and C. Sanchez, *Chem. Mater.*, 2002, **14**, 750.
109. M. Niederberger, M. H. Bartl and G. D. Stucky, *Chem. Mater.*, 2002, **14**, 4364.
110. K. Wada, K. Yano, T. Kondo and T.-a. Mitsudo, *Catal. Today*, 2006, **117**, 242.
111. K. Mori, T. Hara, T. Mizugaki, K. Ebitani and K. Kaneda, *J. Am. Chem. Soc.*, 2004, **126**, 10657.
112. K. Mori, K. Yamaguchi, T. Hara, T. Mizugaki, K. Ebitani and K. Kaneda, *J. Am. Chem. Soc.*, 2002, **124**, 11572.
113. F. Wang, Z. Lu, L. Yang, Y. Zhang, Q. Tang, Y. Guo, X. Ma and Z. Yang, *Chem. Commun.*, 2013, **49**, 6626.
114. A. F. Lee, J. J. Gee and H. J. Theyers, *Green Chem.*, 2000, **2**, 279.
115. M. Wenkin, P. Ruiz, B. Delmon and M. Devillers, *J. Mol. Catal. A: Chem.*, 2002, **180**, 141.
116. F. Alardin, B. Delmon, P. Ruiz and M. Devillers, *Catal. Today*, 2000, **61**, 255.
117. C. Keresszegi, *J. Catal.*, 2004, **225**, 138.
118. H. Pinxt, B. Kuster and G. Marin, *Appl. Catal. A*, 2000, **191**, 45.
119. N. Dimitratos, F. Porta, L. Prati and A. Villa, *Catal. Lett.*, 2005, **99**, 181.
120. L. Prati, A. Villa, C. Campione and P. Spontoni, *Top. Catal.*, 2007, **44**, 319.
121. D. I. Enache, J. K. Edwards, P. Landon, B. Solsona-Espriu, A. F. Carley, A. A. Herzing, M. Watanabe, C. J. Kiely, D. W. Knight and G. J. Hutchings, *Science*, 2006, **311**, 362.
122. J. Long, H. Liu, S. Wu, S. Liao and Y. Li, *ACS Catal.*, 2013, **3**, 647.
123. M. B. Griffin, A. A. Rodriguez, M. M. Montemore, J. R. Monnier, C. T. Williams and J. W. Medlin, *J. Catal.*, 2013, **307**, 111.

124. J. Xu, H. Zhang, Y. Zhao, B. Yu, S. Chen, Y. Li and L. Hao, *Green Chem.*, 2013, **15**, 1520.
125. R. Wang, Z. Wu, C. Chen, Z. Qin, H. Zhu, G. Wang, H. Wang, C. Wu, W. Dong and W. Fan, *Chem. Commun.*, 2013, **49**, 8250.
126. W. Hou, N. A. Dehm and R. W. Scott, *J. Catal.*, 2008, **253**, 22.
127. P. Mertens, S. Corthals, X. Ye, H. Poelman, P. Jacobs, B. Sels, I. Vankelecom and D. De Vos, *J. Mol. Catal. A: Chem.*, 2009, **313**, 14.
128. G. J. Hutchings, *Catal. Today*, 2005, **100**, 55.
129. A. F. Lee, C. V. Ellis, K. Wilson and N. S. Hondow, *Catal. Today*, 2010, **157**, 243.
130. A. F. Lee, S. F. J. Hackett, G. J. Hutchings, S. Lizzit, J. Naughton and K. Wilson, *Catal. Today*, 2009, **145**, 251.
131. R. W. Scott, O. M. Wilson, S.-K. Oh, E. A. Kenik and R. M. Crooks, *J. Am. Chem. Soc.*, 2004, **126**, 15583.
132. R. W. Scott, C. Sivadinarayana, O. M. Wilson, Z. Yan, D. W. Goodman and R. M. Crooks, *J. Am. Chem. Soc.*, 2005, **127**, 1380.
133. W. Hou, N. A. Dehm and R. W. J. Scott, *J. Catal.*, 2008, **253**, 22.
134. P. Dash, T. Bond, C. Fowler, W. Hou, N. Coombs and R. W. J. Scott, *J. Phys. Chem. C*, 2009, **113**, 12719.
135. A. Maclennan, A. Banerjee, Y. Hu, J. T. Miller and R. W. J. Scott, *ACS Catal.*, 2013, **3**, 1411.
136. H. Wu, Q. Zhang and Y. Wang, *Adv. Synth. Catal.*, 2005, **347**, 1356.
137. C. M. A. Parlett, L. J. Durndell, A. Machado, G. Cibin, D. W. Bruce, N. S. Hondow, K. Wilson and A. F. Lee, *Catal. Today*, 2014, **229**, 46.
138. B. Feng, Z. Hou, H. Yang, X. Wang, Y. Hu, H. Li, Y. Qiao, X. Zhao and Q. Huang, *Langmuir*, 2009, **26**, 2505.
139. R. Dun, X. Wang, M. Tan, Z. Huang, X. Huang, W. Ding and X. Lu, *ACS Catal.*, 2013, **3**, 3063.
140. W. C. Ketchie, M. Murayama and R. J. Davis, *J. Catal.*, 2007, **250**, 264.
141. (a) J. Smidt, W. Hafner, R. Jira, J. Sedlmeier, R. Sieber, R. Rüttinger and H. Kojer, *Angew. Chem.*, 1959, **71**, 176; (b) J. Smidt, W. Hafner, R. Jira, R. Sieber, J. Sedlmeier and A. Sabel, *Angew. Chem. Int. Ed.*, 1962, **1**, 80.
142. For a review, see: J. Piera and J. E. Bäckvall, *Angew. Chem. Int. Ed.*, 2008, **47**, 3506.
143. (a) T. Blackburn and J. Schwartz, *J. Chem. Soc., Chem. Commun.*, 1977, 157; (b) T. Blackburn, J. Schwartz, R. J. Gillespie, J. Murray-Rust, P. Murray-Rust, A. E. A. Porter, D. Ayalon-Chass, E. Ehlinger, P. Magnus, A. G. Schultz and I.-C. Chiu, *J. Chem. Soc., Chem. Commun.*, 1978, 632.
144. E. Gómez-Bengoa, P. Noheda and A. M. Echavarren, *Tetrahedron Lett.*, 1994, **35**, 7097.
145. S. Aït-Mohand, F. Hénin and J. Muzart, *Tetrahedron Lett.*, 1995, **36**, 2473.
146. K. P. Peterson and R. C. Larock, *J. Org. Chem.*, 1998, **63**, 3185.
147. T. Nishimura, T. Onoue, K. Ohe and S. Uemura, *Tetrahedron Lett.*, 1998, **39**, 6011.

148. T. Nishimura, T. Onoue, K. Ohe and S. Uemura, *J. Org. Chem.*, 1999, **64**, 6750.
149. R. Bortolo, D. Bianchi, R. D'Aloisio, C. Querci and M. Ricci, *J. Mol. Catal. A: Chem.*, 2000, **153**, 25.
150. G.-J. Brink, I. W. C. E. Arends and R. A. Sheldon, *Science*, 2000, **287**, 1636.
151. G.-J. ten Brink, I. W. C. E. Arends and R. A. Sheldon, *Adv. Synth. Catal.*, 2002, **344**, 355.
152. G.-J. ten Brink, I. W. C. E. Arends, M. Hoogenraad, G. Verspui and R. A. Sheldon, *Adv. Synth. Catal.*, 2003, **345**, 497.
153. G.-J. ten Brink, I. W. C. E. Arends, M. Hoogenraad, G. Verspui and R. A. Sheldon, *Adv. Synth. Catal.*, 2003, **345**, 1341.
154. For an overview of this work, see: I. W. C. E. Arends, G.-J. ten Brink and R. A. Sheldon, *J. Mol. Catal. A: Chem.*, 2006, **251**, 246.
155. N. R. Conley, L. A. Labios, D. M. Pearson, C. C. L. McCrory and R. M. Waymouth, *Organometallics*, 2007, **26**, 5447.
156. D. M. Pearson, N. R. Conley and R. M. Waymouth, *Organometallics*, 2011, **30**, 1445.
157. R. M. Painter, D. M. Pearson and R. M. Waymouth, *Angew. Chem. Int. Ed.*, 2010, **49**, 9456.
158. K. Chung, S. M. Banik, A. G. De Crisci, D. M. Pearson, T. R. Blake, J. V. Olsson, A. J. Ingram, R. N. Zare and R. M. Waymouth, *J. Am. Chem. Soc.*, 2013, **135**, 7593.
159. L. Bettucci, C. Bianchini, J. Filippi, A. Lavacchi and W. Oberhauser, *Eur. J. Inorg. Chem.*, 2011, 1797.
160. D. R. Jensen, M. J. Schultz, J. A. Mueller and M. S. Sigman, *Angew. Chem. Int. Ed.*, 2003, **42**, 3810.
161. V. Jurčík, T. E. Schmid, Q. Dumont, A. M. Z. Slawin and C. S. J. Cazin, *Dalton Trans.*, 2012, **41**, 12619.
162. J. A. Mueller, C. P. Goller and M. S. Sigman, *J. Am. Chem. Soc.*, 2004, **126**, 9724.
163. M. J. Schultz, C. C. Park and M. S. Sigman, *Chem. Commun.*, **2002**, 3034.
164. M. J. Schultz, S. S. Hamilton, D. R. Jensen and M. S. Sigman, *J. Org. Chem.*, 2005, **70**, 3343.
165. (a) D. S. Bailie, G. M. A. Clendenning, L. McNamee and M. J. Muldoon, *Chem. Commun.*, 2010, **46**, 7238; (b) L. M. Dornan, G. M. A. Clendenning, M. B. Pitak, S. J. Coles and M. J. Muldoon, *Catal. Sci. Technol.*, 2014, **4**, 2526.
166. C. Melero, O. N. Shishilov, E. Alvarez, P. Palma and J. Cámpora, *Dalton Trans.*, 2012, **41**, 14087.
167. B. A. Steinhoff and S. S. Stahl, *J. Am. Chem. Soc.*, 2006, **128**, 4348.
168. (a) M. C. Denney, N. A. Smythe, K. L. Cetto, R. A. Kemp and K. I. Goldberg, *J. Am. Chem. Soc.*, 2006, **128**, 2508; (b) J. M. Keith, R. P. Muller, R. A. Kemp, K. I. Goldberg, W. A. Goddard III and J. Oxgaard, *Inorg. Chem.*, 2006, **45**, 9631; (c) J. M. Keith, R. J. Nielsen,

J. Oxgaard and W. A. Goddard III, *J. Am. Chem. Soc.*, 2005, **127**, 13172; (d) R. J. Nielsen and W. A. Goddard III, *J. Am. Chem. Soc.*, 2006, **128**, 9651; (e) M. M. Konnick, N. Decharin, B. V. Popp and S. S. Stahl, *Chem. Sci.*, 2011, **2**, 326.

169. (a) C. R. Landis, C. M. Morales and S. S. Stahl, *J. Am. Chem. Soc.*, 2004, **126**, 16302; (b) B. V. Popp and S. S. Stahl, *J. Am. Chem. Soc.*, 2007, **129**, 4410; (c) M. M. Konnick and S. S. Stahl, *J. Am. Chem. Soc.*, 2008, **130**, 5753; (d) N. Decharin and S. S. Stahl, *J. Am. Chem. Soc.*, 2011, **133**, 5732; (e) B. V. Popp and S. S. Stahl, *Chem. Eur. J.*, 2009, **15**, 2915; (f) N. Decharin, B. V. Popp and S. S. Stahl, *J. Am. Chem. Soc.*, 2011, **133**, 13268; (g) J. M. Keith and W. A. Goddard III, *Organometallics*, 2012, **31**, 545.
170. S. S. Stahl, J. L. Thorman, R. C. Nelson and M. A. Kozee, *J. Am. Chem. Soc.*, 2001, **123**, 7188.
171. M. M. Konnick, I. A. Guzei and S. S. Stahl, *J. Am. Chem. Soc.*, 2004, **126**, 10212.
172. B. A. Steinhoff, I. A. Guzei and S. S. Stahl, *J. Am. Chem. Soc.*, 2004, **126**, 11268.
173. B. V. Popp, J. E. Wendlandt, C. R. Landis and S. S. Stahl, *Angew. Chem. Int. Ed.*, 2007, **46**, 601.
174. T. Iwasawa, M. Tokunaga, Y. Obora and Y. Tsuji, *J. Am. Chem. Soc.*, 2004, **126**, 6554.
175. B. A. Steinhoff, A. E. King and S. S. Stahl, *J. Org. Chem.*, 2006, **71**, 1861.
176. M. Mifsud, K. V. Parkhomenko, I. W. C. E. Arends and R. A. Sheldon, *Tetrahedron*, 2010, **66**, 1040.
177. For examples, see: (a) M. N. Vargaftik, V. P. Zagorodnikov, I. P. Stolarov, I. I. Moiseev, V. A. Likholobov, D. I. Kotchubey, A. L. Chuvilin, V. I. Zaikovsky, K. I. Zamaraev and G. I. Timofeeva, *J. Chem. Soc., Chem. Commun.*, 1985, 937; (b) G. Kovtun, T. Kameneva, S. Hladyi, M. Starchevsky, Y. Pazdersky, I. Stolarov, M. Vargaftik and I. Moiseev, *Adv. Synth. Catal.*, 2002, **344**, 957.
178. K. Ebitani, Y. Fujie and K. Kaneda, *Langmuir*, 1999, **15**, 3557.
179. (a) D. Choudhary, S. Paul, R. Gupta and J. H. Clark, *Green Chem.*, 2006, **8**, 479; (b) B. Karimi, S. Abedi, J. H. Clark and V. Budarin, *Angew. Chem. Int. Ed.*, 2006, **45**, 4776; (c) Z. Hou, N. Theyssena, A. Brinkmanna, K. V. Klementiev, W. Grünert, M. Bühl, W. Schmidt, B. Spliethoff, B. Tesche, C. Weidenthaler and W. Leitner, *J. Catal.*, 2008, **258**, 315; (d) B. Karimi, A. Zamani, S. Abedi and J. H. Clark, *Green Chem.*, 2009, **11**, 109; (e) H. Yang, Z. Ma, Y. Qing, G. Xie, J. Gao, L. Zhang, J. Gao and L. Du, *Appl. Catal. A: Gen.*, 2010, **382**, 312; (f) H. Yang, X. Han, Z. Ma, R. Wang, J. Liu and X. Ji, *Green Chem.*, 2010, **12**, 441.
180. (a) D. R. Jensen, J. S. Pugsley and M. S. Sigman, *J. Am. Chem. Soc.*, 2001, **123**, 7475; (b) J. A. Mueller, D. R. Jensen and M. S. Sigman, *J. Am. Chem. Soc.*, 2002, **124**, 8202; (c) D. R. Jensen and M. S. Sigman, *Org. Lett.*, 2003, **5**, 63; (d) S. K. Mandal, D. R. Jensen, J. S. Pugsley and M. S. Sigman, *J. Org. Chem.*, 2003, **68**, 4600; (e) S. K. Mandal and M. S. Sigman, *J. Org.*

Chem., 2003, **68**, 7535; (f) J. A. Mueller and M. S. Sigman, *J. Am. Chem. Soc.*, 2003, **125**, 7005.

181. (a) E. M. Ferreira and B. M. Stoltz, *J. Am. Chem. Soc.*, 2001, **123**, 7725; (b) J. T. Bagdanoff, E. M. Ferreira and B. M. Stoltz, *Org. Lett.*, 2003, **5**, 835; (c) J. T. Bagdanoff and B. M. Stoltz, *Angew. Chem. Int. Ed.*, 2004, **43**, 353; (d) R. M. Trend and B. M. Stoltz, *J. Am. Chem. Soc.*, 2004, **126**, 4482; (e) R. J. Nielsen, J. M. Keith, B. M. Stoltz and W. A. Goddard III, *J. Am. Chem. Soc.*, 2004, **126**, 7967; (f) D. C. Ebner, R. M. Trend, C. Genet, M. J. McGrath, P. O'Brien and B. M. Stoltz, *Angew. Chem. Int. Ed.*, 2008, **47**, 6367; (g) D. C. Ebner, J. T. Bagdanoff, E. M. Ferreira, R. M. McFadden, D. D. Caspi, R. M. Trend and B. M. Stoltz, *Chem. Eur. J.*, 2009, **15**, 12978.

CHAPTER 5

Gold-Based Catalysts

CRISTINA DELLA PINA, ERMELINDA FALLETTA AND MICHELE ROSSI*

Dipartimento di Chimica e ISTM-CNR, Università degli Studi di Milano, Via Golgi 19, 20133 Milan, Italy
*Email: michele.rossi@unimi.it

5.1 Introduction

Catalysis has been an essential part of many industrial and technological developments that have occurred in the past century and it still plays a central role in the sustainable progress of our society. The discovery and optimization of catalytic systems have often been performed by preparing and testing a large number of samples with different compositions, supporting materials and synthetic methods. In this context, the unique catalytic activity of gold, regarded until recently as a poorly active metal, was discovered. The early pioneering reports on CO oxidation[1–4] and liquid-phase oxidation of diols with gold catalysis[5] opened the way and nowadays several protocols for preparing effective gold-based catalysts are available, which take advantage of the peculiarities of this metal in terms of activity, selectivity, reusability and resistance to poisons.[6–12] The breakthrough was marked since bulk gold was found to be dramatically enhanced when downsized to nanometric particles (AuNPs). Such a behavior has been underpinned by a series of kinetic studies showing how the activity is markedly dependent on the metallic gold particle size. Actually, many investigations on the liquid-phase oxidation of alcohols and carbohydrates demonstrated that only sufficiently small gold particles displayed catalytic activity.[6,11,12] In that respect gold, alone or alloyed with other metals, is often

RSC Green Chemistry No. 28
Transition Metal Catalysis in Aerobic Alcohol Oxidation
Edited by Francesca Cardona and Camilla Parmeggiani

Published by the Royal Society of Chemistry, www.rsc.org

the catalyst of choice for facilitating the conversion of various organic substrates to fine chemicals under eco-friendly conditions. In addition to low-temperature CO oxidation,[1–17] H_2O_2 synthesis from its elements,[18–20] water gas shift reaction,[21,22] C–C coupling reactions[23–25] and the synthesis of *N*- and *O*-heterocycles,[26] the successful conversion of alcohols and carbohydrates to the corresponding carbonylic and carboxylic compounds witnesses the catalytic effectiveness of gold and its ability to discriminate among different functional groups.[5,11,27–60] The selective oxidation of alcohols is of major importance in both organic synthesis and industrial applications. The world production of carbonyl compounds exceeds 10^7 t per year, the traditional route being the oxidation of alcohols using stoichiometric amounts of polluting oxidants. However, the last decade has seen a growing number of papers and patents reporting innovative protocols for conducting the aerobic oxidation of alcohols with the aid of heterogeneous or homogeneous catalysts.[61,62] Among these, supported metal nanoparticles have attracted a great deal of interest as they provide clean, selective and efficient reactions. In addition, these nanoparticulate catalysts can be easily recovered, thus permitting their reuse. Numerous metals, including gold, have been assembled on different supporting materials, aiming to fulfill some fundamental aspects required in green chemistry: using eco-friendly solvents such as water, employing benign oxidizing agents, such as O_2 or air and H_2O_2, limiting or avoiding the addition of activators such as inorganic bases and carrying out the reactions at as close as possible to ambient temperature.

Palladium and platinum had long been studied before the introduction of nanometric gold for the same kind of catalytic reactions, such as alcohol oxidation. Actually, these reactions performed with Pd, Pt and Au share several similarities and a synergistic effect among these metals has even been discovered.[6–12] In this chapter, an overview of the selective oxidation of alcohols performed with supported gold-based catalysts in the liquid phase is presented, with a particular focus on substrates of interest such as glycerol and allyl alcohol. Some gaseous-phase processes worthy of mention are occasionally included when pertinent.

5.2 Preparation and Characterization of Gold-Based Catalysts

Several methods are available for preparing supported gold-based catalysts, and impregnation to incipient wetness imbibition, deposition–precipitation and deposition/immobilization of colloidal gold on supports are the most commonly used.[1–12]

Since the first attempts to produce gold catalysts, it has become evident that their catalytic performance depends strongly on a combination of parameters: morphology, dispersion, metal particle size, temperature, pH and interactions between gold particles and supporting material.[3] A further complication is related to the low melting point of gold, a reason why the traditional catalyst synthetic methods often fail to produce high metal dispersion on the support. One way around this limitation is the immobilization of a preformed metallic colloidal dispersion (sol), a methodology leading to good control of metal particle size as it reduces the influence of the supporting material on metal dispersion.[6,11,12] Before considering the sol immobilization protocol in detail, a summary of the principal preparation methods is presented. In the impregnation method ('incipient wetness'), a solution of the metal precursor is allowed to fill the pores of the supporting material up to incipient wetness imbibition of its volume. The metal dispersion ultimately obtained is markedly affected by the conditions of drying. Since the original procedure generally led to gold particles larger than 5 nm, which show little activity, modified impregnation methods were developed in which the key issue was impregnation with hydrogen tetrachloroaurate followed by an additional step to ion exchange chlorides with the hydroxyls of a base. The resulting material was dried and then activated by reduction with hydrogen gas, not by calcination, in a method that is simple and more suitable for the production of effective gold catalysts, as the liquid-phase oxidation of glucose to gluconic acid demonstrates.[43,44]

Another popular approach for preparing supported gold-based catalysts is the deposition–precipitation (DP) method. This consists in the deposition of the metal hydroxide or hydrated oxide on the surface of a support by gradually increasing the pH of the metal precursor solution in which the support is suspended. The precipitate may be nucleated by surface functional groups, thus allowing the active phase to be more tightly attached to the support. As a gold precursor, hydrogen tetrachloroaurate is the favorite choice also for this kind of preparation. In aqueous media, hydrogen tetrachloroaurate(III) hydrolyzes to release various species with different amounts of chloride, water and hydroxyls {*e.g.*, $[AuCl_4]^-$, $[AuCl_3(H_2O)]$, $[AuCl_3OH]^-$, $[AuCl_2(OH)_2]^-$, $[AuCl(OH)_3]^-$, $[Au(OH)_4]^-$}, whose composition can vary with both pH value and chloride concentration. Typical precipitating agents are either sodium hydroxide (DP NaOH) or urea (DP urea).

An advance in terms of stability, reproducibility and overall catalytic performance was reached when the immobilization of a preformed gold colloidal dispersion (sol) on activated carbons and various oxides was set up.[5,6] This method, in fact, presents great advantages with respect to other procedures, mainly regarding the nanoparticles size control and metal dispersion. The gold nanoparticles (ideal average diameter 1–10 nm) are prepared in the presence of a stabilizing/protecting agent (polymer, surfactant, polar molecule) and subsequently immobilized on a supporting material, this latter step being crucial for achieving optimum metal dispersion. The type and amount of the protecting agent [mainly poly(vinyl alcohol) (PVA) and polyvinylpyrrolidone (PVP), but also polyhydroxylated compounds such as D-glucose], the reducing agent (*i.e.* $NaBH_4$), pH value, gold sol concentration and gold loading on the support deeply affect the final catalytic performance. Also, the surface properties and morphology of the support are key factors, the metal dispersion being enhanced by increasing the functionalities of the supporting material.[63] The generation of a protected gold sol involves the reduction of an Au(III) salt (*i.e.* $HAuCl_4$) to Au(0) by $NaBH_4$ addition, in the presence of the protecting agent at room temperature.[6,28] A typical recipe to yield a brown tea-colored colloidal dispersion of 3–5 nm gold particles employs Au–PVA (molar ratio from 1.5 to 50) or Au–glucose (molar ratio from 30 to 3000) and Au–$NaBH_4$ (1 : 1 w/w). The preparation of bimetallic systems (*i.e.* Au–Pd, Au–Pt, Au–Cu) follows the same protocol, with the addition of a second metal precursor (generally $PdCl_2$, $KPtCl_4$ or $CuCl_2$) to gold. Within a few seconds after metal colloid generation, the sol is immobilized by adding under vigorous stirring the support, typically carbon or metal oxides such as Al_2O_3, SiO_2, TiO_2, Fe_2O_3, ZrO_2 or CeO_2, after adjusting the pH to the isoelectric point (IEP) of the supporting material. After 15–30 min, depending on the support, the slurry is filtered and the catalyst washed thoroughly with warm distilled water, until chloride and protecting agent residues are no longer detected, and finally dried overnight or calcined. It has been found that the kinetics of adsorption depend on the sol stabilizer, and also on the IEP and surface area of the supporting material.[63–65] As far as the protecting agents are concerned, it is unknown exactly how they mediate catalyst activity and stability, but it is a fact that they influence the final performance by creating a shield limiting the contact between nanoparticles and reagents. Since their role as protectors and stabilizers of the intrinsically unstable nanoparticles is fundamental for guaranteeing a long lifetime to the catalyst necessary for industrial applications, finding a proper protective agent is essential. PVA fulfills such requirements as it provides good metal dispersion regardless of the support and it can be easily removed by thermal decomposition (>300 °C) or under milder conditions by solvent washing.[64,66–69] The latter method has the advantage of limiting the growth of metallic nanoparticles that normally occurs during thermal treatment and it is suitable for non-refractory supports.

The nature of the supporting material is even a key factor influencing the final catalytic performance, as is shown in Section 5.3. Interestingly, gold supported on various carbons displayed different performances while catalyzing the same reactions.[70] The most relevant parameter for the choice of the support seems to be the type and distribution of surface groups, as these can favor or inhibit the grafting of the nanoparticles. Hence, before depositing the metal on the surface, acidic or basic treatments of the supporting materials are often required in order to optimize the catalytic performance in terms of activity, selectivity and durability. For instance, the metal dispersion of a platinum catalyst (Pt/C) was found to depend on the carbon pretreatment, in addition to the preparation protocol, during the aerobic oxidation of benzyl alcohol to acid. Thus, the platinum catalyst prepared by impregnation of CO_2-activated carbon allowed the best activity and selectivity to be achieved compared with platinum catalysts prepared on air-activated carbon or by cation exchange.[71]

Furthermore, alloying gold with a second metal (Pt, Pd and Cu) has been found to enhance the catalytic performance of gold in numerous oxidation reactions due to a synergistic effect among the metals.[18,19,42,46–48,72–77] This has also resulted in particularly stable catalysts with respect to the monometallic ones, as reported in the following sections.

A thorough catalyst characterization requires a concert of analytical techniques. Inductively coupled plasma atomic emission spectroscopy (ICP-AES) is employed to determine the real metal content, while a combination of X-ray powder diffraction (XRPD) and transmission electron microscopy (TEM) provides the average value and statistical distribution of metal particle diameters. Scanning transmission electron microscopy–high angle annular dark field (STEM-HAADF) and scanning transmission electron microscopy–X ray energy-dispersive spectroscopy (STEM-XEDS) represent the advanced evolution of such techniques. The elemental composition of the surface (1–10 nm depth) and the electronic state of each element can be determined by X-ray photoelectron spectroscopy (XPS). The presence of different types of gold species can be detected by the temperature-programmed reduction (TPR) technique, because different reduction temperatures are a clear indication for different metal species on an identical supporting material or for diverse interactions of the same species with different materials.

5.3 Selective Oxidation of Alcohols by Gold-Based Catalysis

The wide distribution, easy availability from renewable sources and employment of their derivatives as valuable chemicals make alcohols strategic substrates to be converted into the corresponding aldehydes, ketones or acids. Finding proper catalytic systems able to facilitate their selective oxidation under sustainable conditions is what both academia and industry are aiming for. This has led to a competition among platinum, palladium, ruthenium, copper and gold catalysts, which represent the most promising candidates for substituting the traditional polluting methods. Here, attention is focused on gold compared with other metals, trying to offer a general picture of how properly tuning various parameters can boost the catalytic activity while addressing the selectivity towards a given product.

The effective aerobic oxidation of alcohols to carbonylic and carboxylic compounds employing gold began at the end of the 1990s (Scheme 5.1).[5] The discovery of gold as a powerful catalyst opened the way to more and more sophisticated catalytic systems, where gold, alone or alloyed with other metals, is supported on a wide range of materials from different carbons and oxides to the ultimate carbon nanotubes.[11,12,59] Furthermore, one can choose whether to perform the oxidation by homogeneous or heterogeneous catalysis, in the gaseous or liquid phase, with ordinary solvents, supercritical CO_2 or solventless, while varying the pressure, temperature and

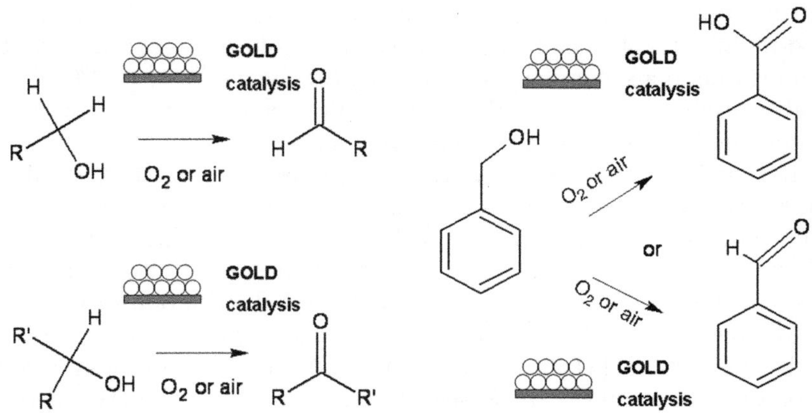

Scheme 5.1 Aldehydes, ketones and carboxylic acids from the aerobic oxidation of alcohols with gold-based catalysis.

catalyst-to-substrate ratio. All these factors dramatically affect the yield, selectivity and productivity while making difficult to compare different literature data and to draw general conclusions.

Several research groups around the world have made notable achievements in this huge field, as detailed throughout this chapter, but Rossi's, Hutchings' and Corma's studies deserve particular mention. Rossi and co-workers were the first to discover the catalytic effectiveness of gold in the liquid-phase oxidation of the C–OH functional group.[5,11,12,31] They even found how to address the synthesis with preference for carbonylic derivatives, when conducting the oxidation of alcohols in the gaseous phase, or to carboxylic acids or esters in the liquid phase under mild conditions in the presence of alkali. The strict synergistic effect between gold nanoparticles and the supporting material, mainly nanometric CeO_2, was particularly emphasized by Corma and co-workers,[78] whereas the bimetallic Au–Pd catalysts synthesized by Hutchings's group highlighted the synergistic effect between the two metals, dramatically enhancing the catalytic performances of the individual metals.[79]

Recently, Fan's group carried out the selective oxidation of alcohols to aldehydes/ketones with O_2 over a series of supported gold catalysts, confirming that the catalytic performance strongly depends on either the support or the preparation method.[51] Among them, Au/CuO catalyst prepared by coprecipitation was particularly active and selective for the oxidation of benzylic, cyclic and unsaturated alcohols to the corresponding aldehydes or ketones by molecular oxygen at atmospheric pressure. Cyclooctanol and cyclododecanol were oxidized to ketones with notable conversion and selectivity higher than 99%. Regarding the catalyst stability, Au/CuO could be recycled for use after regeneration. The addition of alkali metal salts to the reaction mixture did not increase the activity but turned out to improve the catalytic stability.

Two different research groups recently employed magnesium oxide as a supporting material for gold in the selective oxidation of alcohols under solvent-free conditions, thus showing its versatility.[56,57] Choudhary and Dumbre[56] prepared an MgO-supported nano-gold catalyst by the deposition–precipitation technique and tested it in the oxidation of different primary alcohols (Ph–CH_2OH, NO_2–Ph–CH_2OH, MeO–Ph–CH_2OH, Ph–CH_2–CH_2OH) and aldehydes to the corresponding aldehydes and carboxylic acids, respectively, by molecular oxygen (under atmospheric pressure) in the absence of any solvent. This led to high activity/selectivity and excellent reusability. The influence of the catalyst calcination temperature (400–900 °C), reaction temperature (50–120 °C) and different solvents (toluene, *p*-xylene, DMF or DMSO) on the oxidation reaction was also studied. The catalyst displayed inferior performance in the presence of different solvents, whereas in all cases the catalyst could be easily removed from the reaction mixture and reused several times without any significant decrease in activity. One year later, Gusevskaya and co-workers[57] showed gold nanoparticles supported on magnesium oxide to be efficient heterogeneous catalysts for the liquid-phase oxidation of another wide range of alcohols using molecular oxygen as a sole oxidant, employing toluene or methanol as the solvent but in the absence of co-catalysts or additives. The Au/MgO catalyst was also prepared through the deposition–precipitation method and characterized by XRD, XPS, XAS, HRTEM, UV–Vis spectroscopy and N_2 adsorption techniques. Various carbonylic monoterpenoids important for the fragrance and pharmaceutical industries were obtained in good to excellent yields starting from biomass-based monoterpenic alcohols, such as isoborneol, perillyl alcohol and carveol. Moreover, the Au/MgO catalyst allowed the selective one-pot oxidative esterification of benzyl alcohol, in alkali-free methanol solutions and in the absence of any additive, to give methyl benzoate in virtually quantitative yield.[57]

A novel frontier of supported gold catalysts for the selective oxidation of alcohols is the use of carbon nanotubes (CNTs) as the supporting material.

Doris and coworkers prepared and then assessed the catalytic activity of CNT–gold nanohybrids in the aerobic oxidation of alcohols under mild conditions (*i.e.* room temperature, air), first taking benzyl alcohol as a model substrate, and subsequently screening variously substituted primary and secondary alcohols.[59] They found the CNT-supported gold nanohybrid to be active for the aerobic oxidation of diverse alcohols without any added O_2 and no heating. While alcohols and aldehydes were chemoselectively oxidized to the corresponding acids, under anhydrous conditions, primary alcohols could also be converted to the corresponding aldehydes. Finally, they elucidated the role of water in the oxidation process by labeling water with ^{18}O. This CNT-based system compares favorably with other supported gold nanoparticle catalysts in terms of recyclability, catalytic activity, selectivity and mildness of the operating conditions, and with the advantage of no added O_2.

Gold nanoparticles were supported on multi-walled carbon nanotubes (MWCNTs) following different methods by Órfão's group.[66] The catalytic systems were tested in the selective oxidation of glycerol under basic conditions, for evaluating the effect of the preparation technique on the activity and selectivity. As expected, the catalytic performances depended markedly on the gold crystallite size. The sol immobilization method was found to be the most suitable technique to prepare gold supported on CNTs. In contrast to what was observed for the catalytic activity, the diverse preparation methods did not significantly affect the selectivity. The use of MWCNTs as a supporting material for Au nanoparticles resulted in the oxidation of the secondary hydroxyl group and a notable high dihydroxyacetone selectivity of about 60% was attained independently of the preparation protocol. It was also concluded that dihydroxyacetone in the final mixture can be stabilized by lowering the pH to about 3.

A bifunctional, efficient and recyclable heterogeneous catalyst consisting of gold nanoparticles supported on supramolecular ionic liquid-grafted graphene was recently presented by Shaabani *et al.*[54] This catalytic system was employed for the aerobic oxidation of various primary and secondary aliphatic and aromatic alcohols to the corresponding aldehyde and ketone derivatives in water as a green solvent and at room temperature. The innovative technique of grafting a supramolecular ionic liquid to graphene led to a suitable support for the stabilization and encapsulation of gold nanoparticles able to control their size and uniform distribution on the support. Moreover, this heterogeneous graphene-based catalyst displays good dispersity in water and removes the need for an added base due to the presence of basic centers in its structure guaranteeing an environmentally friendly process with excellent yield and selectivity. Also, the catalyst is stable in the reaction media, as proved by investigations on recyclability.

Probably the first example of full conversion of alcohols by aerobic oxidation was the recent study of Kobayashi *et al.*, who engineered an innovative gold-immobilized microchannel flow reactor. The oxidation of various alcohols to the corresponding aldehydes and ketones with good to excellent yields was always achieved, with no gold leaching and steady catalytic activity for 4 days.[80]

It has already been mentioned that the catalytic performance of gold can be boosted when it is alloyed to other metals, mainly palladium and platinum. Hutchings and co-workers are particularly active in this field, resulting in several papers reporting different methods of preparation of bimetallic catalysts and their application in the selective oxidation of various substrates, including alkanes and alcohols. They use benzyl alcohol oxidation as a model reaction aiming at gaining insight into the synergistic effect between gold and the second metal. One of their most recent studies involved Au–Pd nanoalloys supported on Mg–Al mixed metal oxides (MgAl-MMO) prepared using sol-immobilization and tested in the solvent-free oxidation of benzyl alcohol with molecular oxygen at low pressure and temperature.[53] When using this support, alloying Pd with Au resulted in an increase in both activity and selectivity towards benzaldehyde and also an improved resistance to catalyst deactivation in comparison with the monometallic Pd and Au catalysts. The turnover number (TON) for an Au/Pd 1 : 1 molar ratio catalyst reached 13 000 h^{-1} after 240 min and the selectivity towards benzaldehyde was 93%. This high catalytic activity was maintained after three successive cycles. XRD, UV–Vis and HRTEM confirmed the formation of Au–Pd nanoalloys with an average size of 2.7 nm instead of core–shell bimetallic particles and segregated monometallic particles. The ensemble and electronic effects between Pd and Au in the Au–Pd/MgAl-MMO catalyst were determined by IR spectroscopy of CO chemisorption and XPS analysis. It was found that Au atoms draw electron density away from Pd atoms, thus enhancing the interaction of Pd atoms with the substrate and thereby increasing the catalytic activity. In order to investigate the support effect, they also prepared an Au–Pd/MgO catalyst for comparison. They suggested that the promotional effect on the catalytic activity of the MgAl-MMO was connected with the acidic sites of the support, whereas the basic sites were responsible for the high selectivity. A possible mechanism was proposed: the oxidation of benzyl alcohol would proceed through the co-operation between the Au–Pd nanoalloys and the base/acid sites on the surface of the support.

Also in 2013, Hutchings and co-workers took benzyl alcohol oxidation as a model reaction for studying the effect of other supporting materials on Au, Pd and AuPd nanoparticles, namely hierarchical titanium silicalite (TS-1), titania and carbon.[55,58] Hierarchical TS-1 materials were prepared from the crystallization of silanized protozeolitic units and used as supports for gold, palladium and gold–palladium nanoparticles. The choice of these supporting materials was based on their secondary porous system within the supermicro/mesopore range and enhanced surface area over a standard reference TS-1 material. The presence of the secondary porosity not only improved the accessibility to the active sites of the relatively large reactant molecules but also boosted the metal dispersion. Benzyl alcohol was oxidized by an '*in situ*-generated' hydroperoxy species, formed from a dilute mixture of hydrogen and oxygen, under mild conditions. More exactly, the catalysts were evaluated for peroxide production at 2 and 30 °C. At sub-ambient conditions the rate of oxidation was found to be very low, due to a low production rate of H_2O_2. In contrast, increasing the reaction temperature to 30 °C improved the peroxide production rate and, hence, the benzaldehyde production rate. In order to confirm that the formation of peroxide species promotes the reaction, a comparison of reactions in the presence and absence of hydrogen was carried out at 30 °C. This resulted in a fivefold increase in benzaldehyde yield when H_2 was present. Therefore, the authors concluded that the use of highly active zeolitic material with secondary porosity enables this approach to be potentially broadened to a greater number of oxidizable reactants than previously considered.[55]

The parallel work performed by the same group in 2013 considered more traditional materials, namely activated carbon and titania, used as supports for Au–Pt and Pd–Pt. The supported gold–platinum and palladium–platinum nanoalloys were prepared using the sol-immobilization method. In order to obtain core–shell structures besides alloys, a sequential reduction of the metal combinations was carried out to allow the formation of core–shell structures, whereas the simultaneous reduction of both metals led to homogeneous alloys. The materials were then characterized using XPS and aberration-corrected STEM. The supported nanoparticles were tested in the oxidation of benzyl alcohol (and also toluene) using *tert*-butyl hydroperoxide at 80 °C, in order to elucidate any potential structure–activity relationships. Amongst the Au–Pt catalyst variants, the supported nanoalloys prepared using simultaneous reduction gave the best catalytic performance. However, the Pt_{shell}–Au_{core}/C catalyst, prepared by the sequential addition of Pt to Au 'seeds', displayed an improved catalytic activity in comparison with the TiO_2-supported analog. In contrast, both the Au_{shell}–Pt_{core} variant (prepared by the sequential addition of Au to the Pt 'seeds') and the Au–Pt variant (prepared by simultaneous metal reduction) allowed a higher catalytic activity to be achieved for TiO_2-supported nanoparticles. As far as the product distribution is concerned, when Au–Pd catalysts were used a significant amount of toluene was formed, whereas when the Pd component of the alloy was replaced with Pt, toluene was not detected and the main products were benzaldehyde and reaction products derived from the aldehyde. Furthermore, the catalytic performance of the Au–Pt and Au_{shell}–Pt_{core} materials is very different when supported on carbon and TiO_2, since mainly benzoic acid is formed by the TiO_2-supported Au–Pt nanoparticles. In contrast, in the case of Pt_{shell}–Au_{core} nanoparticles, both TiO_2 and carbon led to benzaldehyde as the major oxidation product. Hence the importance of the alloy composition and morphology, in addition to the support identity, is dramatically highlighted.[58]

Interestingly, around the same time, Wang's group prepared innovative monometallic Au, Pd and Pt catalysts supported on a hybrid of natural DNA and MMT (montmorillonite), showing higher activity and selectivity for the oxidation of various primary alcohols than DNA-templated nanoparticles, but also with comparable performance to the bimetallic catalysts reported in the literature.[52] These organic–inorganic hybrids are made up of Na-montmorillonite, which is a type of naturally occurring clay, and the natural DNA from fish sperm as the template of metal nanoparticles. Depending on these catalytic systems, highly efficient formation of the corresponding aldehydes, carboxylic acids and esters was obtained. Both oxidative self-esterification and cross-esterification of various alcohols could be smoothly performed. Although bimetallic nanoparticles as active catalysts for selective oxidation of primary alcohols to the aldehydes, carboxylic acids and esters have been reported, this is probably the first example of such transformations being easily catalyzed by monometallic nanoparticles. Furthermore, taking advantage of the water-soluble reversibility of these catalysts, all the transformations could be carried out effectively in water under mild conditions and the catalysts were recovered by a simple phase separation process and reused.[52]

A key point in the gold-catalyzed conversion of alcohols in the liquid phase is the role of alkali. Only rare cases of conversion were observed in the absence of a base and this has been a matter of study since alkali was found to be compulsory in the oxidation of hydroxyl and carbonyl groups. The base reasonably appears to be involved in the first step of the oxidative process for the creation of the alkoxide anion. Hence the basic strength of the alcoholic group is crucial.[81] Moreover, the presence of a base promotes the desorption of the products from the surface, thus prolonging the catalyst life. Two ways can be chosen for providing a reaction system with a basic environment: continuous addition of alkali (pH-controlled conditions) or an excess of base added at the beginning of the reaction. These conditions allow the formation of an intermediate aldehyde, which is more easily oxidized to carboxylate than the parent alcohol. In some cases, the aldehyde can be isolated in high yield (90%) even in the presence of alkali, as in *o*-hydroxybenzyl oxidation catalyzed by Au/Fe_2O_3 at 50 °C, atmospheric pressure and with Na_2CO_3.[82] However, carrying out the reaction in the presence of a base represents a limit when free carboxylic acids, and not the corresponding carboxylates, are the desired products. The use of bimetallic catalysts seems to offer a valid alternative, since most of the reactions carried out with gold alloyed with palladium or platinum are not base assisted.

The reactions presented here were mostly performed in the presence of a solvent and partly under solvent-free conditions. This is an important point in catalysis: finding a proper solvent that improves catalytic performances while respecting ecological restrictions is not always an easy task. The ideal medium would be water, being eco-friendly, cheap and easy to handle, but unfortunately its use depends on the hydrophobic/hydrophilic properties of both the substrate and the catalytic system. Anyway, water represents the solvent of choice for an wide range of substrates, such as polyhydroxylated compounds, and solventless conditions – when possible – can provide a valid solution for water-insoluble compounds. On the other hand, under solventless conditions drawbacks can occur, principally related to the physical state of the substrate and its thermal stability. However, solventless conditions often emphasize the action of the base, so important in gold-catalyzed alcohols oxidation. For instance, in the solvent-free oxidation of benzyl alcohol conducted at moderate temperature (80 °C) with Au/TiO_2 catalysis and addition of K_2CO_3 (0.2% with respect to the alcohol), a marked effect of the base was evident as the TOF reached 7851 h^{-1}, much higher than the =19 h^{-1} achieved without any alkaline promoter. On changing the type of base, some differences were observed: sodium carbonate and acetate improved the oxidation, whereas stronger bases did not have the same prominent effect.[83] Gold-based catalysts are the ideal choice in water because, compared with more traditional catalysts such as palladium or platinum, they do not suffer from deactivation when employed in the presence of molecular oxygen; moreover, they are highly chemoselective.[5] Regarding the support material, activated carbon and graphite are the most investigated materials for aqueous-phase oxidation by gold due to both the high performance under mild conditions and economic reasons: these materials are, in fact, cheap and allow an easy recovery of the precious metal by burning off the support.[7] However, hydrophilic supports are preferable when water is the solvent, because the dependence on a basic promoter is reduced, as in the case of ethanol oxidation to acetic acid catalyzed by Au/$MgAl_2O_4$. This reaction, conducted under forced conditions (3.5 atm of air, 150–180 °C) with notable yields up to the 92% in acetic acid, represents the first example of a gold-catalyzed oxidation proceeding in water in the absence of a base.[84]

Concerning the oxidizing agent, hydrogen peroxide represents an efficient oxidant for performing environmentally friendly oxidations of organic compounds in addition to air or molecular oxygen. As an example, Kholdeeva *et al.* developed new clean and efficient methods for the production of 2-methyl-1,4-naphthoquinone (MNQ, menadione, vitamin K_3) based on the oxidation of 2-methyl-1-naphthol (MNL) with cheap and environmentally benign oxidants, aqueous hydrogen peroxide, *tert*-butyl hydroperoxide and molecular oxygen, using three types of heterogeneous catalysts, namely mesoporous titanium–silicate Ti-MMM-2, silica-supported iron phthalocyanine and supported gold nanoparticles. Interestingly, on performing blank tests, it was also found that this reaction proceeds fairly well without any catalyst with oxygen at 3 atm in non-polar solvents.[50]

Until a few years ago, catalytic methods for the oxidation of amino alcohols were almost unknown in organic synthesis, owing to the inhibiting effect of the amino group on traditional metal catalysts, but over the last 5 years a number of papers have appeared on this topic.[11,12,32] Rossi and co-workers first found gold to be an exception when compared with palladium and platinum under similar conditions, displaying notable and stable catalytic performance in the aerobic oxidation of amino alcohols.[12,32] By reacting different kinds of amino alcohols at 70 °C, with O_2 under a slight oxygen pressure (3 atm) and a substrate-to-NaOH molar ratio of 1:1, it was found that alkali increased the oxidation rate and alumina was a better supporting material than carbon. In the case of *N*-substituted amino alcohols, the oxidation took place exclusively at the nitrogen atom, leading to the corresponding *N*-oxide without any amino acid formation. Therefore, starting from 3-dimethylamino-1-propanol, 100% regioselectivity to the corresponding *N*-oxide was reached. In principle, the oxidation of the amino group is possible both in the absence and in the presence of alkali: without any alkali, 100% selectivity was observed with different metal catalysts, but only gold-containing catalysts allowed 100% conversion, whereas Pt/C was inert and Rh/C led to only 20% conversion to unidentified compounds.[32]

An effective alternative to liquid-phase reactions, but suitable only for thermally stable and volatile alcohols, is to perform selective oxidation in the gaseous phase. This procedure has the advantages over liquid-phase reactions of being solventless and alkali free and allowing the oxidation to be stopped at the corresponding aldehyde or ketone. Furthermore, gold-based catalysts have displayed good performance in terms of activity and selectivity also in gaseous-phase processes. One of the first examples was the selective oxidation of a series of C_3–C_5 alcohols and benzyl alcohol carried out with Au/SiO_2. With the temperature ranging from 250 to 300 °C, the activity followed the order $C_4 > C_5 > C_3$, whereas secondary alcohols required milder temperatures (100–150 °C). Regarding benzyl alcohol, a high selectivity (>99.5%) could be obtained only with a restricted range of conversion (50–75%).[27] However, the need to produce benzaldehyde in high yields, avoiding expensive and polluting purification techniques, prompted the authors to extend this research. As a strategy, in order to stabilize the almost total selectivity achieved with the gold on silica catalyst while ensuring the complete conversion of the reagent, a series of mono- and bimetallic catalysts were investigated. Among them, the gold–copper combination supported on silica was particularly effective. This study emphasizes the key-role of gold while highlighting a synergistic effect with copper. Although gold alone led to useful values of selectivity and conversion, the most promising achievements were obtained with a bimetallic Cu–Au/SiO_2 catalyst, producing over 99% selectivity at 98% conversion.[85]

In the wide range of alcohols so far examined, glycerol is at present the most intensively investigated, with numerous papers, patents and reviews that definitely confirm supported gold to be a valid alternative to the traditional platinum- and palladium-based catalysts.

Another substrate of interest is allyl alcohol, being a potential starting material for the synthesis of 3-hydroxypropionic acid, which occupies the third position in the top-12 list of chemical intermediates to be converted into a number of high-value materials.[86]

The following sections are therefore focused on the oxidation of glycerol (Section 5.3.1) and allyl alcohol (5.3.2).

5.3.1 Glycerol Oxidation

The necessity to find eco-friendly routes for replacing petroleum derivatives prompted both academia and industry to search for potential starting materials in biomass. The synthesis of biodiesel is probably the most promising solution, being a fuel comprised of monoalkyl esters of long-chain fatty acids produced by chemically reacting a vegetable oil or animal fat with an alcohol such as methanol (transesterification reaction). However, the growing overproduction of glycerol, the main by-product of biodiesel synthesis, has meanwhile introduced another challenge for scientists, who are now called upon to find effective routes for converting glycerol into valuable chemicals (Scheme 5.2).[34,35]

The catalytic approach employing gold supported on different materials seems to meet expectations, according to two research teams particularly involved in the application of gold catalysis to glycerol oxidation: Hutchings' group at Cardiff University and Prati group at Milan University. Gold-based catalysts are reported to guarantee high performance, especially in terms of selectivity, which is so important in the entangled net of possible products. Accordingly, in 2002, Hutchings and co-workers achieved 100% selectivity

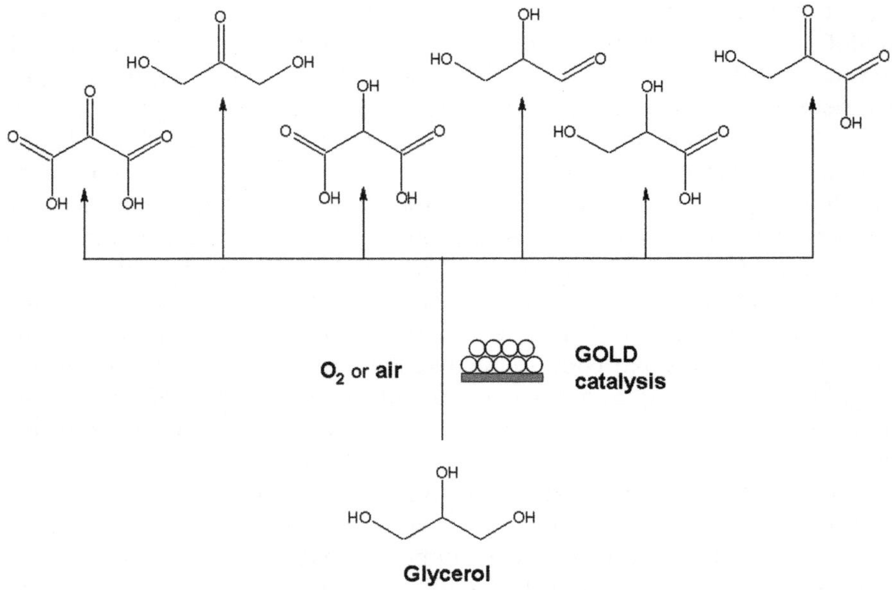

Scheme 5.2 Schematic pathway of the potential products from glycerol aerobic oxidation with gold-based catalysis.

with respect to sodium glycerate at 50–60% conversion when gold was supported on graphite in aqueous solution at 100 °C and in the presence of NaOH.[87] They showed how to convert glycerol to lactate *via* preliminary reduction to 1,2-propanediol followed by oxidation of the primary alcohol group to carboxylate. Some years later, they confirmed that a second metal alloyed with gold can lead to boosted catalyst performance compared with the monometallic version.[88] Gold–palladium supported on two different materials, titania and carbon, prepared following wet impregnation and sol immobilization, markedly improved the catalytic activity while retaining the high selectivity towards lactate (up to 96% selectivity at 94% conversion). The sol immobilization protocol led to catalysts allowing the highest activity for lactate formation, whereas on comparing a series of C_3 alcohols with glycerol the reactivity trend observed was glycerol > 1,2-propanediol > 1,3-propanediol > 1-propanol > 2-propanol.

Parallel studies on the oxidation to glycerate carried out by Prati *et al.* were aimed at evaluating gold nanoparticles, either unsupported or supported on carbon and titania, in terms of selectivity and activity, the role of the base added in the reacting mixture and also hydrogen peroxide formed during the reaction.[89] It was found that the C–C bond fission of glycerol increased with the reaction time and, partially, with the rate of degradation of the by-product H_2O_2, thus lowering the selectivity to glycerate. The authors also adopted the strategy of alloying gold with a second metal (Pt and Pd) in order to overcome such limits. The catalytic activity could be enhanced by employing the bimetallic Au–Pt and Au–Pd systems, whereas the selectivity to the desired product could be tuned by the composition and texture of the catalyst (particle size, kind of support and alloyed phases) and also by the experimental conditions (pH, temperature, reaction time). In particular, pure gold was found to allow glyceric acid formation, while palladium addition led to further transformation to tartronic acid and platinum addition favored glycolic acid formation as a consequence of C–C bond fission. In addition to the metal particle size, the atomic ratio between the metals was shown to affect dramatically the catalytic activity and selectivity. As expected, the key role of the supporting materials (carbon, graphite, TiO_2, Ti/SiO_2, SiO_2) was also highlighted.[75,76,89] A marked advance was reached when the need for a base could be overcome while maintaining valuable performance (80% glyceric acid selectivity at full conversion).[76] As already mentioned, the pH of the reaction medium and the type of catalyst (Pt-, Pd- and Au-based catalysts) significantly affect the glycerol oxidation reaction. Whereas platinum or palladium do not distinguish between acidic or basic conditions, gold activity is almost exclusively limited to basic media where sodium hydroxide as a base is generally the first choice. The basicity is thought to favor the first step of the oxidation mechanism, that is, hydrogen abstraction from the hydroxyl groups.[76,87] Therefore, the main product is always the sodium salt of glyceric acid with yields depending on the NaOH-to-glycerol ratio, as an increased ratio was found to decrease the tartronic acid formed in favor of glyceric acid. This represents a drawback, since working in acidic media would prevent the salt formation, whose disposal cost is always an issue in the design of industrial processes.

The different catalytic behaviors of palladium, platinum and gold induced by pH during glycerol oxidation are now briefly reviewed. We intentionally avoid the bimetallic formulations where the synergistic effect between the metals is so strict that it would be impossible to decouple the pH effect on the various metals. Monometallic palladium, platinum and gold supported on carbon are taken as model catalysts.

Concerning Pd/C catalyst, the experiments were usually performed at pH 11 in order to increase the activity of the catalysts, with glyceric acid being found as the prevalent oxidation product resulting from the activation of the primary hydroxyl group of glycerol. Some workers also found tartronic acid and oxalic acid to be formed in minor amounts.[46,75,76,90–92] A systematic study of the relation between pH and the catalytic activity, employing 5% Pd/C as a catalyst, was carried out by Gallezot's group.[91] As a general trend, increasing pH improved the catalytic activity, as pH 11 allowed full conversion of glycerol to be achieved after a reaction time of $\sim$4 h, whereas lower pH values resulted in incomplete conversion even after a doubled reaction time. Regarding the selectivity at pH 11, glyceric acid was the main product but, reaching full conversion, no more than 67% could be registered as due to oxalic acid formation. Since the selectivity of the co-produced tartronic acid and dihydroxyacetone remained unchanged, even at high conversion, the decomposition of glyceric acid to oxalic acid by a decarboxylation mechanism ($-CO_2$) was suggested. Confirmation that C_1

products can be formed came from a more recent study by Prati and co-workers, who used carbon-supported palladium (1% Pd/C) under acidic conditions (pH 5).[46] Even though the activity was low (5% conversion of glycerol after 2 h), it is worth mentioning that C_1 compounds were the principal products, as a consequence of C–C bond cleavage induced by hydrogen peroxide, a known by-product under these reaction conditions.[46]

Platinum has also been employed for selectively oxidizing glycerol under acidic or basic conditions. A study in the mid-1990s examined the catalytic activity of carbon-supported platinum catalysts at pH 6 and 11. The catalytic activity was significantly lower in acidic media than under basic conditions and even the overall yield of the identified products (glyceric acid, dihydroxyacetone and tartronic acid) was favored by higher pH, leading to 35% under acidic and 70% under basic conditions. As in the case of palladium, the principal product was glyceric acid but with lower yield (from 70 to 55%) while the dihydroxyacetone yield increased from 8 to 12%.[91,93–95] More recent studies using carbon-supported platinum under acidic media were reported by Liang *et al.*[96] and Prati and co-workers.[46,76] Whereas two decades ago low activity was observed for this kind of catalyst at low pH, Liang *et al.* claimed a 50% conversion and Prati and co-workers even 78%. Owing to the different experimental conditions, a clear cross-evaluation of such results is impossible, but both groups reported an important selectivity towards C1 compounds, namely carbon dioxide and formic acid. An older earlier paper from by Hutchings and co-workers showed that passing from acidic to basic conditions doubled the glycerol conversion by on increasing the NaOH/- to-glycerol molar ratio in the range 0–2 with 5 wt% Pt/C at 60 °C (conversion from 27 to 56% after 3 h of reaction). At the same time, the selectivity towards glyceric acid and glyceraldehyde increased from 28 to 69% and from 0 to 30%, respectively, whereas oxalic acid was detected only under acidic conditions. Interestingly, the nature of metal hydroxides was found to markedly affect the catalytic performance following this trend related to conversion: NaOH (63%) > CsOH (52%) > LiOH (49%) > RbOH (37%) > KOH (35%).[92]

Coming back to gold after this short excursion on platinum and palladium, research on the selective oxidation of glycerol with gold catalysis is still in progress, as the most recent papers witness.[97,98] Efforts are being made to improve mainly the experimental conditions and catalyst design, even searching for novel supporting materials, such as hydrotalcite,[97] or boosting more traditional supports, such as alumina,[98] with a particular interest in bimetallic formulations as the only ones allowing alkali-free conditions.

5.3.2 Allyl Alcohol Oxidation

The interest of the scientific community in 3-hydroxypropionic acid (3-HP) is related to the manufacture of high-performance polymers where 3-HP is an essential starting material. Despite all the efforts so far to find effective biological and eco-friendly chemical routes, a large-scale process for industrial application is still lacking. Allyl alcohol could be a potential starting material for achieving 3-HP, as Rossi and co-workers reported.[49] They presented an innovative route to 3-HP by oxidizing allyl alcohol in aqueous alkali solution with O_2 in the presence of a gold catalyst, namely Au/C and Au/TiO_2. The selectivity to 3-HP could be tuned by regulating the amount of alkali and the temperature, thus depressing acrylate and glycerate formation. Au/C was shown to be superior, allowing a 79% yield of 3-HP at full conversion compared with a 53% yield to at 97% conversion using a titania-supported gold catalyst after 1 day of reaction. Both tests were carried out using NaOH in excess (3 : 1), oxygen pressure 3 atm and 50 °C. A possible mechanism was proposed in which primarily formed acrolein undergoes a Michael-type addition of water, finally leading to 3-HP.[49]

Although elegant, this process presents two main challenging issues. The first is the control of selectivity among the possible products, acrylate, acrolein and glycerate. The second, probably the major, limitation is related to Au/C recycling since the selectivity to 3-HP dropped after the second reaction cycle, not depending on leaching. In order to overcome such drawbacks, Rossi and co-workers decided to exploit the power of bimetallic formulations to boost the gold performance, by preparing and testing a series of bimetallic Au–M nanoparticles (M = Pt, Pd, Cu or Ag) supported on carbon.[77] The conversion, selectivity and lifetime of the bimetallic catalysts could be tuned and markedly improved by optimizing the experimental conditions and the preparation protocols. This strategy was successful as notable performances were maintained even after four reaction cycles, especially for the 0.3% $Au_{99}Cu_1/C$ system. Thus the selectivity towards 3-HP changed from 95% (first cycle) to 78% (fourth cycle) while maintaining the same conversion (around 98%). Owing to the extremely dilute amount of the second metal, it was not possible to detect it despite using sophisticated STEM analysis. However, as the synergistic effect between gold and the alloying metal is evident, the authors justified the stabilization of the catalytic activity as being due to electronic modifications in the gold nanoparticles. A significant particle size effect, in this case, was considered to be unlikely, as STEM analyses indicated sintering of the metal particles. Therefore, the authors suggested that the negative effect of particle growth could be outweighed by the low redox potential of copper, thus allowing the sustained high performance.[77]

A recent paper from Schüth and co-workers advanced a bio-renewable pathway towards allyl alcohol. They converted glycerol to allyl alcohol through dehydration and consecutive hydrogen transfer with an iron oxide catalyst.[99] This important finding, together with the enhanced bimetallic catalysts successfully employed by Rossi and co-workers for selectively oxidizing allyl alcohol, could open the way to a green route towards 3-hydroxypropionic acid (Scheme 5.3).

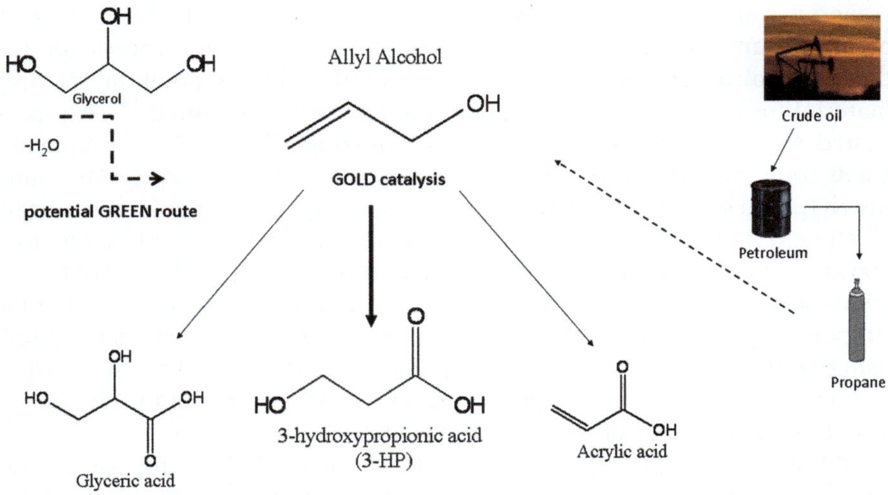

Scheme 5.3 Schematic pathway of the two possible routes towards allyl alcohol (from glycerol or crude oil) and the products from allyl alcohol aerobic oxidation with gold-based catalysis.

5.4 Conclusion

Two decades have passed by since the discovery of the catalytic power of gold. This marked a breakthrough in chemistry, especially in organic synthesis, allowing catalyzed selective oxidations of a wide number of substrates to be carried out to produce important chemicals under eco-friendly conditions. The keystone is the nanometric scale: only when gold could be applied as nanoparticles with diameters in the range 1–10 nm did its extraordinary catalytic activity definitely come out and it still continues to show more of this peculiarity. Gold, alone or alloyed with a second metal, has turned out to be particularly effective in the selective oxidation of various alcohols, as highlighted throughout this chapter. The progressive advances in gold catalyst design, including effective bimetallic formulations, and the proper tuning of the experimental conditions are increasingly shortening the distance between bench experiments and industrial processes.

References

1. M. Haruta, T. Kobayashi, H. Sano and N. Yamada, *Chem. Lett.*, 1987, **16**, 405.
2. M. Haruta, N. Yamada, T. Kobayashi and S. Iijima, *J. Catal.*, 1989, **115**, 301.
3. M. Haruta, *Catal. Today*, 1997, **36**, 153.
4. M. Haruta, *Stud. Surf. Sci. Catal.*, 1997, **110**, 123.
5. L. Prati and M. Rossi, *J. Catal.*, 1998, **176**, 552.
6. *Catalysis by Gold, Catalytic Science Series*, ed. G. C. Bond, C. Louis and D. T. Thompson, vol. 6, Imperial College Press, London, 2006.

7. M. C. Daniel and D. Astruc, *Chem. Rev.*, 2004, **104**, 293.
8. A. S. K. Hashmi and G. J. Hutchings, *Angew. Chem. Int. Ed.*, 2006, **45**, 7896.
9. A. S. K. Hashmi, *Chem. Rev.*, 2007, **107**, 3180.
10. S. Carrettin, P. Concepcion, A. Corma, N. Lopez, M. Jose and V. F. Puntes, *Angew. Chem. Int. Ed.*, 2004, **43**, 2538.
11. C. Della Pina, E. Falletta, L. Prati and M. Rossi, *Chem. Soc. Rev.*, 2008, **37**, 2077.
12. C. Della Pina, E. Falletta and M. Rossi, *Chem. Soc. Rev.*, 2012, **41**, 350.
13. G. C. Bond and D. T. Thomson, *Gold Bull.*, 2000, **33**, 41.
14. M. S. Chen and D. W. Goodman, *Science*, 2004, **306**, 252.
15. G. J. Hutchings, *Catal. Today*, 2005, **100**, 55.
16. N. Weiher, A. M. Beesley, N. Tsapatsaris, L. Delannoy, C. Louis, J. A. van Bokhoven and S. L. M. Schroeder, *J. Am. Chem. Soc.*, 2007, **129**, 2240.
17. T. V. W. Janssens, B. S. Clausen, B. Hvolbaek, H. Falsig, C. H. Christensen, T. Bligaard and J. K. Nørskov, *Top. Catal.*, 2007, **44**, 15.
18. G. J. Hutchings, *Chem. Commun.*, 2008, 1148.
19. J. K. Edwards, A. F. Carley, A. A. Herzing, C. J. Kiely and G. J. Hutchings, *Faraday Discuss.*, 2008, **138**, 225.
20. J. K. Edwards, B. Solsona, N. N. Edwin, A. F. Carley, A. A. Herzing, C. J. Kiely and G. J. Hutchings, *Science*, 2009, **323**, 1037.
21. Q. Fu, H. Saltsburg and M. Flytzani-Stephanopoulos, *Science*, 2003, **301**, 935.
22. Q. Fu, S. Kudriavtseva, H. Saltsburg and M. Flytzani-Stephanopoulos, *Chem. Eng. J.*, 2003, **93**, 41.
23. A. S. K. Hashmi, J. P. Weyrauch, M. Rudolph and E. Kurpejovic, *Angew. Chem. Int. Ed.*, 2004, **43**, 6545.
24. J. Xie, H. Li, J. Zhou, Y. Cheng and C. Zhu, *Angew. Chem. Int. Ed.*, 2012, **51**, 1252.
25. C. González-Arellano, A. Abad, A. Corma, H. García, M. Iglesias and F. Sánchez, *Angew. Chem. Int. Ed.*, 2007, **46**, 1536.
26. A. S. K. Hashmi, *Pure Appl. Chem.*, 2010, **82**, 1517.
27. S. Biella and M. Rossi, *Chem. Commun.*, 2003, 378.
28. M. Comotti, C. Della Pina, R. Matarrese and M. Rossi, *Angew. Chem. Int. Ed.*, 2004, **43**, 5812.
29. S. D. Pollington, D. I. Enache, P. Landon, S. Meenakshisundaram, N. Dimitratos, A. Wagland, G. J. Hutchings and E. H. Stitt, *Catal. Today*, 2009, **145**, 169.
30. M. Lanfranchi, L. Prati, M. Rossi and A. Tiripicchio, *Chem. Commun.*, 1993, 1698.
31. L. Prati and M. Rossi, *Stud. Surf. Sci. Catal.*, 1997, **110**, 509.
32. S. Biella, G. Castiglioni, C. Fumagalli, L. Prati and M. Rossi, *Catal. Today*, 2002, **72**, 43.
33. S. Biella, L. Prati and M. Rossi, *J. Catal.*, 2003, **197**, 207.
34. M. Pagliaro and M. Rossi, *The Future of Glycerol: New Usages for a Versatile Raw Material*, RSC Green Chemistry Book Series, RSC Publishing, Cambridge, UK, 2008.

35. M. Pagliaro, R. Ciriminna, H. Kimura, M. Rossi and C. Della Pina, *Angew. Chem. Int. Ed.*, 2007, **46**, 4434.
36. M. Comotti, C. Della Pina, E. Falletta and M. Rossi, *J. Catal.*, 2006, **244**, 122.
37. Y. Önal, S. Schimpf and P. Claus, *J. Catal.*, 2004, **223**, 122.
38. P. Beltrame, M. Comotti, C. Della Pina and M. Rossi, *Appl. Catal. A*, 2006, **297**, 1.
39. P. Beltrame, M. Comotti, C. Della Pina and M. Rossi, *J. Catal.*, 2004, **228**, 282.
40. M. Comotti, C. Della Pina, E. Falletta and M. Rossi, *Adv. Synth. Catal.*, 2006, **348**, 313.
41. C. Della Pina, E. Falletta, M. Rossi and A. Sacco, *J. Catal.*, 2009, **263**, 92.
42. M. Comotti, C. Della Pina and M. Rossi, *J. Mol. Catal. A*, 2006, **251**, 89.
43. C. Baatz, N. Decker and U. Pruesse, *J. Catal.*, 2008, **258**, 165.
44. C. Baatz and U. Pruesse, *J. Catal.*, 2007, **249**, 34.
45. F. Porta and L. Prati, *J. Catal.*, 2004, **224**, 397.
46. C. Bianchi, P. Canton, N. Dimitratos, F. Porta and L. Prati, *Catal. Today*, 2005, **102**, 203.
47. N. Dimitratos, F. Porta, L. Prati and A. Villa, *Catal. Lett.*, 2005, **99**, 181.
48. A. Villa, D. Wang, P. Spontoni, R. Arrigo, D. Sub and L. Prati, *Catal. Today*, 2010, **15**, 89.
49. C. Della Pina, E. Falletta and M. Rossi, *ChemSusChem.*, 2009, **2**, 57.
50. O. A. Kholdeeva, O. V. Zalomaeva, A. B. Sorokin, I. D. Ivanchikova, C. Della Pina and M. Rossi, *Catal. Today*, 2007, **121**, 58.
51. H. Wang, W. Fan, Y. He, J. Wang, J. N. Kondo and T. Tatsumi, *J. Catal.*, 2013, **299**, 10.
52. L. Tang, X. Guo, Y. Li, S. Zhang, Z. Zha and Z. Wang, *Chem. Commun.*, 2013, **49**, 5213.
53. J. Feng, C. Ma, P. J. Miedziak, J. K. Edwards, G. L. Brett, D. Li, Y. Du, D. J. Morgan and G. J. Hutchings, *Dalton Trans.*, 2013, **42**, 14498.
54. M. Mahyari, A. Shaabani and Y. Bide, *RSC Adv.*, 2013, **3**, 22509.
55. I. Moreno, N. F. Dummer, J. K. Edwards, M. Alhumaimess, M. Sankar, R. Sanz, P. Pizarro, D. P. Serrano and G. J. Hutchings, *Catal. Sci. Technol.*, 2013, **3**, 2425.
56. V. R. Choudhary and D. K. Dumbre, *Catal. Commun.*, 2011, **13**, 82.
57. V. V. Costa, M. Estrada, Y. Demidova, I. Prosvirin, V. Kriventsov, R. F. Cotta, S. Fuentes, A. Simakov and E. V. Gusevskaya, *J. Catal.*, 2012, **292**, 148.
58. V. Peneau, Q. He, G. Shaw, S. A. Kondrat, T. E. Davies, P. Miedziak, M. Forde, N. Dimitratos, C. J. Kiely and G. J. Hutchings, *Phys. Chem. Chem. Phys.*, 2013, **15**, 10636.
59. R. Kumar, E. Gravel, A. Hagege, H. Li, D. V. Jawale, D. Verma, I. N. N. Namboothiri and E. Doris, *Nanoscale*, 2013, **5**, 6491.
60. W. Wang, Y. Cong, L. Zhang, Y. Huang, X. Wang and T. Zhang, *Tetrahedron Lett.*, 2014, **55**, 124.
61. K. Weissermel and H.-J. Arpe, in *Industrial Organic Chemistry*, ed. C. R. Lindley, VCH, New York, 3rd edn, 1997, pp. 167–168.

62. C. Parmeggiani and F. Cardona, *Green Chem.*, 2012, **14**, 547.
63. L. Prati, A. Villa, C. E. Chan-Thaw, R. Arrigo, D. Wang and D. S. Su, *Faraday Discuss.*, 2011, **152**, 353.
64. S. Coluccia, G. Martra, F. Porta, L. Prati and M. Rossi, *Catal. Today*, 2000, **61**, 165.
65. A. Villa, D. Wang, D. Su, G. M. Veith and L. Prati, *Phys. Chem. Chem. Phys.*, 2010, **12**, 2183.
66. E. G. Rodrigues, S. A. C. Carabineiro, J. J. Delgado, X. Chen, M. F. R. Pereira and J. M. Órfão, *J. Catal.*, 2012, **285**, 83.
67. N. E. Kolli, L. Delannoy and C. Louis, *J. Catal.*, 2013, **297**, 79.
68. P. G. N. Mertens, M. Bulut, L. E. M. Gevers, I. F. J. Vankelecom, P. A. Jacobs and D. E. De Vos, *Catal. Lett.*, 2005, **102**, 57.
69. J. A. Lopez-Sanchez, N. Dimitratos, C. Hammond, G. L. Brett, L. Kesavan, S. White, P. Miedziak, R. Tiruvalam, R. L. Jenkins, A. F. Carley, D. Knight, C. J. Kiely and G. J. Hutchings, *Nat. Chem.*, 2011, **3**, 551.
70. L. Prati and F. Porta, *Appl. Catal. A*, 2005, **291**, 199.
71. P. Korovchenko, C. Donze, P. Gallezot and M. Besson, *Catal. Today*, 2007, **121**, 13.
72. L. Kesavan, R. Tiruvalam, M. H. Ab Rahim, M. Izham bin Saiman, D. I. Enache, R. L. Jenkins, N. Dimitratos, J. A. Lopez-Sanchez, S. H. Taylor, D. W. Knight, C. J. Kiely and G. J. Hutchings, *Science*, 2011, **331**, 195.
73. X. Wang, N. S. Venkataramanan, H. Kawanami and Y. Ikushima, *Green Chem.*, 2007, **9**, 1352.
74. N. Dimitratos, J. A. Lopez-Sanchez, S. Meenakshisundaram, J. M. Anthonykutty, G. Brett, A. F. Carley, S. H. Taylor, D. W. Knight and G. J. Hutchings, *Green Chem.*, 2009, **11**, 1209.
75. B. Wang, A. Villa, F. Porta, L. Prati and D. Su, *J. Phys. Chem. C*, 2008, **112**, 8617.
76. A. Villa, A. Gaiassi, M. Veith and L. Prati, *Angew. Chem. Int. Ed.*, 2010, **49**, 4499.
77. E. Falletta, C. Della Pina, M. Rossi, Q. He, C. J. Kiely and G. J. Hutchings, *Faraday Discuss.*, 2011, **152**, 367.
78. A. Abad, P. Conception, A. Corma and H. Garcia, *Angew. Chem. Int. Ed.*, 2005, **44**, 4066.
79. D. Enache, J. Edwards, P. Landon, B. Solsona-Espriu, A. Carley, A. Herzing, M. Watanabe, C. Kiely, D. Knight and G. Hutchings, *Science*, 2006, **311**, 362.
80. N. Wang, T. Matsumoto, M. Ueno, H. Miyamura and S. Kobayashi, *Angew. Chem. Int. Ed.*, 2009, **121**, 4838.
81. W. C. Ketchie, M. Murayama and R. J. Davis, *Top. Catal.*, 2007, **44**, 307.
82. C. Milone, R. Ingoglia, G. Neri and S. Galvagno, *Appl. Catal. A*, 2001, **211**, 251.
83. N. Zheng and G. D. Stucky, *Chem. Commun.*, 2007, 3862.
84. C. H. Christensen, B. Jorgensen, J. Rass-Hansen, K. Egeblad, R. Madsen, S. K. Klitgaard, S. M. Hansen, M. R. Hansen, H. C. Andersen and A. Riisager, *Angew. Chem. Int. Ed.*, 2006, **45**, 4648.

85. C. Della Pina, E. Falletta and M. Rossi, *J. Catal.*, 2008, **260**, 384.
86. T. Werpy, J. Holladay and J. White, *Top Value Added Chemicals from Biomass*, Pacific Northwest National Laboratory, Richland, WA, PNNL-14808, 2004, https://www1.eere.energy.gov/bioenergy/pdfs/35523.pdf.
87. S. Carrettin, P. McMorn, P. Johnston, K. Griffin and G. Hutchings, *Chem. Commun.*, 2002, 696.
88. N. Dimitratos, J. A. Lopez-Sanchez, S. Meenakshisundaram, J. M. Anthonykutty, G. Brett, A. F. Carley, S. H. Taylor, D. W. Knight and G. J. Hutchings, *Green Chem.*, 2009, **11**, 1209.
89. L. Prati, P. Spontoni and A. Gaiassi, *Top. Catal.*, 2009, **52**, 288.
90. B. Katryniok, H. Kimura, E. Skrzyńska, J.-S. Girardon, P. Fongarland, M. Capron, R. Ducoulombier, N. Mimura, S. Paul and F. Dumeignil, *Green Chem.*, 2011, **13**, 1960.
91. R. Garcia, M. Besson and P. Gallezot, *Appl. Catal. A*, 1995, **127**, 165.
92. S. Carrettin, P. McMorn, P. Johnston, K. Griffin, C. Kiely and G. J. Hutchings, *Phys. Chem. Chem. Phys.*, 2003, **5**, 1329.
93. P. Gallezot, *Catal. Tod.*, 1997, **37**, 405.
94. P. Fordham, M. Besson and P. Gallezot, *Catal. Lett.*, 1997, **46**, 195.
95. H. Kimura and K. Tsuto, *Appl. Catal. A*, 1993, **96**, 217.
96. D. Liang, J. Gao, J. Wang, P. Chen, Z. Hou and X. Zheng, *Catal. Commun.*, 2009, **10**, 1586.
97. D. Tongsakul, S. Nishimura and K. Ebitani, *ACS Catal.*, 2013, **3**, 2199.
98. E. Skrzynska, J. Ftouni, A.-S. Mamede, A. Addad, M. Trentesaux, J.-S. Girardon, M. Capron and F. Dumeignil, *J. Mol. Catal. A*, 2014, **382**, 71.
99. Y. Liu, H. Tüysüz, C.-J. Jia, M. Schwickardi, R. Rinaldi, A. H. Lu, W. Schmidt and F. Schüth, *Chem. Commun.*, 2010, **46**, 1238.

CHAPTER 6

Aerobic Oxidation of Alcohols Catalyzed by Iron or Vanadium Compounds

MASAYUKI KIRIHARA

Department of Materials and Life Science, Shizuoka Institute of Science and Technology, 2200-2 Toyosawa, Fukuroi, Shizuoka, 437-8555, Japan
Email: kirihara@ms.sist.ac.jp

6.1 Oxidations Catalyzed by Iron Compounds

6.1.1 $Fe(NO_3)_3$–$FeBr_3$-Catalyzed Aerobic Oxidation

The first practical aerobic oxidation of an alcohol catalyzed by iron compounds was reported by Martin and Suárez in 2002 (Scheme 6.1).[1] Based on their earlier report of the efficient and selective aerobic oxidation of sulfides to sulfoxides with a binary system of $Fe(NO_3)_3$ and $FeBr_3$,[2] they successfully attempted the aerobic oxidation of alcohols using the same catalytic system. Several secondary and benzylic alcohols were efficiently oxidized in the presence of catalytic amounts of the binary Fe(III) system. It is interesting that primary alcohols are inert under the reaction conditions. Therefore, a secondary alcohol, even in the presence of a primary alcohol, can be selectively oxidized to produce a hydroxy ketone using this method. Furthermore, the presence of an amino group in the substrate did not interfere with the oxidation process. Two active species may be at work in the Fe(III)-catalyzed reaction, either oxidation by NO_2 or oxidation by a bromide–bromine couple controlled by Fe(III).[1]

RSC Green Chemistry No. 28
Transition Metal Catalysis in Aerobic Alcohol Oxidation
Edited by Francesca Cardona and Camilla Parmeggiani

Published by the Royal Society of Chemistry, www.rsc.org

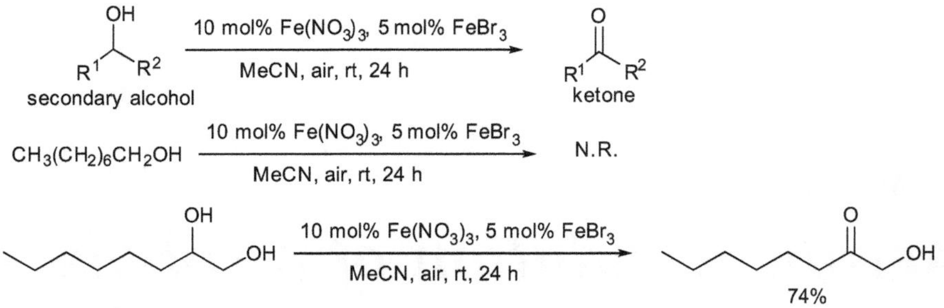

Scheme 6.1 Aerobic oxidation catalyzed by $Fe(NO_3)_3$–$FeBr_3$.

Scheme 6.2 Aerobic oxidation catalyzed by $FeCl_3$–TEMPO–$NaNO_2$.

6.1.2 Fe(III)-TEMPO-Catalyzed Aerobic Oxidation

TEMPO (2,2,6,6-tetramethylpiperidine-1-oxyl)[3] and related compounds[4] have been used as catalysts for the oxidation of organic compounds for many years. Recently, catalytic systems for alcohol oxidation involving transition metals and TEMPO have attracted a significant amount of attention owing to the use of molecular oxygen as the terminal oxidant. In 2005, Liang and co-workers reported that $FeCl_3$–TEMPO–$NaNO_2$ catalyzed the selective and mild aerobic oxidation of a broad range of alcohols to the corresponding aldehydes and ketones (Scheme 6.2).[5] The ideal conditions for the oxidation used 5 mol% $FeCl_3$, 2 mol% TEMPO and 5 mol% $NaNO_2$ as catalysts in (trifluoromethyl)benzene ($PhCF_3$) at room temperature under ambient air pressure. Primary and secondary benzylic alcohols, cinnamyl alcohol and secondary aliphatic alcohols were efficiently oxidized to the corresponding aldehydes and ketones in high yields with excellent selectivity. On the other hand, aliphatic primary alcohols were converted to the corresponding aldehydes with modest selectivity accompanied by both acids and esters. It is notable that sulfide did not interfere with the alcohol and was not oxidized under the reaction conditions. In fact, when a mixture of alcohols and methyl phenyl sulfide was reacted with air under these reaction conditions, the alcohols were selectively oxidized to aldehydes, whereas methyl phenyl sulfide remained unreacted.

A plausible mechanism for this catalytic oxidation can be described as a cascade of redox reactions involving two cycles as depicted in Scheme 6.3.

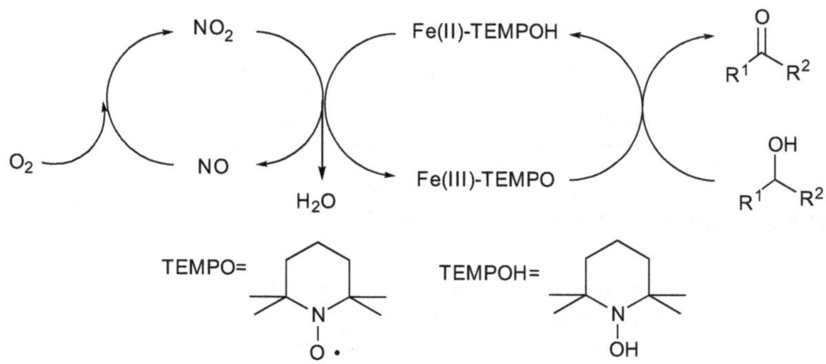

Scheme 6.3 A plausible reaction mechanism for $FeCl_3$–TEMPO–$NaNO_2$-catalyzed aerobic oxidation.

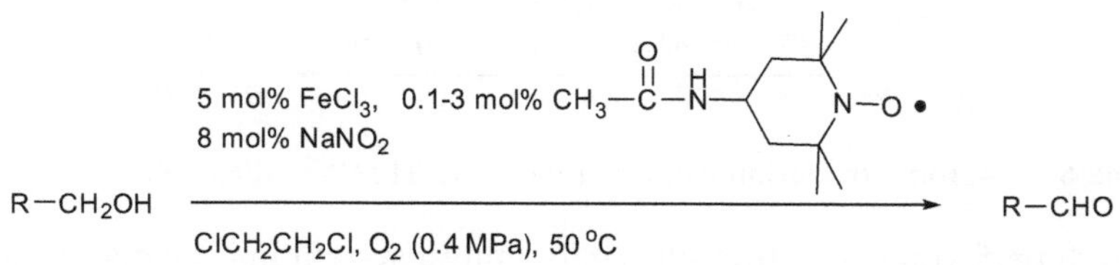

Scheme 6.4 Aerobic oxidation catalyzed by $FeCl_3$-4-acetamido-TEMPO–$NaNO_2$.

In order to improve the selectivity of the oxidation of aliphatic primary alcohols to aldehydes, Liang and co-workers further examined the reaction conditions and found that using 4-acetamido-TEMPO instead of TEMPO provided the best results (Scheme 6.4).[6] Several primary alcohols were selectively oxidized to the corresponding aldehydes in high yields in the presence of 5 mol% $FeCl_3$, 0.1–3 mol% TEMPO and 8 mol% $NaNO_2$ in 1,2-dichloroethane under an oxygen atmosphere at 50 °C. This reaction system was applicable to a broad range of primary and secondary alcohols, which could contain carbon–carbon double bonds, and also N, O and S heteroatoms.

Wang and Liang also reported that a combination of $Fe(NO_3)_3$ and 4-OH-TEMPO efficiently catalyzed the oxidation of alcohols under an ambient air atmosphere in acetonitrile at room temperature (Scheme 6.5).[7]

Recently, Ma and co-workers reported an $Fe(NO_3)_3$–TEMPO-catalyzed aerobic oxidation in the presence of a catalytic amount of NaCl in 1,2-dichloroethane, which efficiently converted essentially all types of alcohols, including compounds bearing different unsaturated carbon–carbon bonds, within a couple of hours at room temperature (Scheme 6.6).[8–11] The exact role of NaCl is not clear. This reaction system is applicable to allenols,[8] propargylic alcohols,[9] allylic alcohols[10] and indole carbinols.[11]

Gao *et al.* recently reported that silica gel not only enhances the catalytic reaction rate but also increases selectivity for the products in the aerobic oxidation of alcohols catalyzed by $FeCl_3\cdot 6H_2O$–TEMPO (Scheme 6.7).[12]

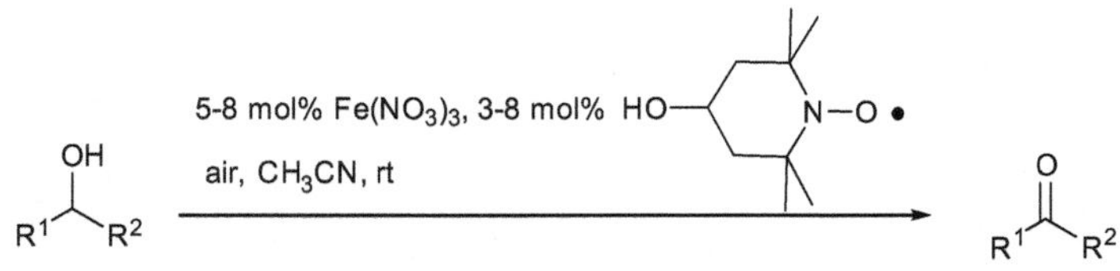

Scheme 6.5 Aerobic oxidation catalyzed by $Fe(NO_3)_3$-4-OH-TEMPO.

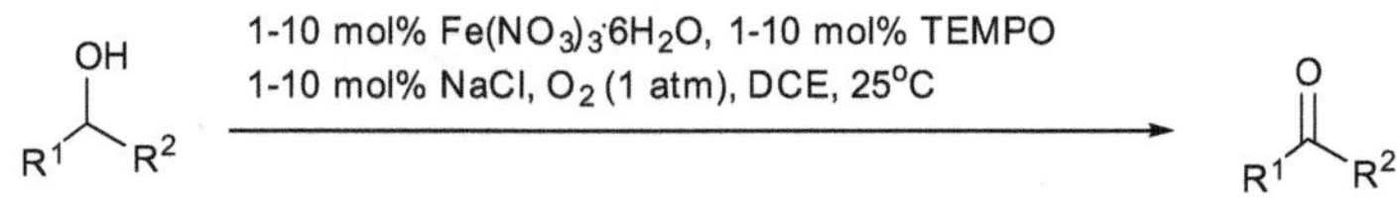

Scheme 6.6 Aerobic oxidation catalyzed by $Fe(NO_3)_3$–TEMPO–NaCl.

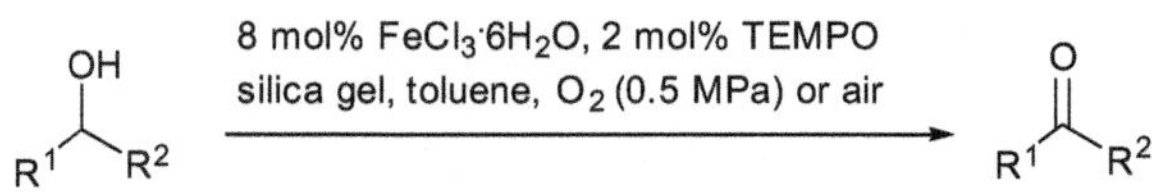

Scheme 6.7 Aerobic oxidation catalyzed by $FeCl_3$–TEMPO–silica gel.

The high performance of this silica gel–catalyst system may be due to surface silanol groups. Primary and secondary benzylic alcohols were oxidized efficiently to the corresponding aldehydes and ketones in high yields. Conversely, aliphatic alcohols were less reactive under the reaction conditions.

6.1.3 Aerobic Oxidation Catalyzed by Recyclable Fe(III)–TEMPO Derivatives

The first bimagnetic catalyst, Fe and TEMPO supported in an ionic liquid (IL), was developed by Zhang and co-workers in 2011.[13] A task-specific bimagnetic imidazolium salt, [Imim-TEMPO][$FeCl_4$], containing cooperative functionalities was easily synthesized and applied to the selective aerobic oxidation of aromatic alcohols (Scheme 6.8). Quantitative yields with excellent selectivity were achieved under mild conditions without the use of any organic solvent. It is worth noting that the reaction also performed well even when using air as an oxidant instead of pure oxygen. The catalyst [Imim-TEMPO][$FeCl_4$] was readily recovered by diethyl ether extraction and could be used repeatedly by charging with fresh substrate and $NaNO_2$. The catalyst could be reused at least five times with retention of high activity and selectivity. At least in principle, the use of a magnet to recover this magnetic IL is also possible.

Gao and co-workers selected a representative mesoporous silica, SBA-15, as the support for TEMPO and found that $FeCl_3 \cdot 6H_2O$–SBA-15–TEMPO/$NaNO_2$ efficiently catalyzed the oxidation of alcohols with either O_2 or air as the terminal oxidant (Scheme 6.9).[14] Various alcohols were oxidized at a low catalyst loading (0.1–1 mol%) to the corresponding aldehydes and ketones in

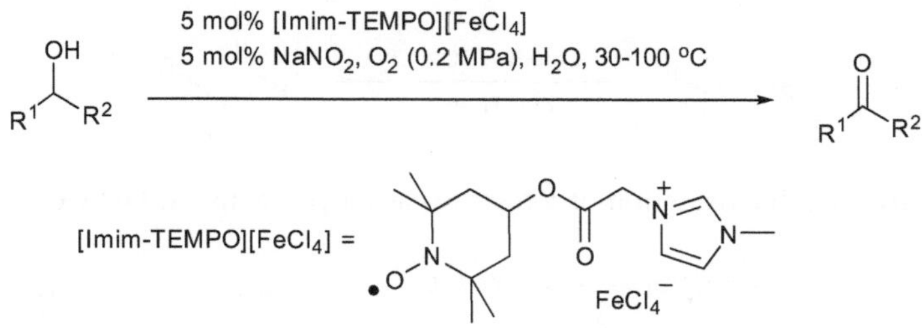

Scheme 6.8 Aerobic oxidation catalyzed by Fe and TEMPO supported in IL.

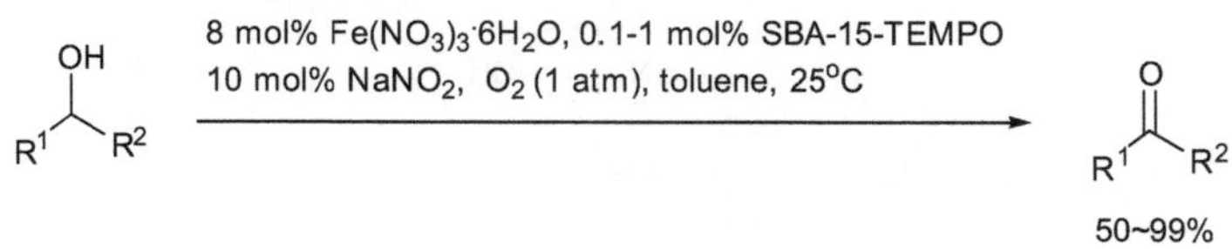

Scheme 6.9 Aerobic oxidation catalyzed by $FeCl_3 \cdot 6H_2O$–SBA-15–TEMPO/$NaNO_2$.

high yields. The catalyst SBA-15–TEMPO can be reused for 10 reaction cycles without a significant loss of catalytic activity.

6.2 Oxidations Catalyzed by Vanadium Compounds

6.2.1 Aerobic Oxidation of α-Hydroxycarbonyls and Related Compounds

In 1999, Kirihara *et al.* reported that α-hydroxycarbonyls were efficiently oxidized to α-dicarbonyls using a catalytic amount of $VOCl_3$ in acetonitrile under an oxygen atmosphere (Scheme 6.10).[15] Various α-hydroxy ketones and α-hydroxy esters afforded the desired α-dicarbonyls in almost quantitative yield. On the other hand, alcohols lacking a carbonyl moiety at the α-position did not react under the same conditions.

Toste and co-workers developed the oxidative kinetic resolution of α-hydroxy esters by their aerobic asymmetric oxidation catalyzed by $VO(O\textit{i}Pr)_3$ and chiral *O,N,O*-chelating tridentate ligands (Scheme 6.11).[16] Several racemic aryl-, vinyl- and alkyl-substituted α-hydroxy esters were efficiently resolved with selectivity factors (k_{rel}) ranging from 12 to >50.

Chen and co-workers synthesized a series of chiral vanadyl carboxylates derived from *N*-salicylidene-L-α-amino acids and V(V).[17,18] These chiral catalysts were effective in the aerobic oxidative kinetic resolution of a variety of racemic α-hydroxy esters,[17] amides[17] and phosphonates[18] (Scheme 6.12).

During the course of their investigations, Chen and co-workers found that a pentanuclear C_4-symmetric cluster complex was formed for the 3,5-di-*tert*-butyl analog when potassium salts were employed.[19] The cluster complex was utilized to facilitate the asymmetric aerobic oxidations of various

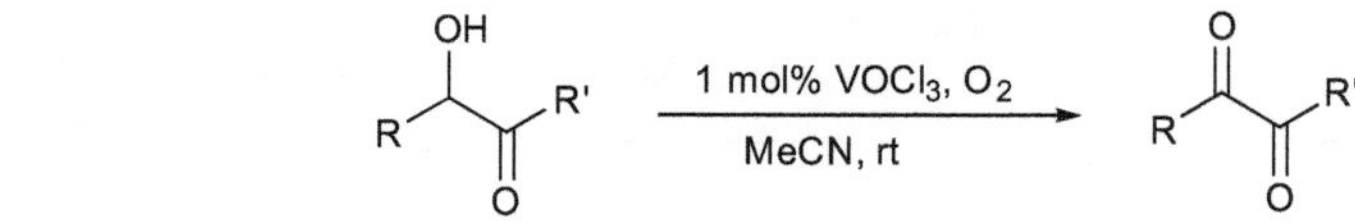

Scheme 6.10 Aerobic oxidation of α-hydroxycarbonyls catalyzed by $VOCl_3$.

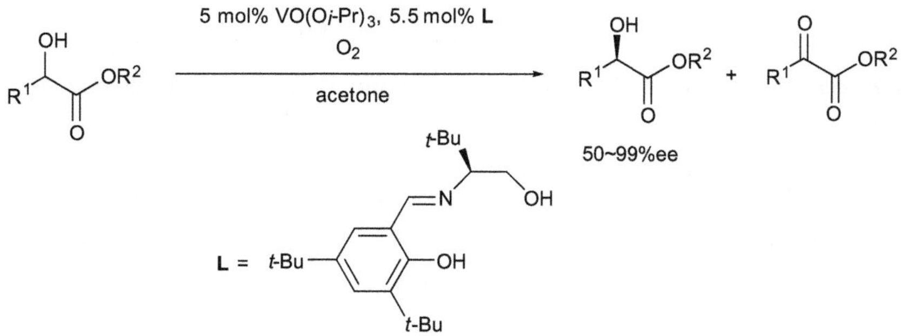

Scheme 6.11 Oxidative kinetic resolution of α-hydroxy esters catalyzed by VO(O*i*Pr)$_3$ and chiral *O*,*N*,*O*-chelating tridentate ligands.

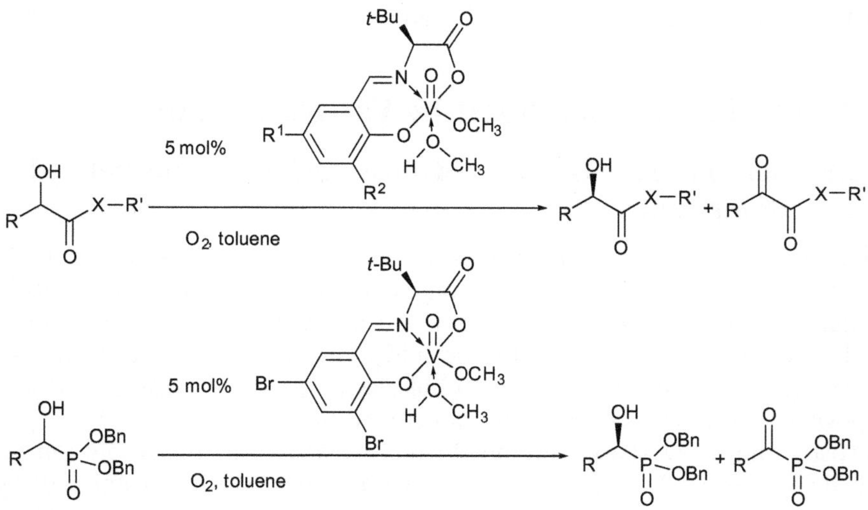

Scheme 6.12 Oxidative kinetic resolution of α-hydroxy esters, -amides and -phosphonates catalyzed by chiral vanadyl carboxylates.

racemic α-hydroxy esters, -amides and thioesters with excellent selectivity factors ($k_{rel} = 40$ to >500) (Scheme 6.13).

Subsequently, Chen's group prepared polystyrene-supported chiral vanadyl catalysts through the click chemistry approach.[20] These catalysts promoted the asymmetric aerobic oxidation of α-hydroxy (thio)esters and -amides with enantioselectivities of up to 99% *ee* in chloroform (Scheme 6.14). These catalysts can be readily recovered by filtration and

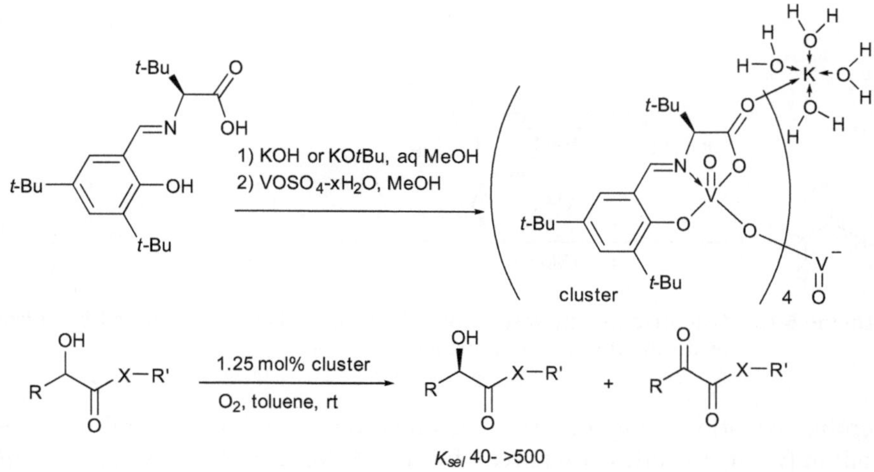

Scheme 6.13 Oxidative kinetic resolution of α-hydroxycarbonyls catalyzed by a pentanuclear C_4-symmetric cluster complex.

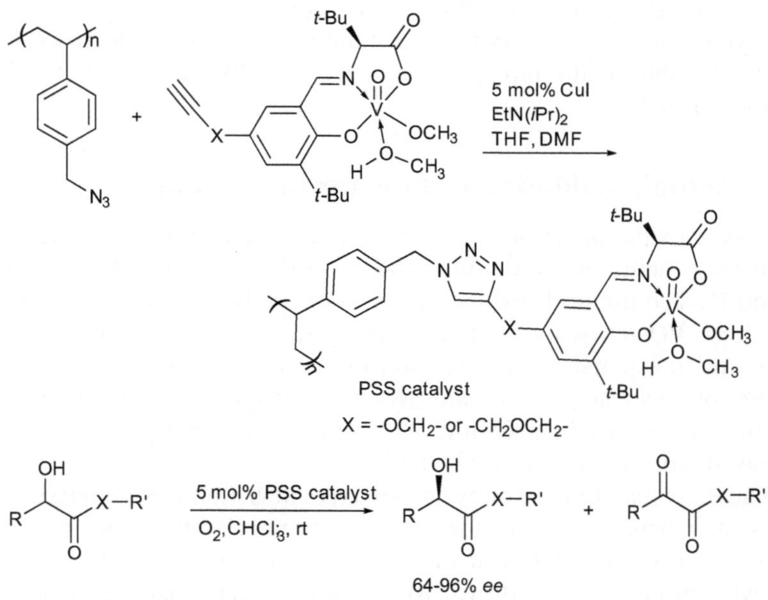

Scheme 6.14 Oxidative kinetic resolution of α-hydroxycarbonyls catalyzed by recyclable vanadium complexes.

reused at least four times without a decrease in reactivity or enantioselectivity.

Unfortunately, asymmetric oxidation of α-hydroxy ketones provided products with only moderate selectivity using the chiral vanadyl complexes developed by Chen and co-workers.[20] Therefore, they undertook further investigations into the effective kinetic resolution of α-hydroxy ketones using

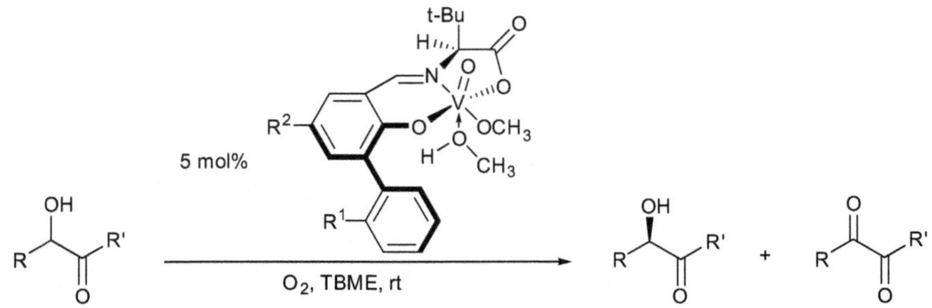

Scheme 6.15 Oxidative kinetic resolution of α-hydroxy ketones catalyzed by chiral oxidovanadium(v) methoxides.

aerobic oxidation catalyzed by chiral vanadium complexes. Chiral oxidovanadium(v) methoxides prepared from 3,5-diphenyl-*N*-salicylidene-L-*tert*-butylglycine (or 3-*o*-biphenyl-5-nitro-*N*-salicylidene-L-*tert*-butylglycine) and vanadyl sulfate in air-saturated MeOH showed the best results for the aerobic oxidative kinetic resolution of alkyl, aryl and heteroaryl α-hydroxy ketones with different α-substituents at room temperature in toluene or *tert*-butyl methyl ether (TBME). The kinetic resolution selectivity factors for the aerobic oxidation process were in the range 12 to >1000 (Scheme 6.15).[21]

6.2.2 Aerobic Oxidation of Propargylic Alcohols

The aerobic oxidation of propargylic alcohols catalyzed by an oxovanadium complex was achieved by Uemura and co-workers (Scheme 6.16).[22] Several V(v) and V(IV) compounds exhibited excellent catalytic activity, but not V_2O_5. They chose $VO(acac)_2$ as a suitable catalyst, because it is stable, ease to handle and inexpensive. The reaction of several propargylic alcohols with 1 mol% of $VO(acac)_2$ under an oxygen atmosphere in the presence of molecular sieves 3A in acetonitrile provided the corresponding carbonyl compounds in moderate to high yields.

A plausible reaction pathway is depicted in Scheme 6.17. First, a V(IV) species (**A**) reacts with O_2 to form a V(v) (**B**). Species **B** further reacts with propargylic alcohol to afford a vanadium alcoholate (**C**). Next, the product carbonyl compound and a dihydroxyvanadium species (**D**) are formed *via* β-hydrogen elimination from **C**. The subsequent dehydration of **D** by oxidation with oxygen re-produces species **B**.

Subsequently, Uemura and co-workers[23] attempted to heterogenize this vanadium system to allow reuse of the catalyst (for green/sustainable chemistry) and eventually they discovered that calcium phosphate–vanadate apatite (CPVAP)[24] could be used as a recyclable catalyst for the aerobic oxidation of propargylic alcohols (Scheme 6.18). A variety of propargylic alcohols were oxidized to the corresponding carbonyl compounds in the presence of a catalytic amount of CPVAP in *n*-butyronitrile at 100 °C under an

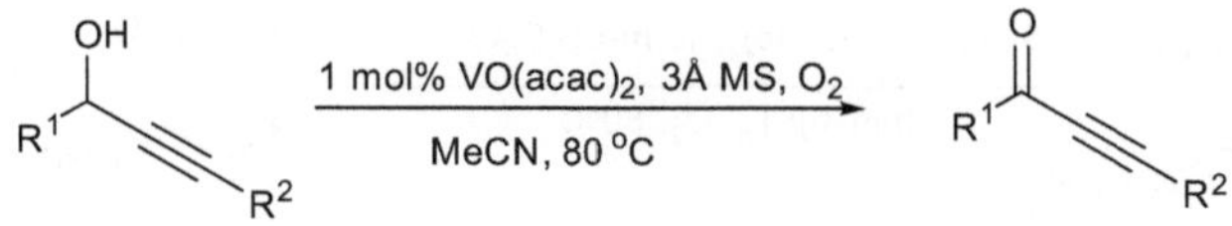

Scheme 6.16 Aerobic oxidation of propargylic alcohols catalyzed by an oxovanadium complex.

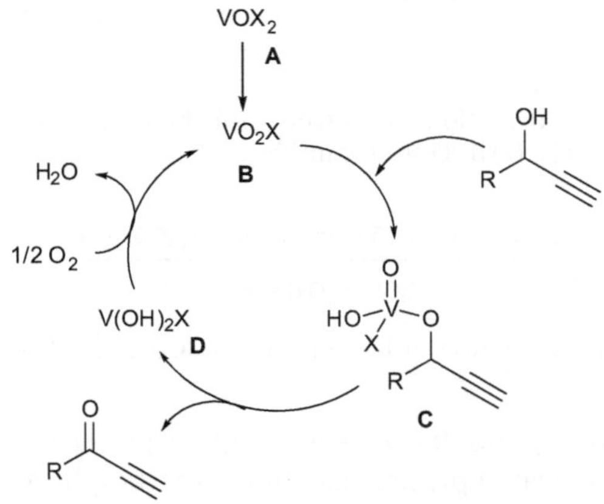

Scheme 6.17 Plausible reaction mechanism for oxovanadium-catalyzed aerobic oxidation of propargylic alcohols.

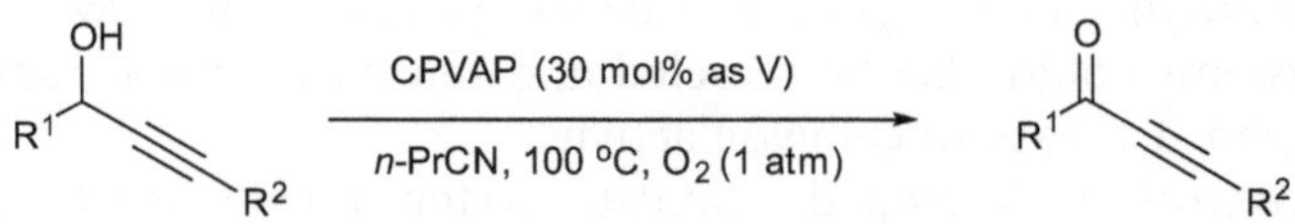

Scheme 6.18 Aerobic oxidation of propargylic alcohols catalyzed by a recyclable vanadium complexes.

oxygen atmosphere. CPVAP can be readily separated by filtration and reused at least 10 times without an appreciable decrease in catalytic activity.

6.2.3 Aerobic Oxidation of Allylic and Benzylic Alcohols

In 2007, Jiang and Ragauskas reported the vanadium-catalyzed selective aerobic oxidation of activated alcohols (allylic and benzylic) in an ionic liquid solvent. The optimal conditions were treatment of the activated alcohols with 5 mol% $VO(acac)_2$, 10 mol% DABCO (1,4-diazabicyclo[2.2.2]octane) in [bmim][PF_6] (1-butyl-3-methylimidazolium hexafluorophosphate) at 80 °C under an oxygen atmosphere (Scheme 6.19).[25] Several primary and secondary activated alcohols provided the corresponding aldehydes and ketones in high yield. On the other hand, unactivated alcohols provided the carbonyl compounds in poor yields. The catalysts can be recycled and reused for three runs without any significant decrease in catalytic activity.

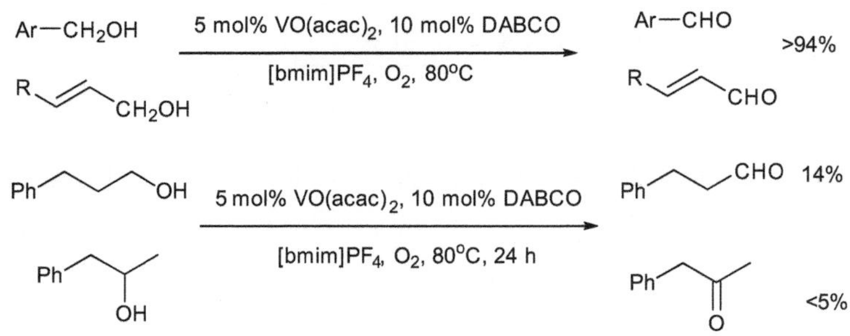

Scheme 6.19 Aerobic oxidation of allylic and benzylic alcohols catalyzed by $VO(acac)_2$–DABCO in $[bmim]PF_6$.

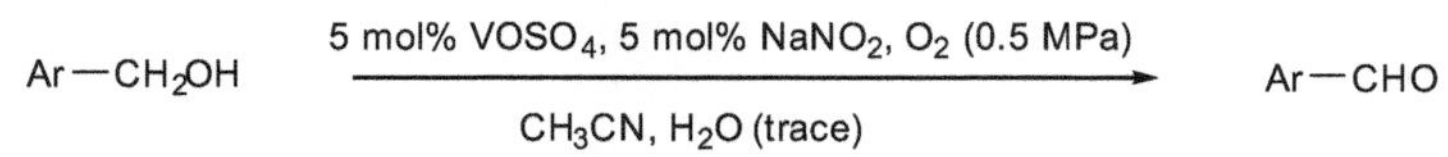

Scheme 6.20 Aerobic oxidation of benzylic alcohols catalyzed by $VOSO_4$–$NaNO_2$.

Subsequently, Jiang and Ragauskas sought the reaction conditions necessary to oxidize activated primary alcohols to carboxylic acids. The reaction of activated primary alcohols with 2 mol% $VO(acac)_2$, 2 mol% Cu(II) 2-ethylhexanoate, 6 mol% DABCO in [hmim][OTf] (1-hexyl-3-methylimidazolium triflate) at 95 °C under an oxygen atmosphere afforded the corresponding carboxylic acids in good to excellent yields.[26] This newly developed catalytic systems could also be recycled and reused for three runs without any significant decrease in catalytic activity.

Xu and co-workers developed a catalytic system for the aerobic oxidation of benzylic alcohols consisting of vanadyl sulfate and sodium nitrite (Scheme 6.20).[27] Benzyl alcohols with various substituents were efficiently converted to their corresponding aldehydes with high conversion and selectivity under 0.5 MPa of O_2 at 80 °C. Addition of a trace amount of water to this system before the reaction was crucial for high efficiency.

Xu and co-workers also developed a method for the oxidation of benzylic alcohols with O_2 in water catalyzed by $VOPO_4$–TEMPO (Scheme 6.21).[28] The presence of the phosphorus species was essential to ensure high efficiency.

In 2011, Ogawa's group reported that a vanadium complex, $VOSO_4$, combined with 4.4′-di-*tert*-butyl-2,2′-bipyridyl efficiently catalyzed the oxidation of benzyl alcohols by O_2 (or air) using water as a solvent (Scheme 6.22).[29] Since they observed the presence of a V(V) species in the reaction as evidenced by ^{51}V NMR spectroscopy, there is the possibility that the V(V) species acts as the active species in the oxidation. This reaction might not involve a radical pathway, because a radical inhibitor (2,6-di-*tert*-butylphenol) did not interfere with the oxidation.

Hanson and co-workers in 2011 developed the catalytic vanadium complex (HQ)2V(O)(O*i*Pr) (HQ = 8-quinolinate) (2 mol%) and NEt_3 (10 mol%) for the oxidation of benzylic, allylic and propargylic alcohols with air

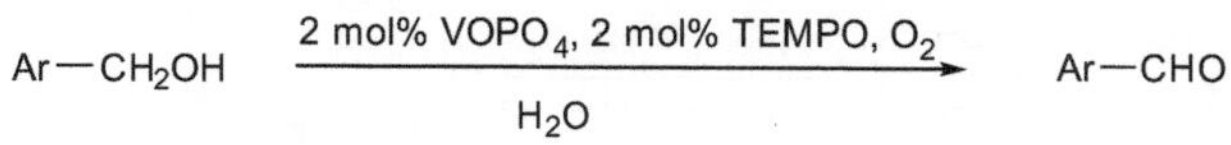

Scheme 6.21 Aerobic oxidation of benzylic alcohols catalyzed by $VOPO_4$–TEMPO.

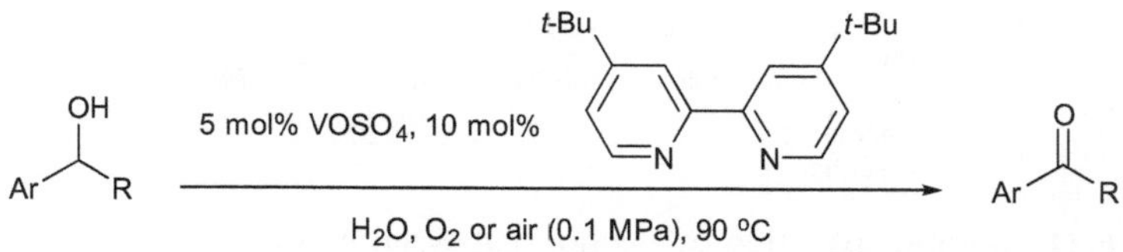

Scheme 6.22 Aerobic oxidation of benzylic alcohols catalyzed by vanadium complex $VOSO_4$ combined with 4.4′-di-*tert*-butyl-2,2′-bipyridyl.

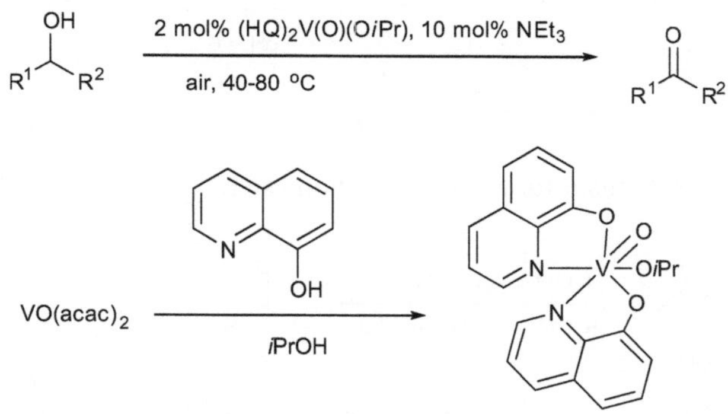

Scheme 6.23 Aerobic oxidation of benzylic, allylic and propargylic alcohols catalyzed by (HQ)2V(O)(O*i*Pr)–NEt_3.

(Scheme 6.23).[30] The activated alcohols produced the desired aldehydes and ketones in high yields. Conversely, simple aliphatic alcohols underwent little or no oxidation, even when heated at 100 °C. The catalyst can be easily prepared in air from the reaction of $VO(acac)_2$ with 8-quinolinol in 2-propanol.

6.2.4 Aerobic Oxidation of Alcohols Including Unactivated Alcohols

Velusamy and Punniyamurthy found that a variety of secondary alcohols, including unactivated aliphatic alcohols, were oxidized to the corresponding ketones under an oxygen atmosphere in the presence of 5 mol% V_2O_5 in toluene at 100 °C. In the case of a primary alcohol, the ester was obtained instead of the corresponding aldehyde (Scheme 6.24).[31] The ester was produced from the subsequent oxidation of a hemiacetal derived from the aldehyde and the unchanged alcohol. The formation of the hemiacetal is due to the weakly acidic nature of the reaction medium (pH 6) (Scheme 6.25).

Addition of K_2CO_3 to the reaction with primary alcohols increased the pH to 9, inhibited hemiacetal formation and afforded the corresponding

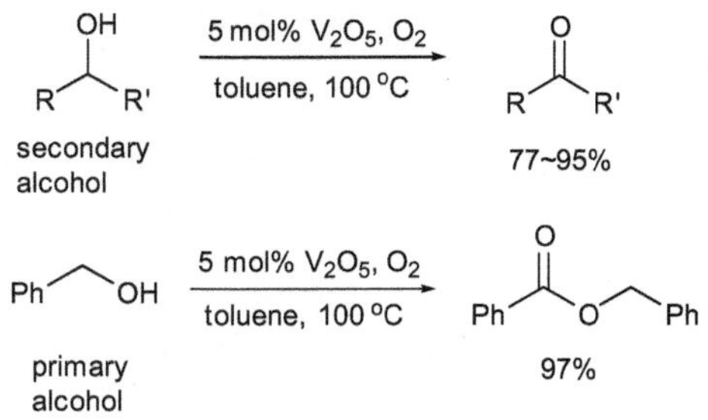

Scheme 6.24 Aerobic oxidation of alcohols catalyzed by V_2O_5.

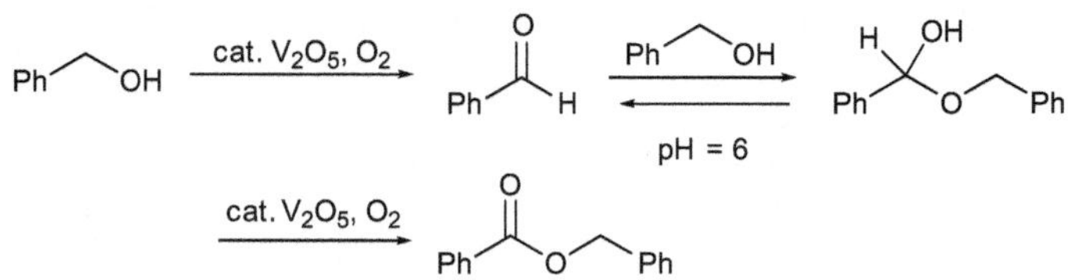

Scheme 6.25 Mechanism for production of an ester.

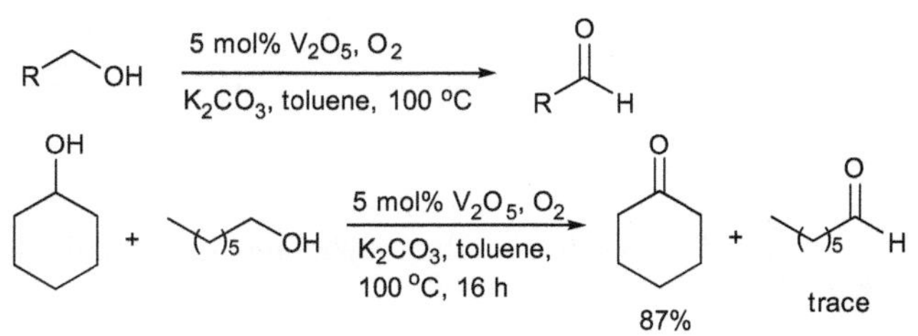

Scheme 6.26 Aerobic oxidation of alcohols catalyzed by V_2O_5 in the presence of K_2CO_3.

aldehydes in moderate to high yields (Scheme 6.26). Interestingly, secondary alcohols can be chemoselectively converted to ketones in high yields in the presence of primary hydroxyl groups.[31]

Punniyamurthy's group also reported that a polyaniline-supported vanadium complex efficiently catalyzes the oxidation of alcohols to aldehydes and ketones in high yields under O_2 in toluene at 100 °C (Scheme 6.27).[32] The catalyst was easily prepared by stirring a 1 : 1 (w/w) mixture of VO(acac)$_2$ and polyaniline in CH_2Cl_2 at room temperature for 48 h.

This catalyst catalyzed the aerobic oxidation of benzyl alcohol in toluene at 100 °C, producing benzaldehyde in 98% yield. On the other hand, the aerobic oxidation of benzyl alcohol catalyzed by unsupported VO(acac)$_2$ was less selective and provided a 2.5 : 1 : 1 mixture of benzaldehyde, benzoic acid and benzyl benzoate (Scheme 6.28). The catalyst is effective for the aerobic oxidation of aliphatic alcohols and can be recycled for three runs without a decrease in activity.[32]

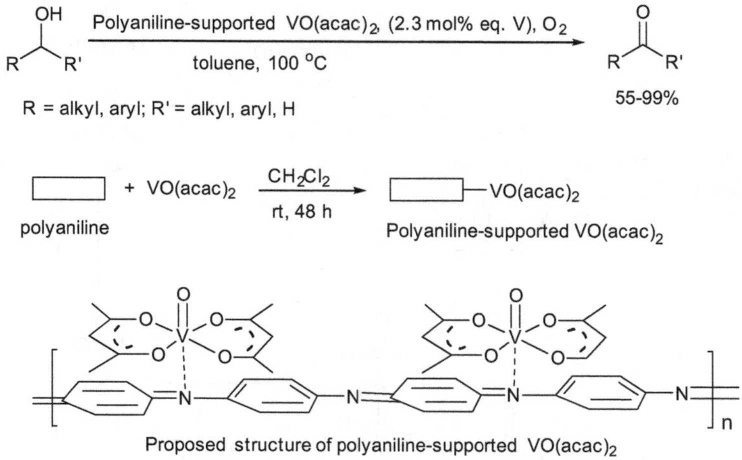

Scheme 6.27 Aerobic oxidation of alcohols catalyzed by recyclable polyaniline-supported $VO(acac)_2$.

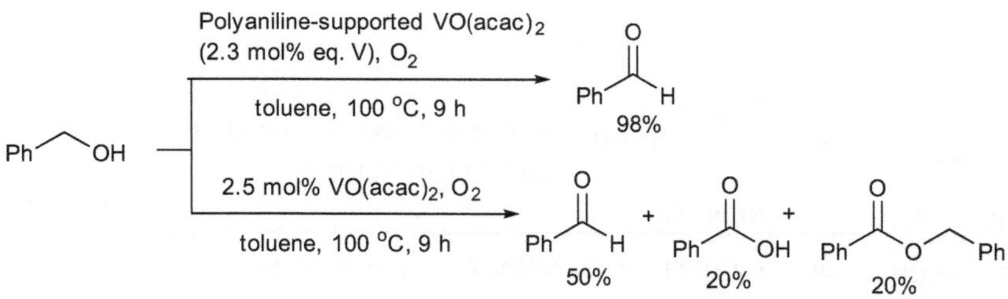

Scheme 6.28 Oxidation of benzyl alcohol catalyzed by polyaniline-supported and -unsupported $VO(acac)_2$.

The proposed reaction mechanism is depicted in Scheme 6.29.[32] The alcohol, followed by O_2, may coordinate with the vanadium catalyst to generate a superoxo complex (**a**) that can readily undergo intramolecular hydrogen abstraction to give an intermediate (**b**), followed by elimination of H_2O_2 to complete the catalytic cycle.

Sobczak and co-workers developed a procedure for the oxidation of alcohols with O_2 catalyzed by *N*-hydroxyphthalimide (NHPI) and vanadium compounds (Scheme 6.30).[33] Various alcohols (primary and secondary) were selectively oxidized to the corresponding carboxylic acids and ketones in high yields under an oxygen atmosphere in the presence of 5 mol% NHPI, 0.2 mol% $VO(acac)_2$ and 0.2 mol% LiCl. Two other catalytic systems based on NHPI–$VO(acac)_2$–Bn_4NCl and NHPI–$(Bn_4N)VO_3$ were also shown to be effective.

A plausible mechanism for the oxidation of cyclohexanol to cyclohexanone was shown to involve both a radical and a molecular pathway (cycles A and B, respectively) (Scheme 6.31).[34]

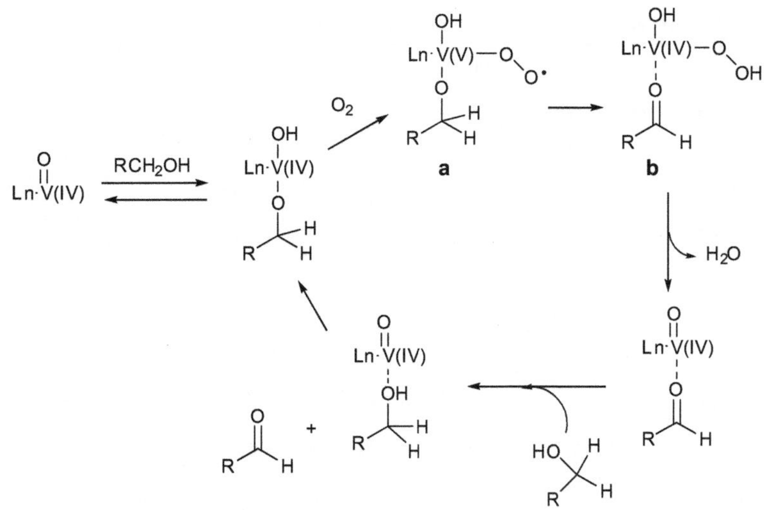

Scheme 6.29 Plausible reaction mechanism for oxidation catalyzed by polyaniline-supported $VO(acac)_2$.

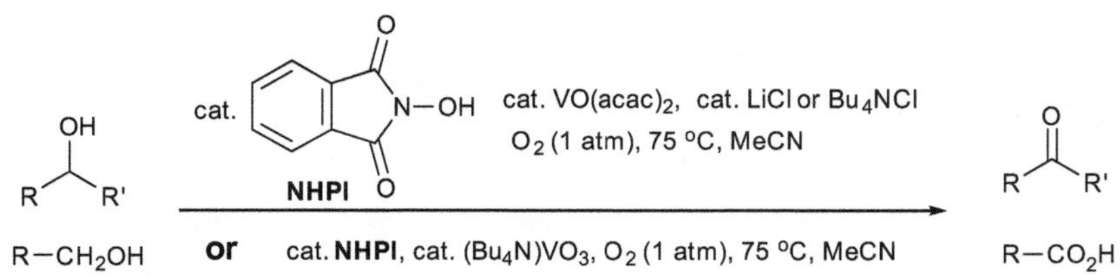

Scheme 6.30 Aerobic oxidation of alcohols catalyzed by NHPI and vanadium compounds.

6.2.5 Aerobic Oxidation of Hydroquinones to Quinones

Uang's group reported that hydroquinone and its derivatives were oxidized to quinines in moderate to high yields with O_2 in the presence of a catalytic amount of $VO(acac)_2$ (Scheme 6.32).[35]

A plausible mechanism for the reaction is depicted in Scheme 6.33. The oxidation may start with a partial disproportionation of the oxovanadium(IV) complex to the corresponding vanadium(III) and vanadium(V) complexes. The vanadium(V) complex then oxidizes hydroquinone to the corresponding quinone and becomes vanadium(III). Then, O_2 oxidizes vanadium(III) to vanadium(V) and generates 1 mol of H_2O_2.

6.2.6 Aerobic Cleavage of Glycol and Related Compounds

α-Hydroxy ketones (α-ketols) can be cleaved chemoselectively with a catalytic amount of $VO(OEt)Cl_2$ under an oxygen atmosphere (Scheme 6.34).[36] Interestingly, *cis*- and *trans*-cyclohexane-1,2-diols are inert under the reaction

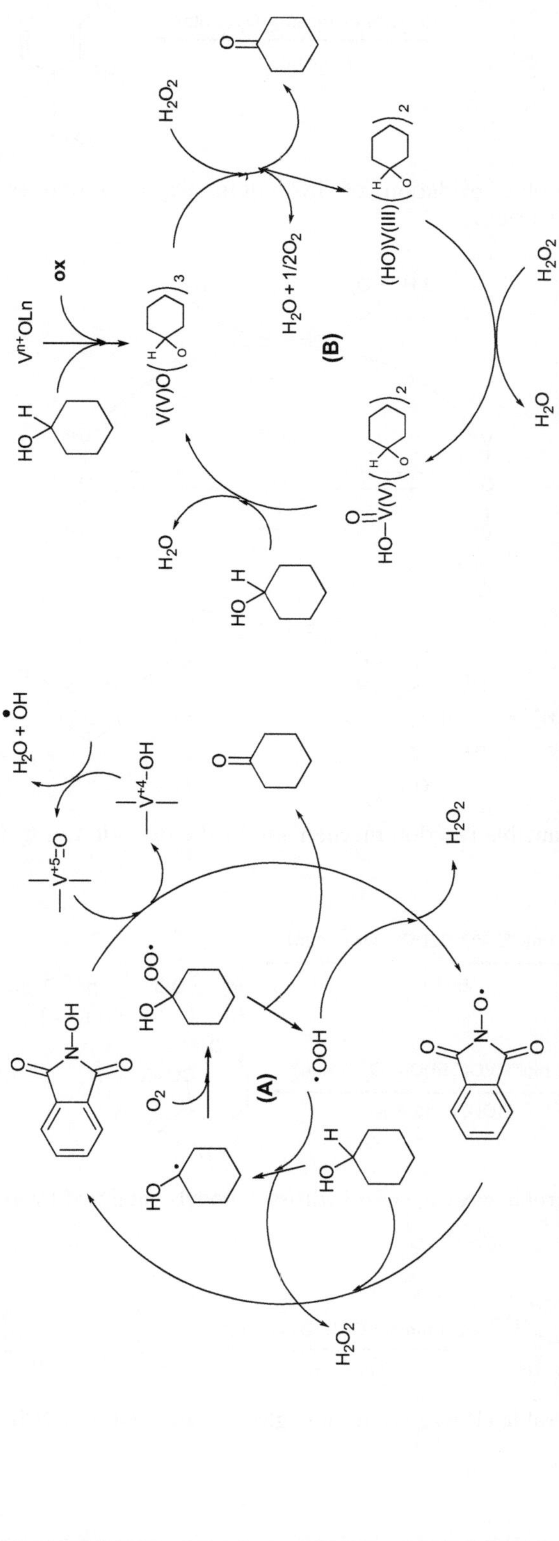

Scheme 6.31 Plausible reaction mechanisms for oxidation catalyzed by NHPI and vanadium compounds.

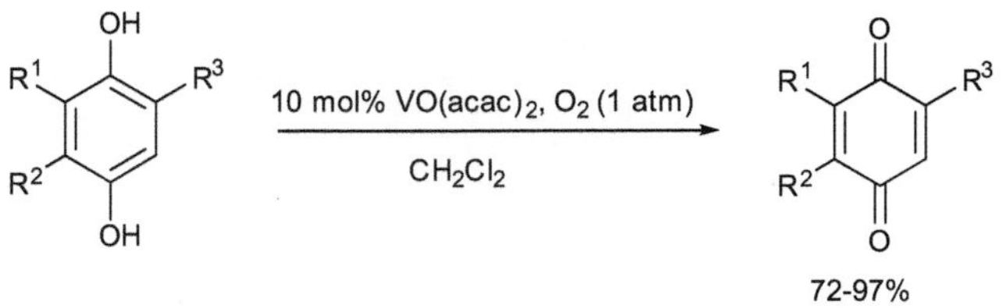

Scheme 6.32 Aerobic oxidation of hydroquinones to quinones catalyzed by $VO(acac)_2$.

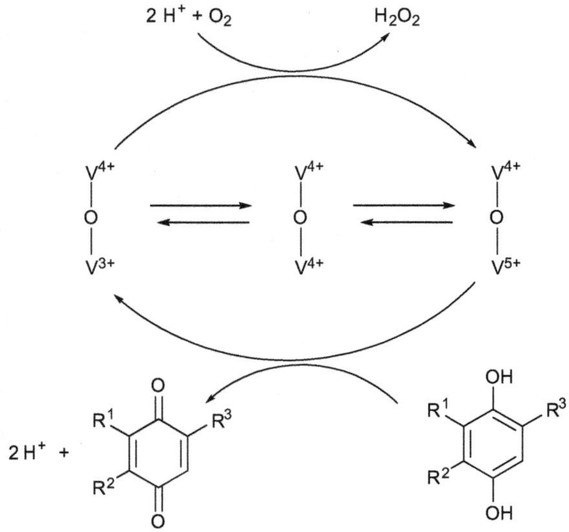

Scheme 6.33 Plausible reaction mechanism for the oxidation of hydroquinones.

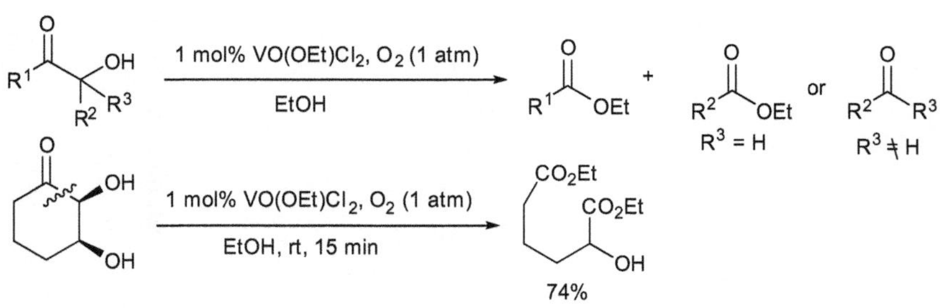

Scheme 6.34 Aerobic cleavage of α-hydroxycarbonyls catalyzed by $VO(OEt)Cl_2$.

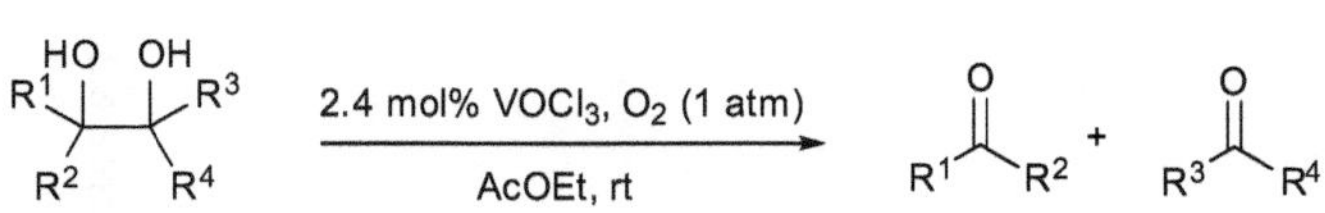

Scheme 6.35 Aerobic cleavage of tertiary glycols catalyzed by $VOCl_3$.

conditions. In fact, only α-cleavage was found to occur in the case of an α,β-dihydroxy ketones.

As described above, secondary 1,2-diols did not react at all under the reaction conditions. However, tertiary 1,2-diols reacted with O_2 in the presence of a catalytic amount of $VOCl_3$, resulting in carbon–carbon bond cleavage and efficient production of the corresponding ketones (Scheme 6.35).[37]

References

1. S. E. Martin and D. F. Suárez, *Tetrahedron Lett.*, 2002, **43**, 4475.
2. S. E. Martin and L. I. Rossi, *Tetrahedron Lett.*, 2001, **42**, 7147.
3. For reviews, see: (a) A. E. J. de Nooy, A. C. Besemer and H. van Bekkum, *Synthesis*, 1996, 1153; (b) R. A. Sheldon, I. W. C. E. Arends, G.-J. ten Brink and A. Dijksman, *Acc. Chem. Res.*, 2002, **35**, 774.
4. (a) M. Shibuya, M. Tomizawa, I. Suzuki and Y. Iwabuchi, *J. Am. Chem. Soc.*, 2006, **128**, 8412; (b) Y. Iwabuchi, *J. Synth. Org. Chem. Jpn.*, 2008, **66**, 1076.
5. N. W. R. Liu, J. Chen and X. Liang, *Chem. Commun.*, 2005, 5322.
6. W. Yin, C. Chu, Q. Lu, J. Tao, X. Liang and R. Liua, *Adv. Synth. Catal.*, 2010, **352**, 113.
7. X. Wang and X. Liang, *Chin. J. Catal.*, 2008, **29**, 935.
8. S. Ma, J. Liu, S. Li, B. Chen, J. Cheng, J. Kuang, Y. Liu, B. Wan, Y. Wang, J. Ye, Q. Yu, W. Yuan and S. Yub, *Adv. Synth. Catal.*, 2011, **353**, 1005.
9. (a) J. Liu, X. Xie and S. Ma, *Synthesis*, 2012, 1569; (b) J. Liu and S. Ma, *Synthesis*, 2013, 1624.
10. J. Liu and S. Ma, *Org. Biomol. Chem.*, 2013, **11**, 4186.
11. J. Liu and S. Ma, *Org. Lett.*, 2013, **15**, 5150.
12. L. Wang, J. Li, Y. Lv, G. Zhao and S. Gao, *Appl. Organomet. Chem.*, 2012, **26**, 37.
13. C.-X. Miao, J.-Q. Wang, B. Yu, W.-G. Cheng, J. Sun, S. Chanfreau, L.-N. He and S.-J. Zhang, *Chem. Commun.*, 2011, **47**, 2697.
14. L. Wang, J. Li, X. Zhao, Y. Lv, H. Zhang and S. Gao, *Tetrahedron*, 2013, **69**, 6041.
15. M. Kirihara, Y. Ochiai, S. Takizawa, H. Takahata and H. Nemoto, *Chem. Commun.*, 1999, 1387.
16. A. T. Radosevich, C. Musich and F. D. Toste, *J. Am. Chem. Soc.*, 2005, **127**, 1090.
17. S.-S. Weng, M.-W. Shen, J.-Q. Kao, Y. S. Munot and C.-T. Chen, *Proc. Natl. Acad. Sci. U. S. A.*, 2006, **103**, 3522.
18. V. D. Pawar, S. Bettigeri, S.-S. Weng, J.-Q. Kao and C.-T. Chen, *J. Am. Chem. Soc.*, 2006, **128**, 6308.
19. C.-T. Chen, S. Bettigeri, S.-S. Weng, V. D. Pawar, Y.-H. Lin, C.-Y. Liu and W.-Z. Lee, *J. Org. Chem.*, 2007, **72**, 8175.
20. S. B. Salunke, N. S. Babu and C.-T. Chen, *Adv. Synth. Catal.*, 2011, **353**, 1234.

21. C.-T. Chen, J.-Q. Kao, S. B. Salunke and Y.-H. Lin, *Org. Lett.*, 2011, **13**, 26.
22. (a) Y. Maeda, N. Kakiuchi, S. Matsumura, T. Nishimura and S. Uemura, *Tetrahedron Lett.*, 2001, **42**, 8877; (b) Y. Maeda, N. Kakiuchi, S. Matsumura, T. Nishimura, T. Kawamura and S. Uemura, *J. Org. Chem.*, 2002, **67**, 6718.
23. Y. Maeda, Y. Washitake, T. Nishimra, K. Iwai, T. Yamauchi and S. Uemura, *Tetrahedron*, 2004, **60**, 9031.
24. C. B. Boechat, J. G. Eon, A. M. Rossi, C. A. C. Perez and R. A. S. SanGil, *Phys. Chem. Chem. Phys.*, 2000, **2**, 4225.
25. N. Jiang and A. J. Ragauskas, *Tetrahedron Lett.*, 2007, **48**, 273.
26. N. Jiang and A. J. Ragauskas, *J. Org. Chem.*, 2007, **72**, 7030.
27. Z. Du, H. Miao, H. Ma, Z. Sun, J. Ma and J. Xu, *Adv. Synth. Catal.*, 2009, **351**, 558.
28. Z. Du, J. Ma, H. Ma, J. Gao and J. Xu, *Green Chem.*, 2010, **12**, 590.
29. S. Kodama, S. Hashidate, A. Nomoto, S. Yano, M. Ueshima and A. Ogawa, *Chem. Lett.*, 2011, **40**, 495.
30. (a) S. K. Hanson, R. Wu and L. A. P. Silks, *Org. Lett.*, 2011, **13**, 1908; (b) B. N. Wigington, M. L. Drummond, T. R. Cundari, D. L. Thorn, S. K. Hanson and S. L. Scott, *Chem. Eur. J.*, 2012, **18**, 14981.
31. S. Velusamy and T. Punniyamurthy, *Org. Lett.*, 2004, **6**, 217.
32. S. R. Reddy, D. Da and T. Punniyamurthy, *Tetrahedron Lett.*, 2004, **45**, 3561.
33. (a) P. J. Figiel, J. M. Sobczak and J. J. Ziolkowski, *Chem. Commun.*, 2004, 244; (b) P. J. Figiel and J. M. Sobczak, *New J. Chem.*, 2007, **31**, 1668.
34. P. J. Figiel and J. M. Sobczak, *J. Catal.*, 2009, **263**, 167.
35. D.-R. Hwang, C.-Y. Chu, S.-K. Wang and B.-J. Uang, *Synlett*, 1999, 77.
36. M. Kirihara, S. Takizawa and T. Momose, *J. Chem. Soc., Perkin Trans. 1*, 1998, 7.
37. M. Kirihara, K. Yoshida, T. Noguchi, S. Naito, N. Matsumoto, Y. Ema, M. Torii, Y. Ishizuka and I. Souta, *Tetrahedron Lett.*, 2010, **51**, 3619.

CHAPTER 7

The Concept of Multicatalysis in Aerobic Oxidation of Alcohols

MICHAEL DAVI[a] AND HÉLÈNE LEBEL*[b]

[a] Equipe Electrochimie et Synthèse Organique, Institut de Chimie et des Matériaux Paris-Est, UMR 7182 CNRS–Université Paris-Est Créteil Val de Marne, 2 rue Henri Dunant, BP 28, 94320 Thiais, France; [b] Department of Chemistry and Centre in Green Chemistry and Catalysis (CGCC), Université de Montréal, Pavillon Roger Gaudry, CP 6128, Succ. Centre-ville, Montréal, Québec, Canada, H3C 3J7
*Email: helene.lebel@umontreal.ca

7.1 Introduction

During recent decades, impressive efforts have been made by the organic chemistry community to synthesize ever more complex molecular structures. Unfortunately, these synthetic routes are often time consuming and expensive and produce large amounts of wastes. Consequently, producing high-value-added fine chemicals, such as pharmaceutical, agrochemical, dye or monomeric compounds is still economically and environmentally challenging. Organic synthesis could be more effective if multiple catalysts act simultaneously or alternately in a single reactor with multiple reagents or reactants. These 'multicatalytic' reactions would maximize yield and minimize time, effort and waste associated to the synthesis of high-value-added molecules.

RSC Green Chemistry No. 28
Transition Metal Catalysis in Aerobic Alcohol Oxidation
Edited by Francesca Cardona and Camilla Parmeggiani

Published by the Royal Society of Chemistry, www.rsc.org

These reactions, named tandem, domino, cascade, *etc.*, introduce molecular complexity by creating several new chemical bonds in a one-pot manner without the isolation of potentially sensitive intermediates.[1] These processes take advantage of the catalyst compatibility and one or more catalysts could execute different chemical reactions in a cooperative, synergistic or sequential way. From a practical point of view, these processes essentially diverge in their experimental protocols and, in this chapter, we prefer to use the general term 'multicatalysis' to denote these reactions.[2]

The transition metal-catalyzed aerobic oxidation of alcohols is an impressive example of a green process. Not only is molecular dioxygen (or air) an inexpensive, renewable and the simplest available oxidant, but also the only by-product of the reaction, water, is also a non-toxic and safe chemical.[3] Consequently, these reactions are an excellent starting point in the development of novel multicatalytic reactions that will address the new environmental challenges encountered by the organic chemistry community.

In this chapter, we present a comprehensive discussion of multicatalytic systems based on the aerobic oxidation of alcohols. A biomimetic multicomponent approach using coupled catalytic systems and electron-transfer mediators is presented. Examples of synergetic catalysis into bimetallic nanoclusters is also mentioned. Finally, catalytic multistep reactions for C–C and/or C–N bond formation are also discussed. This review covers the literature survey up to late 2013.

7.2 Biomimetic Catalytic Oxidation of Alcohols

A number of methods for the oxidation of alcohols to produce the corresponding carbonyl compounds have been developed, as this transformation is central in organic synthesis.[4] Among them, transition metal-catalyzed aerobic oxidations of alcohols are sustainable and environmentally friendly processes. Alcohols are oxidized in the presence of a metal complex (M^{n+2}) producing the corresponding carbonyl compound and the reduced metal complex (M^n); the latter is then reoxidized in the presence of molecular oxygen (or air) affording the active catalytic species while generating water as by-product (Scheme 7.1). The direct reoxidation of reduced catalyst M^n by O_2

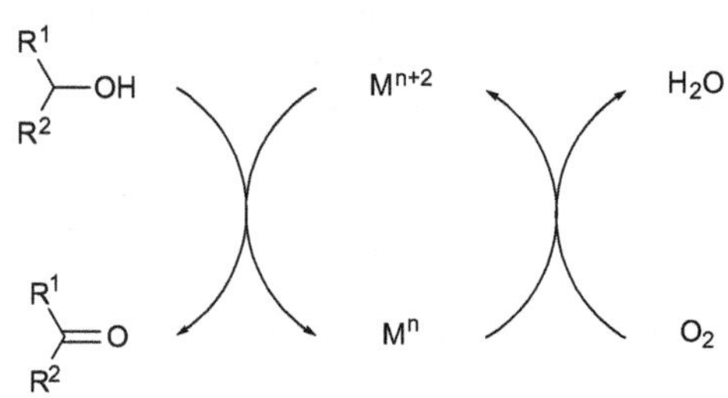

Scheme 7.1 Transition metal-catalyzed aerobic oxidation of alcohols.

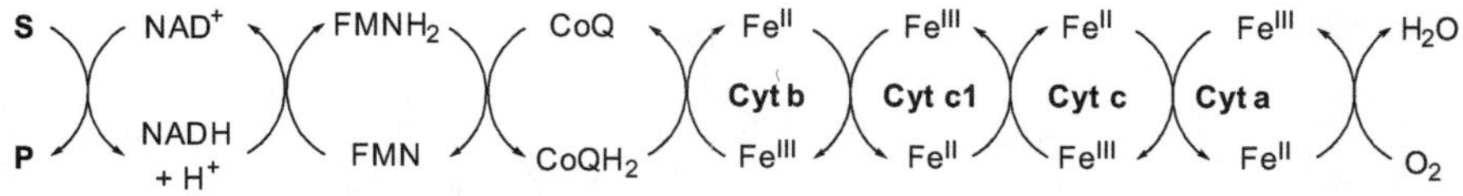

Scheme 7.2 Electron transport chain in mitochondria.

remains challenging because of the high energy barrier associated with this step. As a result, several drawbacks are still associated with these processes including high pressure (O_2 or air), temperature or catalyst loading as well as the use of costly and/or toxic additives.[3,5]

Nature has elaborated a clever and complex system to make these transformations mild and selective. Electrons go through several redox systems (the electron transport chain of mitochondria) from substrate **S** to oxygen as illustrated in Scheme 7.2. Initially, the coenzyme NAD^+ dehydrogenates the substrate (*i.e.* alcohol) and electrons are successively accepted by the flavin mononucleotide (FMN) and ubiquinone (CoQ) cofactors which are acting as electron-transfer mediators (ETM). Finally, electrons go through various cytochrome complexes (**Cyt**) to reduce molecular oxygen to water in presence of protons.[6]

By analogy, the development of a biomimetic multicatalytic systems including transition metal catalyst as dehydrogenation agent, hydroquinone as ETM and metalloporphyrins as oxygen-activating agent should be possible. In 1987, the feasibility of this approach was demonstrated by the Bäckvall's group.[7] Mild and selective 1,4-diacetoxylation of diene **1** was accomplished by using a three-component catalytic system of Pd^{II}–hydroquinone–Co^{II}(TPP) **3** (Equation 7.1).

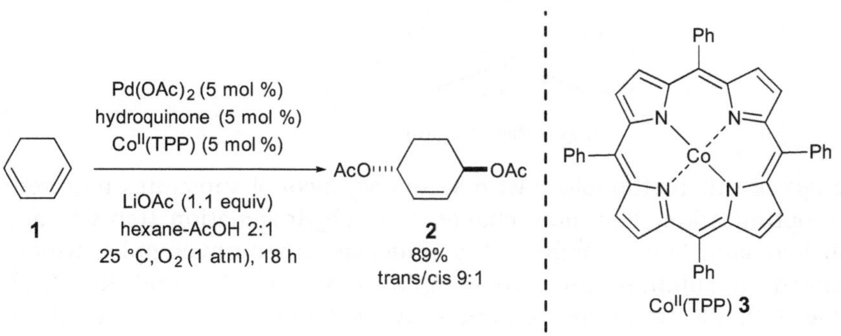

(7.1)

Both the hydroquinone (as an ETM) and the metal–macrocyclic complex, Co^{II}(TPP) **3** (as oxygen-activating agent), were essential to lower the energy barrier for the reoxidation of Pd^0 to Pd^{II} (Scheme 7.3). An overview of biomimetic catalytic oxidation of organic substrates can be found in the excellent review by Piera and Bäckvall.[8]

It is possible to use such a biomimetic approach for the oxidation of alcohols in the presence of low-valent ruthenium complexes known to dehydrogenate alcohols.[9] Benzyl alcohol (**4**) in the presence of

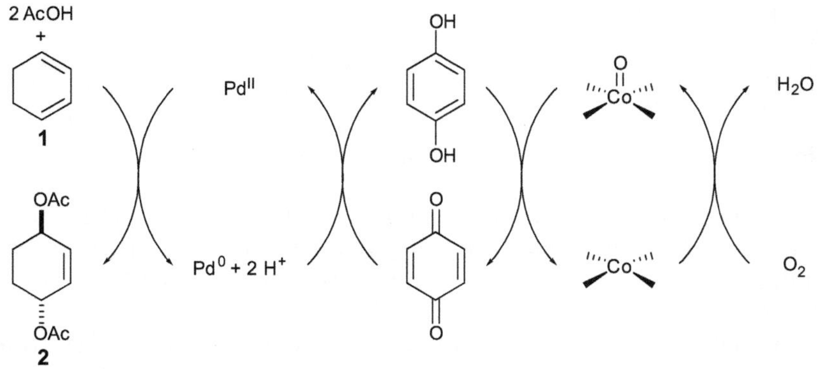

Scheme 7.3 Catalytic cycle of palladium-catalyzed biomimetic aerobic oxidation of 1,3-dienes.

RuCl(OAc)(PPh_3)$_3$–hydroquinone–Co(salophen)(PPh_3) **6** at room temperature under air afforded benzaldehyde (**5**) in 60% yield (Equation 7.2). Turnover numbers (TONs) up to 70 were obtained and the system was efficient with benzylic and allylic primary alcohols.[10]

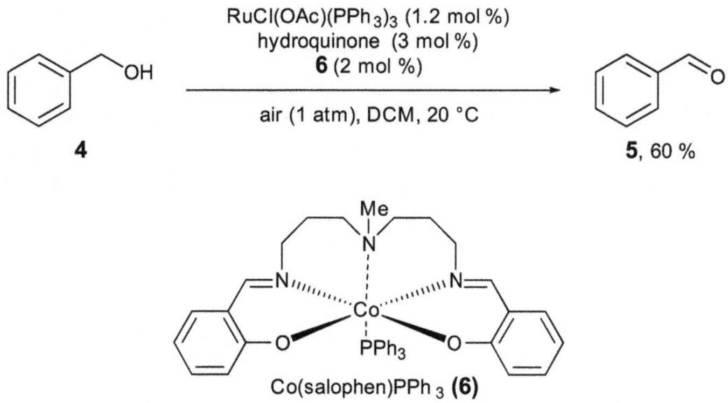

(7.2)

The use of this methodology with secondary alcohol substrates required careful optimization. The more challenging dehydrogenation step was accomplished using Shvo's catalyst 7,[11] a binuclear ruthenium complex, which dissociated in solution into two complexes of Ru^{II} (**7a**) and Ru^0 (**7b**) (Scheme 7.4). Electron-rich 2,6-dimethoxybenzoquinone (**8**) was used to

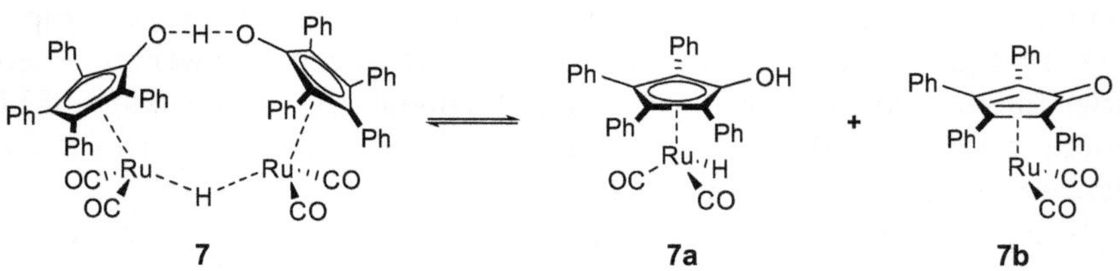

Scheme 7.4 Shvo catalyst 7.

promote a faster electron transfer between Shvo's catalyst **7** and the oxygen-activating agent, Co(salophen) **9**.

A variety of aliphatic and benzylic secondary substrates (including sterically hindered alcohols) were reacted with this three-component catalytic system (**7**–**9**). The reaction proceeded in excellent yields and with high turnover frequencies (TOFs) (Table 7.1).[12] For example, TOFs >4000 h^{-1} were observed for the dehydrogenation of 1-phenylethanol (Table 7.1, entry 1). Reactive *Z* carbon–carbon double bonds were inert under these mild reaction conditions (Table 7.1, entry 4). No epoxide formation was observed, or the reduction of the double bond, despite the presence of ruthenium-hydride **7a**.

Table 7.1 Ruthenium-catalyzed biomimetic aerobic oxidation of secondary alcohols.

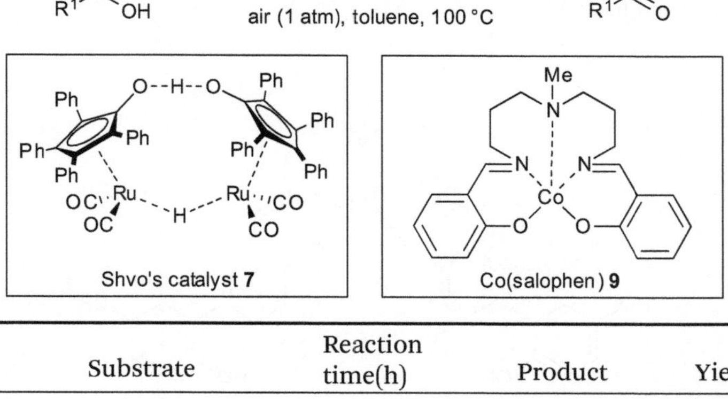

Entry	Substrate		Reaction time(h)	Product	Yield[a,b] (%)
1	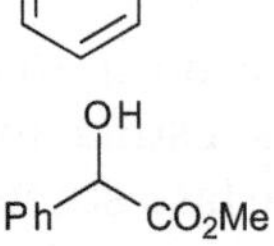		1	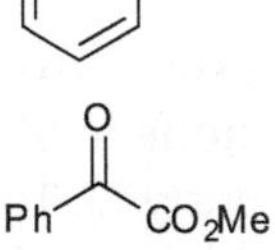	89
2	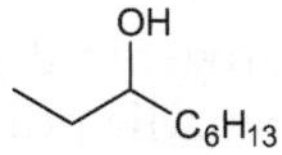		2	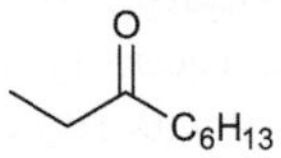	81[b]
3	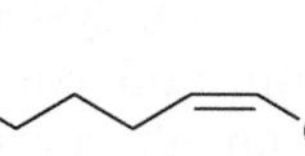		1	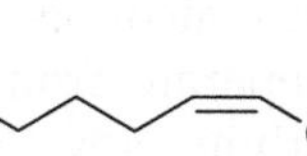	92
4	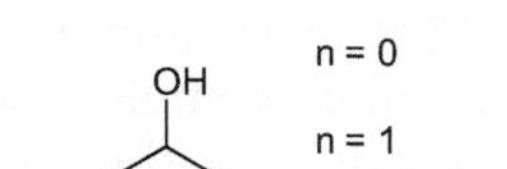		2.5	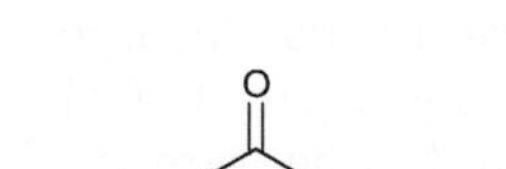	92
5	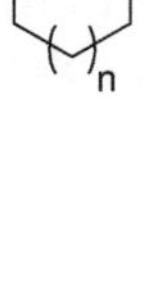	n = 0	1	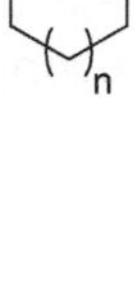	88
6		n = 1	1		92
7		n = 1	1		95[c]
8		n = 2	1		92

Table 7.1 (*Continued*)

Entry	Substrate	Reaction time(h)	Product	Yield[a,b] (%)
9	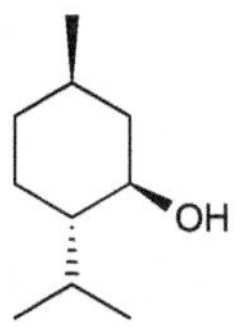	2	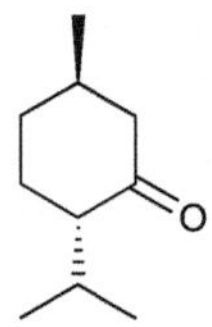	80
10	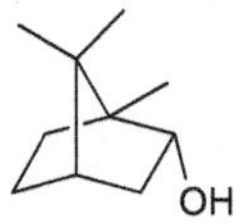	2	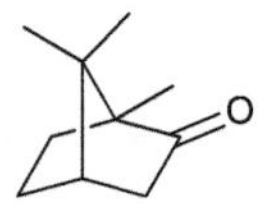	73[d]

[a]Isolated yield.
[b]Reaction performed on 1 mmol scale;
[c]Reaction performed on 0.1 mol scale;
[d]Shvo's catalyst (1 mol%).

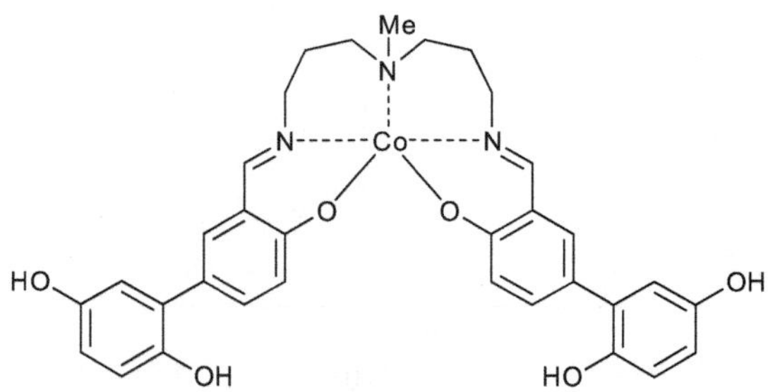

hydroquinone/Co(salophen) hybrid catalyst **10**

Figure 7.1 Hybrid hydroquinone–Schiff-base Co catalyst.

Moreover, the aerobic oxidation of cyclohexanol on a 100 mmol scale provided cyclohexanone in 95% yield when using a substrate-to-catalyst ratio of 1000 (Table 7.1, entry 7). To date, this is one of the fastest catalytic systems reported for the oxidation of secondary alcohols.

To facilitate the recovery of the catalyst, a heterogenized encapsulated zeolite of complex **9** was reported with allylic and benzylic primary alcohols and also secondary alcohols as substrates.[13] Such a catalyst was easier to handle and to separate from the reaction mixture and could be reused several times without any decrease in activity. An alternative improved hybrid catalyst **10** that incorporate both the ETM hydroquinone and the oxygen-activating agent Co(salophen) in the same complex, was subsequently found to perform the biomimetic catalytic oxidation of alcohols (Figure 7.1).[14] Faster electron transfer from the substrate to molecular oxygen became possible owing to the closest proximity of the cobalt center and

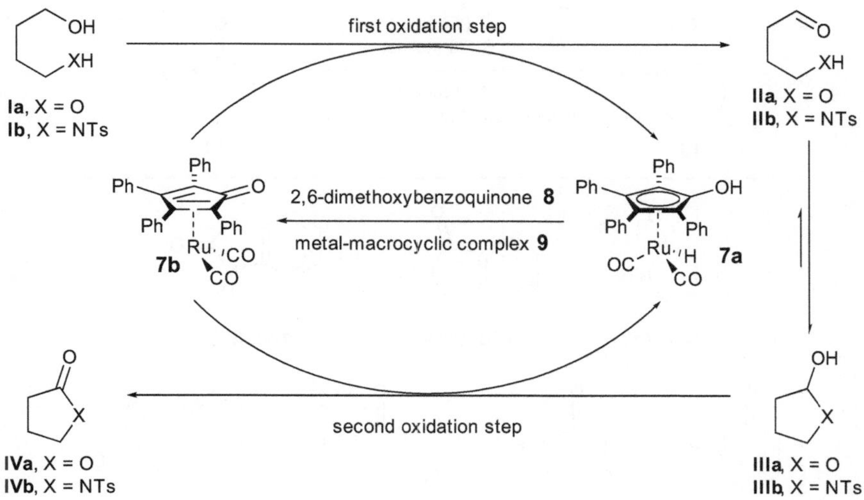

Scheme 7.5 Biomimetic aerobic oxidation of diols and amino alcohols.

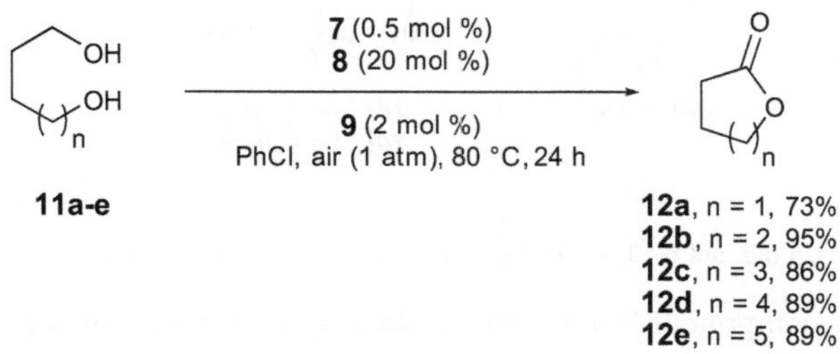

(7.3)

the hydroquinone moiety. As a result, milder reaction conditions with a lower catalyst loading were efficient for the oxidation of secondary alcohols.

Bäckvall's group also reported the biomimetic aerobic oxidation of diols and amino alcohols to produce lactones and lactams, respectively, *via* a two oxidation step mechanism (Scheme 7.5).[15,16] The first dehydrogenation of substrate **I** provided hydroxy- or aminoaldehyde intermediate **II**, which underwent an intramolecular cyclization affording hemi-acetal or -aminal **III**, respectively; the second oxidation produced desired lactone or lactam products **IV**.

The reaction of diols **11a–e** provided corresponding five- to nine-membered lactones **12a–e** in good to excellent yields (Equation 7.3).[17] Conversely, the formation of 2-oxetanones, derived from 1,3-diols, was not observed under these reaction conditions. For the formation of larger rings, diluted reaction conditions were required to avoid the formation of oligomers from intermolecular esterification.

In addition, biologically significant lactams such as 2-piperidinones, 3-morpholinones, 2-piperazinones, indolinones and isoindolinones (**14a–g**) were produced in good to excellent yields from the corresponding

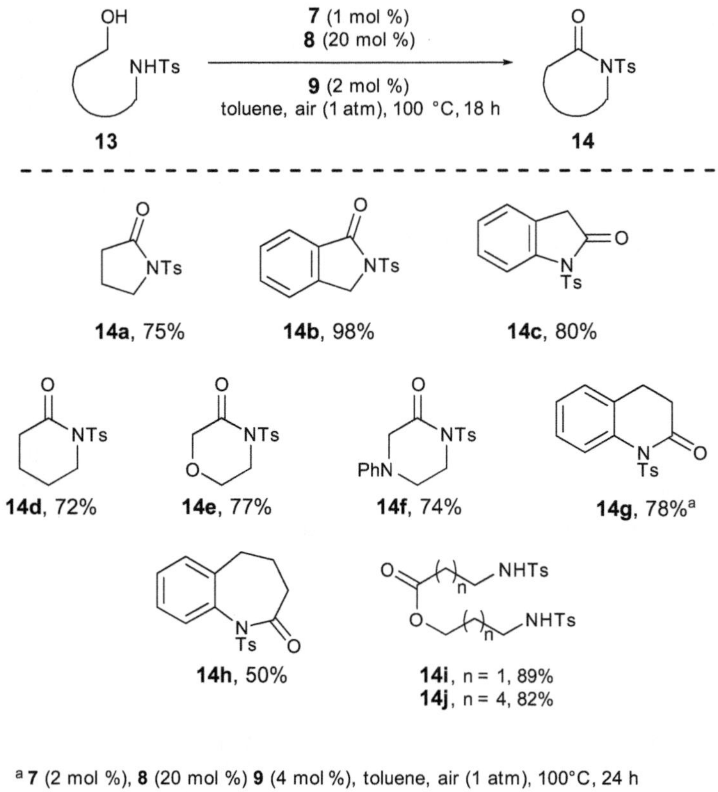

Scheme 7.6 Three-component biomimetic oxidation of *N*-tosylamino alcohols.

N-protected amino alcohols using such a three-component biomimetic oxidation process (Scheme 7.6).[16]

Furthermore, benzo-fused seven-membered lactam **14h** was successfully isolated, albeit only in moderate yields. Conversely, *N*-tosylamino-1-hexanol **13i** did not produce the expected simple seven-membered cyclic lactam. In this case, the first oxidation step led to aminoaldehyde intermediate **IIb** (Scheme 7.5), but the latter did not cyclize to form hemiaminal **IIIb.** Instead intermolecular *O*-nucleophilc attack of **Ib** was observed, providing a hemiacetal intermediate which underwent the second oxidation step affording dimer **14i.** Similarly, β-lactams were not formed from substrate **13j,** which also produced dimer **14j.**

Hirao and co-workers reported a heterogeneous biomimetic multicatalytic process for the oxidation of secondary alcohols using water as a solvent (Scheme 7.7).[17] Gold NPs supported on PMAS catalyzed the dehydrogenation of alcohols; the polymer not only stabilized the Au NPs but also played the role of a redox-active mediator (*i.e.* ETM and oxygen-activating). This heterogeneous system can be reused for four runs without any significant decrease in catalytic activity.

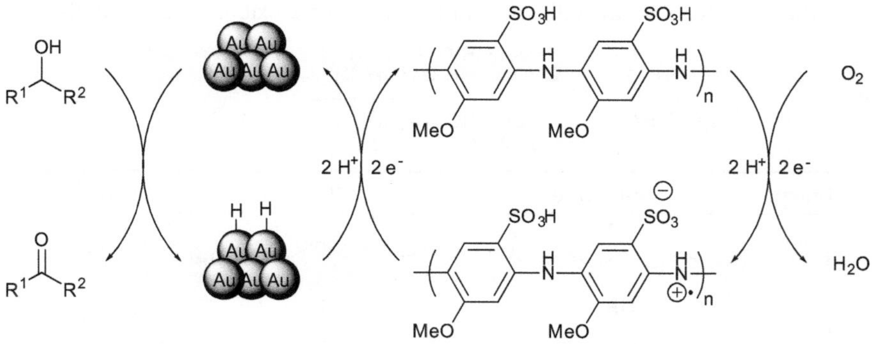

Scheme 7.7 Redox-active PMAS in multicatalytic aerobic oxidation of alcohols.

7.3 Synergistic Bimetallic Systems

Synergistic activities with bimetallic systems have been reported for the catalytic aerobic oxidation of alcohols. In most cases, the first transition metal was the oxidant whereas the second one activated molecular oxygen to regenerate the catalyst in the good oxidation state (Scheme 7.8).

Murahashi's group reported a dual ruthenium–cobalt catalytic system for the aerobic oxidation of alcohols in presence of acetaldehyde (Table 7.2).[18] Secondary alcohols were converted to ketones in good to excellent yields, whereas primary alcohols produced the corresponding carboxylic acid. The proposed mechanism can be rationalized by the formation of a peracid intermediate from acetaldehyde and the Co^{III} complex in presence of molecular oxygen, *via* a radical pathway. The peracid intermediate oxidized the Ru^{III} complex to form the catalytically active low-valent oxoruthenium complex that performed the oxidation of the alcohol substrate (Scheme 7.9).

Other cooperative systems, namely homogeneous Os–Cu[19] and Os–Cr[20] bimetallic catalysts, have also been developed. The Os–Cu bimetallic complexes were effective for the oxidation of primary and secondary allylic and benzylic alcohols to produce aldehydes and ketones (Equations 7.4[19a] and 7.5[19b]). The exact dual activity between both transition metals is unknown

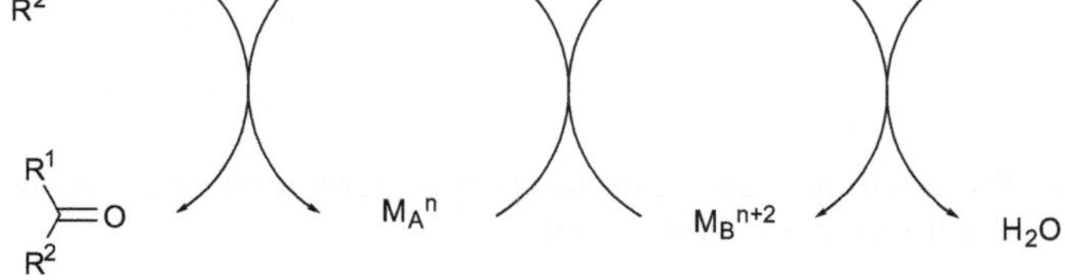

Scheme 7.8 Synergistic bimetallic systems for catalytic aerobic oxidations of alcohols.

Table 7.2 Ru–Co bimetallic system for the aerobic oxidation of alcohols.

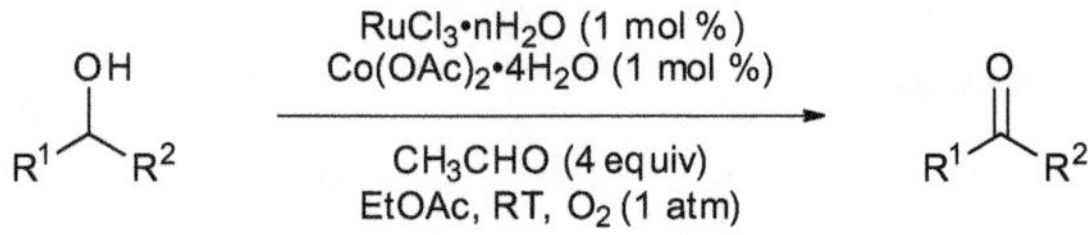

Entry	Substrate	Product	Yield[a] (%)
1	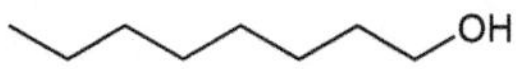		96
2	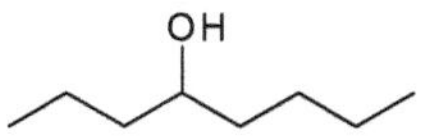	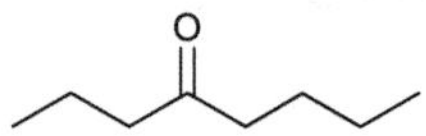	89
3	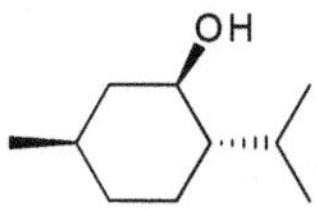	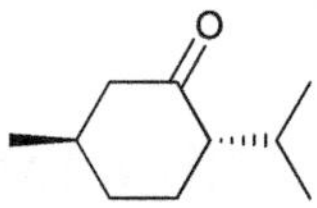	95
4	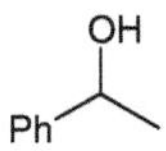	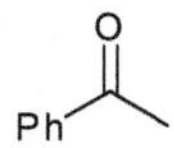	94
5	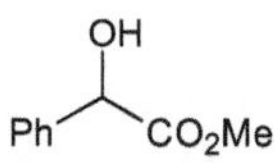	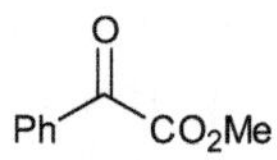	78

[a]Isolated yield.

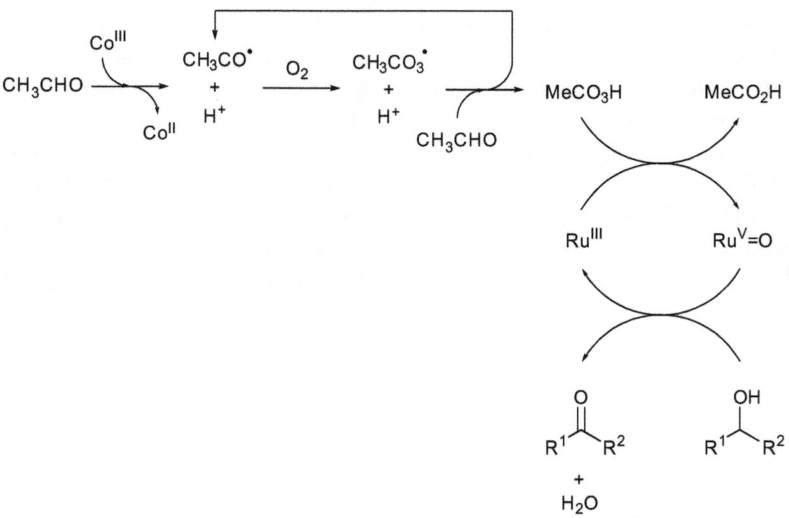

Scheme 7.9 Proposed mechanism of the catalytic aerobic oxidation of alcohols with the Ru–Co bimetallic system.

but the authors suggested that osmium is probably responsible for the oxidation step whereas copper could act as an electron-transfer agent between molecular oxygen and the reduced osmium complex.[19a]

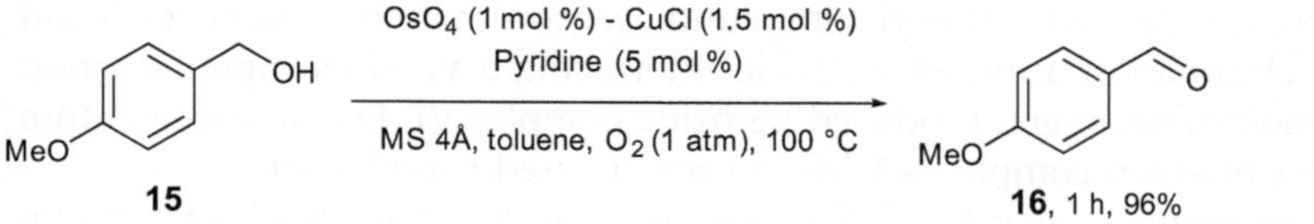

(7.4)

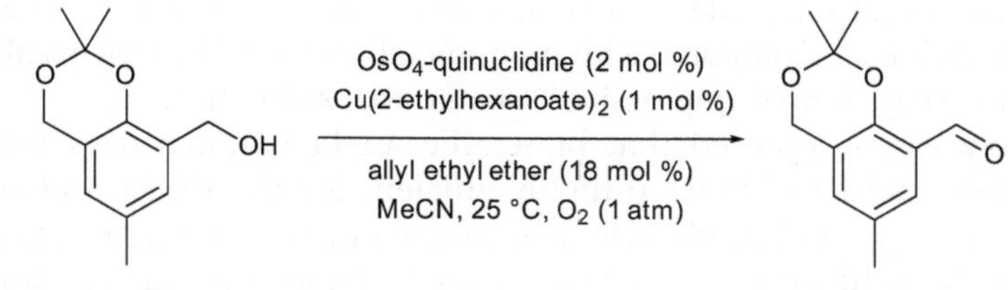

(7.5)

In the case of the Os-Cu bimetallic complex **19**, mechanistic studies, including kinetic and labeling studies, were conducted on the oxidation of benzyl alcohol (**4**) to benzaldehyde (**5**) (Scheme 7.10).[20] The first step involved the coordination of alcohol **4** to osmium complex **19**, followed by proton transfer from the coordinated alcohol to the bridging oxo group. A concerted β-hydrogen elimination oxidized the alkoxide intermediate to

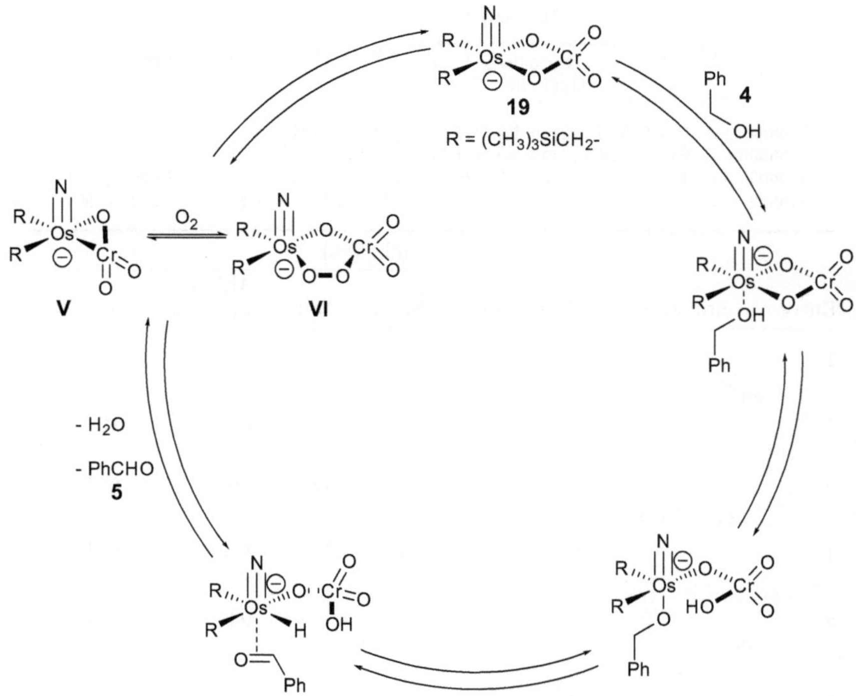

Scheme 7.10 Proposed mechanism of the catalytic aerobic oxidation of alcohols with the Os–Cr bimetallic catalyst.

produce the desired carbonyl compound. Decoordination of aldehyde **5** and water elimination afforded unstable intermediate **V**, which upon reaction with molecular oxygen produced peroxide complex **VI.** Peroxo oxygen atom transfer between complexes **V** and **VI** regenerated complex **19.**

The cooperative activity of bimetallic nanoclusters has also been demonstrated. For instance, a bimetallic Au–Pd/TiO_2 catalyst catalyzed the selective oxidation of alcohols (primary alkyl alcohols) to aldehydes under mild solvent-free reaction conditions.[21] Compared with monometallic Au or Pd nanocrystals, bimetallic Au–Pd/TiO_2 showed impressive TOFs up to 269 000 h^{-1}.

Prati and co-workers reported that bimetallic Au–Pt nanoparticles supported on H-mordenite zeolite were highly efficient for the direct and selective oxidation of glycerol to glyceric acid under practical acidic reaction conditions.[22] Kobayashi's group synthesized Au–Pt bimetallic clusters supported on a cross-linked polystyrene for the aerobic oxidation of alcohols at room temperature without a base in aqueous media.[23] Such a heterogeneous catalyst can be reused several times without a decrease in activity. The same group also reported carbon-stabilized polymer-incarcerated bimetallic nanocluster catalysts (PI-CB catalysts) for the oxidation of alcohols (Table 7.3).[24] Interestingly with these catalysts, the aldehyde *versus* carboxylic acid (or ester) selectivity was strongly dependent on the content of Au

Table 7.3 Carbon-stabilized polymer-incarcerated bimetallic nanocluster-catalyzed aerobic oxidation of alcohols.

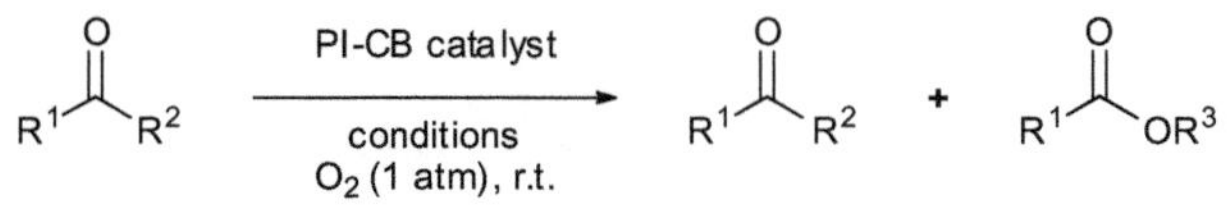

Condition A : PI-CB/Au-Pt (2 mol %), BTF/H_2O (1/1), 0.25 M, 9 h
Condition B : PI-CB/Au-Pt (2 mol %), BTF/H_2O (1/1), 0.125 M, 24 h
Condition C : PI-CB/Au-Pd (1 mol %), MeOH/H_2O (1/2), 0.125 M, K_2CO_3 (3 equiv), 24 h
Condition D : PI-CB/Au-Pd (1 mol %), MeOH/H_2O (2/1), 0.125 M, K_2CO_3 (1 equiv), 24 h

			Yield[a](%)			
Entry	Substrate	Conditions	Ester	Acid	Aldehyde or ketone	Alcohol
1		A	---	8	90	1
2		C	81	5	3	10
3		A	---	n.d.	>99	n.d.
4		D	98	2	n.d.	n.d.
5	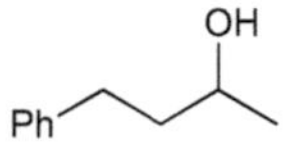	B	---	-	94	n.d.
6	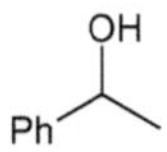	B	---	-	96	n.d.

[a]Yield determined by GC. n.d.: not detected.

versus the other metal. For example, PI-CB/Au–Pt nanocluster, containing a gold-to-platinum ratio of ~1 : 1, afforded the aldehyde as the major product from a primary alcohol. Conversely, esters were mainly produced when PI-CB/Au–Pd catalyst with a gold-to-platinum ratio of ~4 : 1 to 3 : 1 was used.

Kobayashi and co-workers studied the application of PI-CB bimetallic nanoclusters in tandem reactions.[25] They first reported the sequential aerobic oxidation–Michael addition of 1,3-dicarbonyl substrates **20** with secondary allylic alcohols **21** in the presence of such a catalyst, producing desired adducts **22** in good to excellent yields (Scheme 7.11). With such a process, the purification of potentially sensitive α,β-unsaturated aldehydes was avoided.

To develop an asymmetric version of this tandem reaction, they developed a novel heterogeneous bifunctional chiral catalyst by coating a polymer-supported proline organocatalyst with a styrene-based copolymer that can immobilize and incarcerate Au–Pd alloy nanoparticles [PI(Au/Pd)-OC] (Scheme 7.12).

This catalyst performed very well in the enantioselective aerobic oxidation–Michael addition of 1,3-dicarbonyl compounds to primary allylic alcohols affording the desired products in good yields and high enantiomeric ratio (Equation 7.6).[26] Undesired interactions between the organocatalyst and nanoparticles were avoided using this approach.

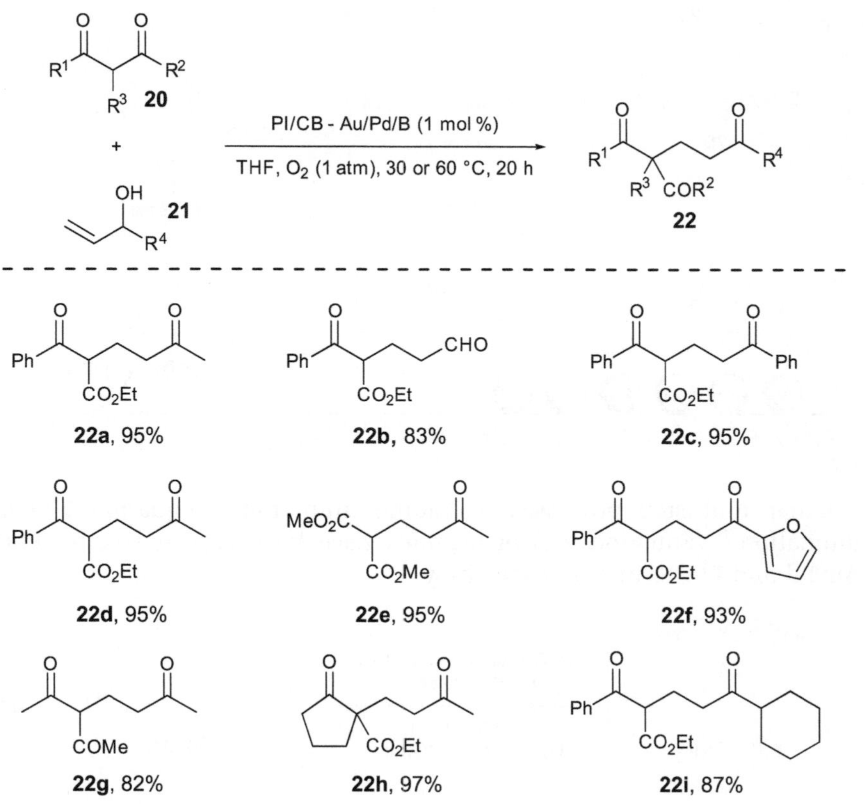

Scheme 7.11 One-pot aerobic oxidation–Michael addition sequence.

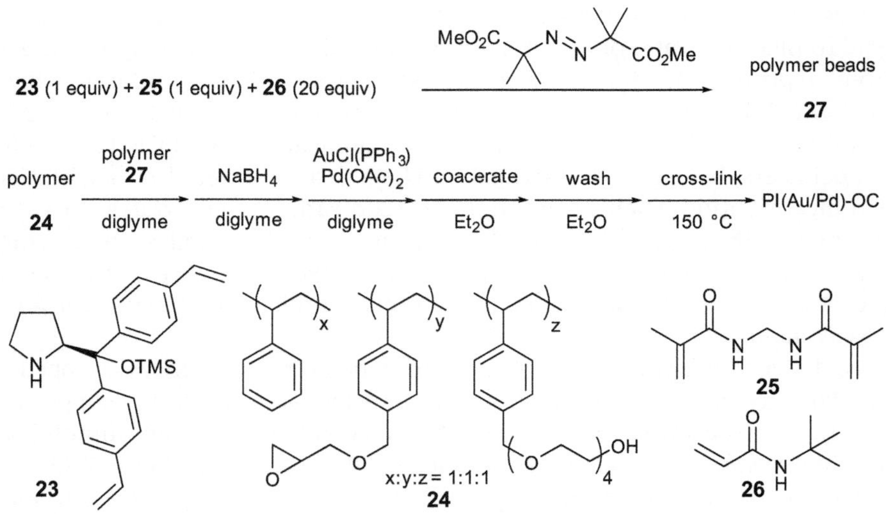

Scheme 7.12 Synthesis of heterogeneous bifunctional chiral catalyst PI(Au/Pd)-OC.

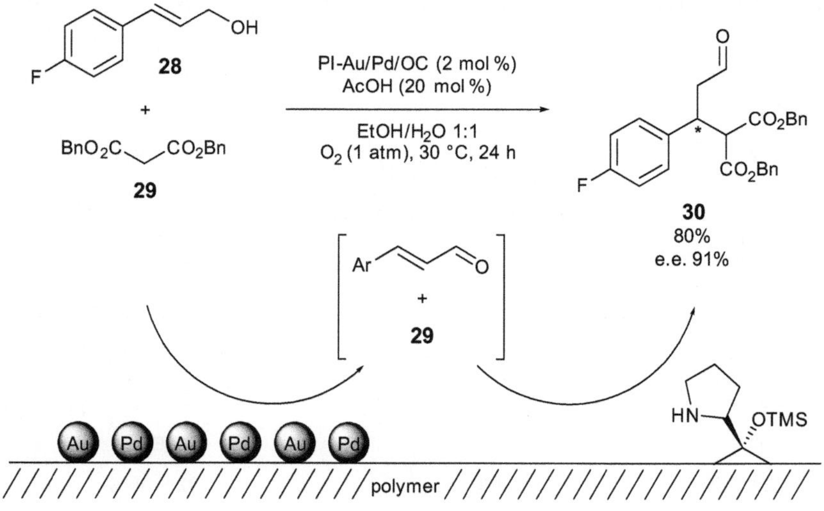

(7.6)

Other multistep processes, including imination[27] (Equation 7.7) and amidation[28] (Equation 7.8) of alcohols have been reported using PI-CB/Au–Pd and PI-CB/Au–Co, respectively.

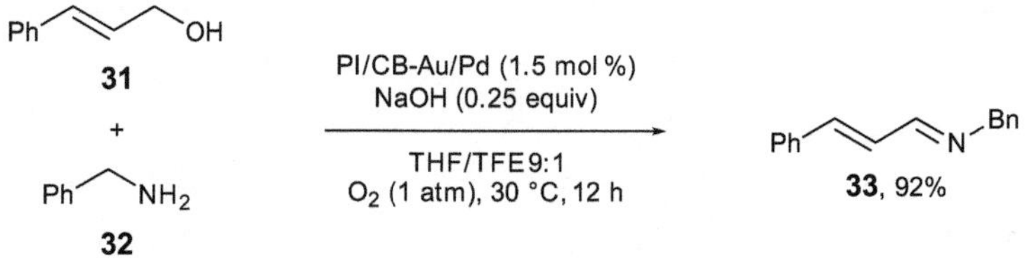

(7.7)

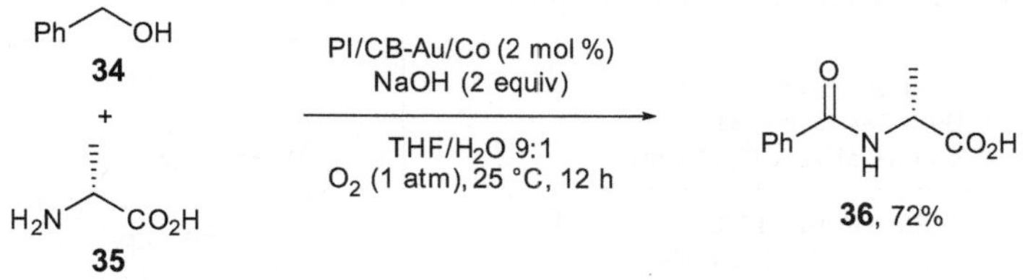

(7.8)

The direct lactamization of amino alcohols catalyzed by PI-CB/Au–Co has also been reported. This strategy is particularly effective for the synthesis of isoindolin-1-one and 3,4-dihydroisoquinolin-1(*2H*)-one ring systems (Equation 7.9).[29]

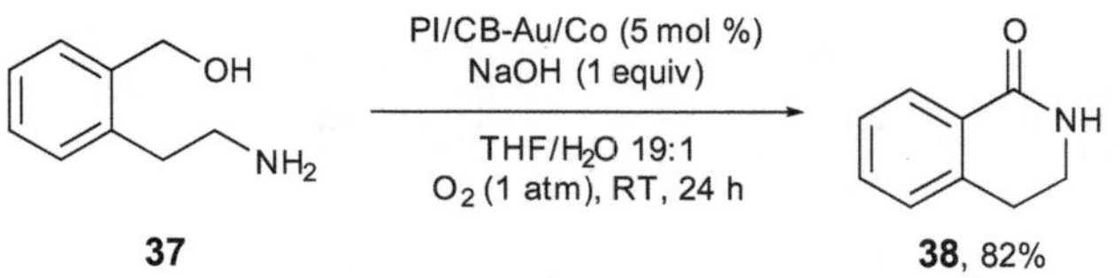

(7.9)

7.4 Catalytic Aerobic Oxidation of Alcohols in Multistep Processes

As mentioned in the last section, catalytic multistep reactions can be an elegant tool to build molecular complexity in a single vessel without purification of intermediates. This section describes multicatalytic processes in which two or more catalysts execute different chemical reactions in tandem or sequentially. Some examples of multicatalysis take advantage of the versatile properties of a transition metal to promote more efficient monocatalytic multistep reactions. In all cases, at least one stage is a catalytic aerobic oxidation step. Only examples in which two or more distinct catalytic cycles are achieved in a single vessel will be presented here.[30]

Lebel and Paquet reported the combination of a palladium and a rhodium catalyst for the olefination of alcohols.[31] Alcohols were treated with a palladium catalyst **39** (which was known to catalyze aerobic oxidation)[32] under an atmosphere of oxygen to produce the corresponding carbonyl derivative. These carbonyl intermediates were further reacted in the same vessel under Wilkinson's catalyst methylenation reaction conditions, affording terminal alkenes in good to excellent yields (Table 7.4). Up to four individual transformations can be achieved starting with the diol substrate **53** (Table 7.4, entry 7).

In the same study, an intramolecular tandem methylenation–ring closing metathesis was accomplished using the metathesis catalyst **55** developed by Nolan's group.[33] Lebel and Paquet demonstrated that both multicatalytic oxidation–methylenation and methylenation–ring closing metathesis can be combined in the same vessel (Equation 7.10). This reaction provided an impressive example of catalyst compatibility as the three different transition metal catalysts coexisted and did not interfere in the reaction mixture.

Table 7.4 Multicatalytic oxidation–olefination reaction.

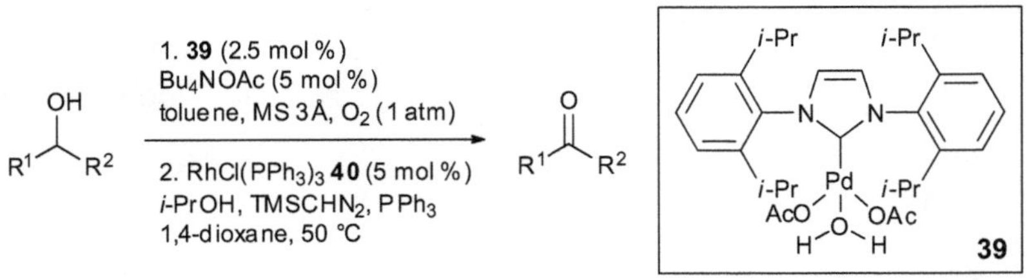

Entry	Substrate	Product	Yield (%)
1	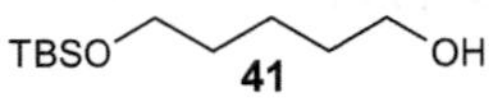 **41**	 **42**	78
2	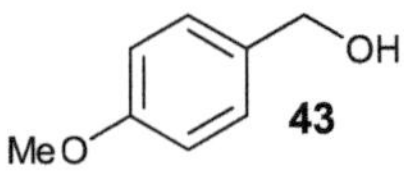 **43**	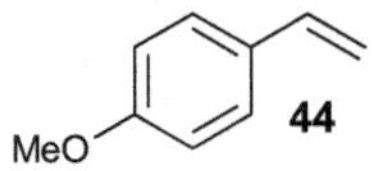 **44**	82
3	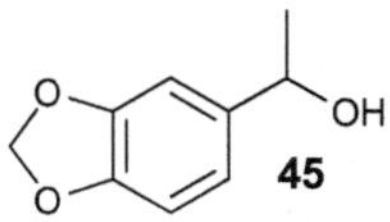 **45**	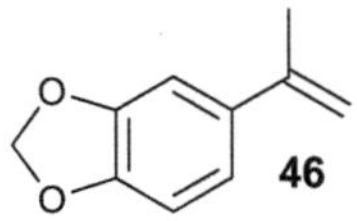 **46**	92
4	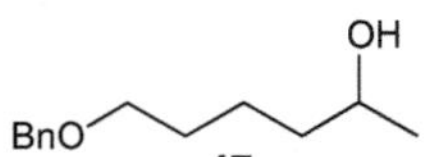 **47**	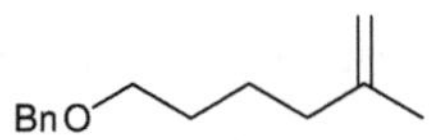 **48**	84
5	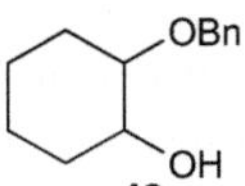 **49**	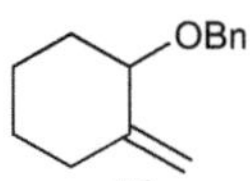 **50**	65
6	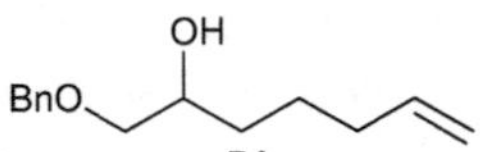 **51**	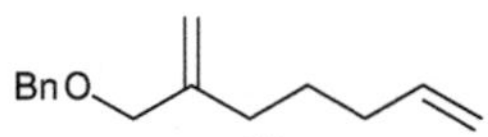 **52**	87
7	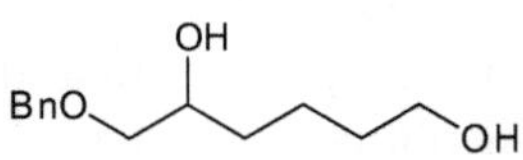 **53**	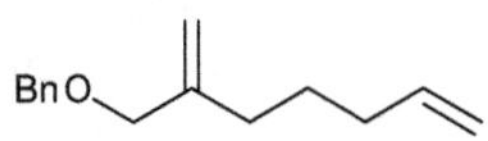 **54**	54

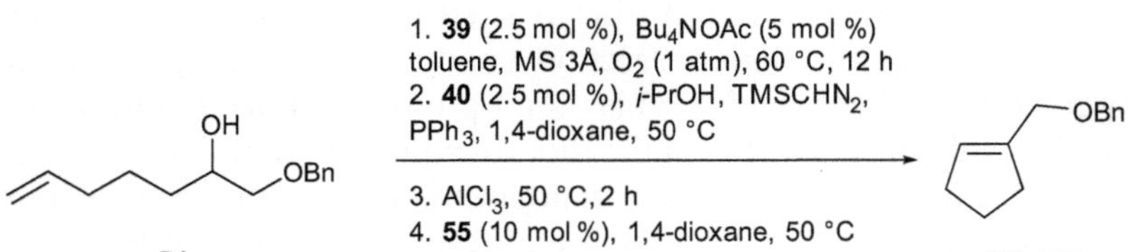

51 **56**, 70% (7.10)

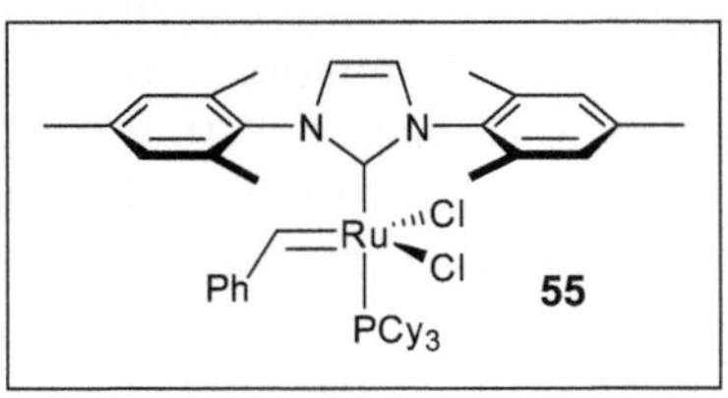

Some multicatalytic processes take advantage of the multifaceted reactivity of a transition metal catalyst to make more effective chemical transformations. Using such an approach, Lebel's group exploited the dual activity of copper complexes in both aerobic oxidation[34] and olefination Wittig-type reaction[35] to develop an improved tandem oxidation–olefination process using a single and inexpensive copper catalyst.[36] Primary and secondary alcohols can be used in combination with trimethylsilyldiazomethane to synthesize a large variety of terminal alkenes (Scheme 7.13). Importantly, the functional group tolerance was excellent and non-basic reaction conditions allowed the olefination of α-chiral alcohols without racemization.

The reactivity of copper salts was not limited to trimethylsilyldiazomethane and various diazo compounds were compatible in the catalytic Wittig-type reaction, providing a large variety of disubstituted olefins.[35c] Not only could one access the terminal alkene from the corresponding alcohol, but also the α,β-unsaturated ester, ketone and amide (Table 7.5).

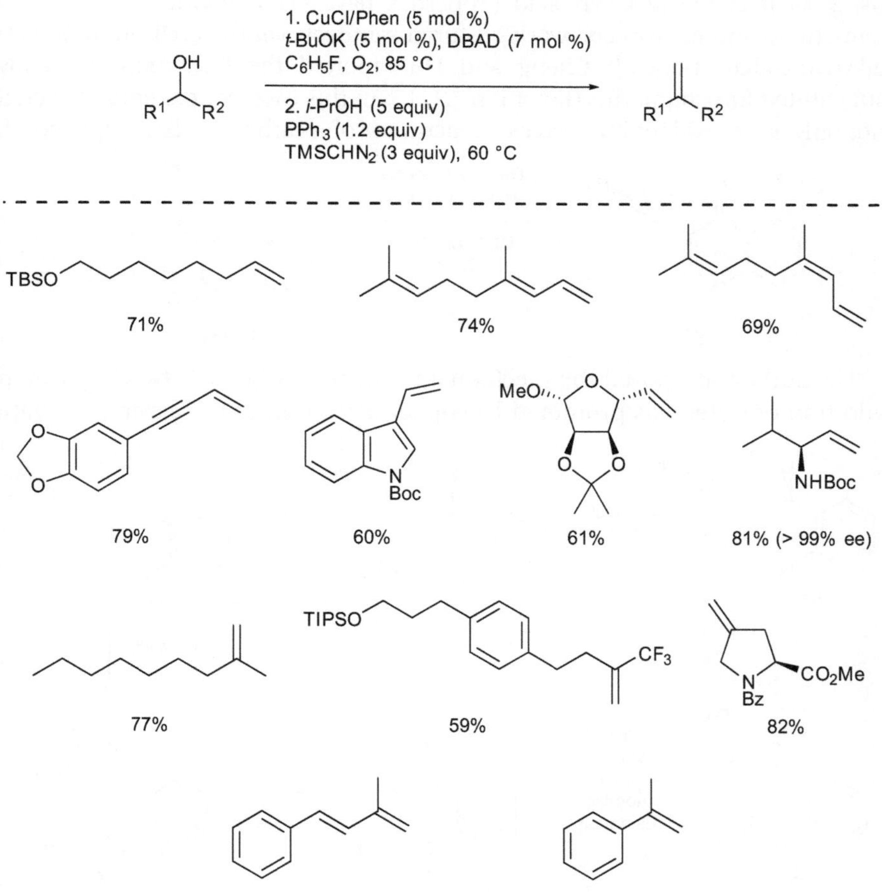

Scheme 7.13 Copper-catalyzed tandem oxidation–olefination.

Table 7.5 Multicatalytic oxidation–olefination reactions with various diazo reagents.

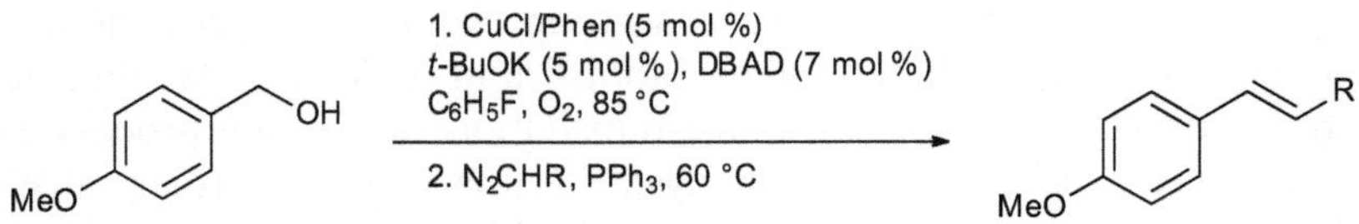

Entry	Diazo compound	Yield[a](%)	E/Z[b]
1	$TMSCHN_2$/*i*-PrOH	85	---
2	N_2CHCO_2Et	87	94:6
3	$N_2CHCOPh$	70	91:9
4	$N_2CHCONMe_2$	74	95:5
5	$N_2CHCON(OMe)Me$	81	97:3

[a]Isolated yield.
[b]Determined by ^1H NMR of the crude product.

Other multicatalytic reactions involving copper salts as single catalysts and using their exquisite Lewis acid properties have been reported. Inspired by Yamamoto and co-workers' work on furan synthesis *via* the cyclization of 2-(1-alkynyl)-2-alken-1-ones,[37] Cheng and Hu reported the formation of highly substituted furocoumarins (Equation 7.11).[38] In this case, copper chloride acted not only as an oxidant but also as a π-activator of a carbon–carbon triple bond.

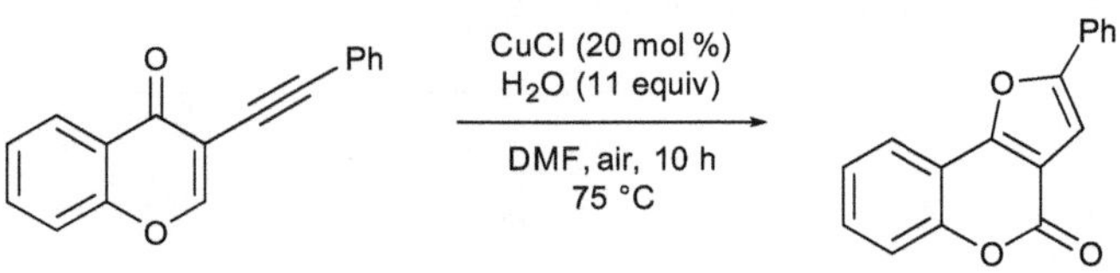

57 58, 86% (7.11)

The authors proposed the mechanism shown in Scheme 7.14. Conjugated addition of water was promoted by copper activation of the carbonyl group.

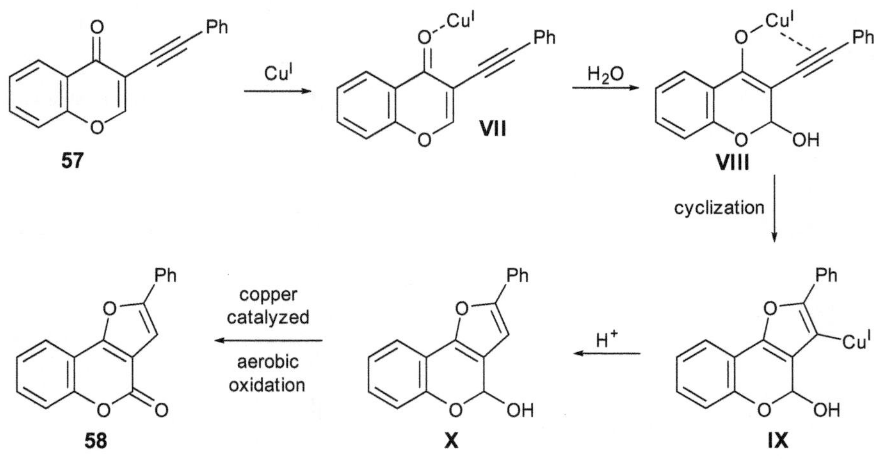

Scheme 7.14 Proposed mechanism of the multicatalytic furocoumarin synthesis.

Coordination of the copper enolate with the alkynyl moiety onto **VIII** induced the cyclization, followed by the protonation of the organocopper intermediate **IX**. The copper-catalyzed aerobic oxidation of the hemiacetal **X** afforded the furanocoumarin product **58**.

When the same reaction was performed using CuBr as catalyst and stoichiometric amounts of $CuCl_2$, chlorofuro[3,2-*c*]coumarin was obtained (Equation 7.12). This product can be further engaged in cross-coupling reactions such as Suzuki cross-coupling (Equation 7.13).

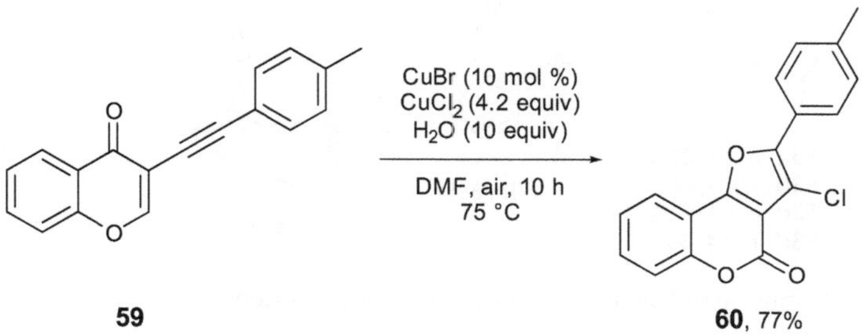

(7.12)

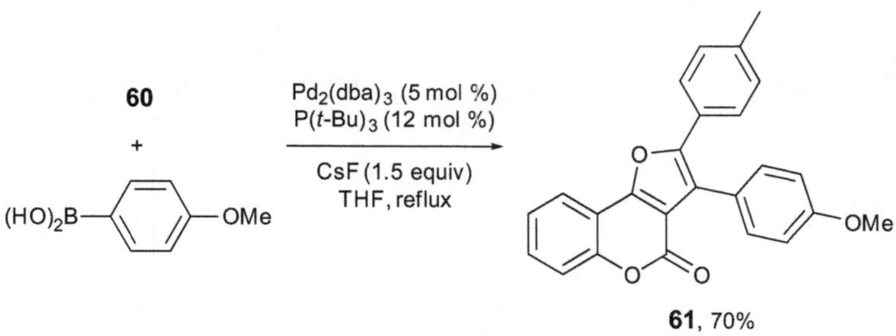

(7.13)

Following the work by Cheng and Hu, a similar one-pot cyclization–oxidation reaction was reported for the synthesis of substituted 2-benzoylfurans **63** from β-alkynyl-α,β-unsaturated aldehydes **62** (Scheme 7.15).[39]

The proposed mechanism for this copper-catalyzed tandem cyclization–aerobic oxidation is different, however (Scheme 7.16). The coordination of the Cu^I with the alkyne group into **62** leads to an intramolecular attack of the aldehyde, providing an organocopper intermediate **XII**. Protonation, followed by the conjugate addition of water, gave substituted furfuryl alcohol **XIV**. Finally, the copper salts catalyzed the aerobic oxidation of **XIV** to give substituted 2-benzoylfurans **63**.

Other transition metal complexes, in addition to copper, are known for their dual activity in different transformations. Recently, He and co-workers reported the synthesis of carbon-supported Ag nanoparticles.[40] These were investigated as catalysts for the oxidation of benzyl alcohols under an air atmosphere and basic DMSO reaction conditions, with the goal of producing benzaldehydes. Unexpectedly, they observed instead the formation of styryl ethers **65** in excellent yields from a variety of benzyl alcohols (Equation 7.14).

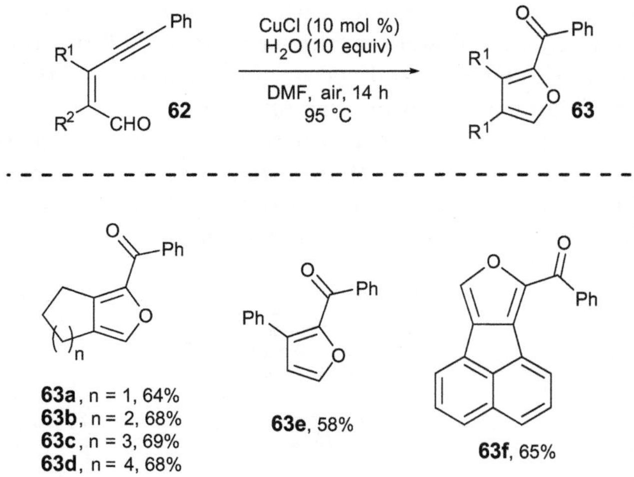

Scheme 7.15 Copper-catalyzed cyclization–oxidation tandem.

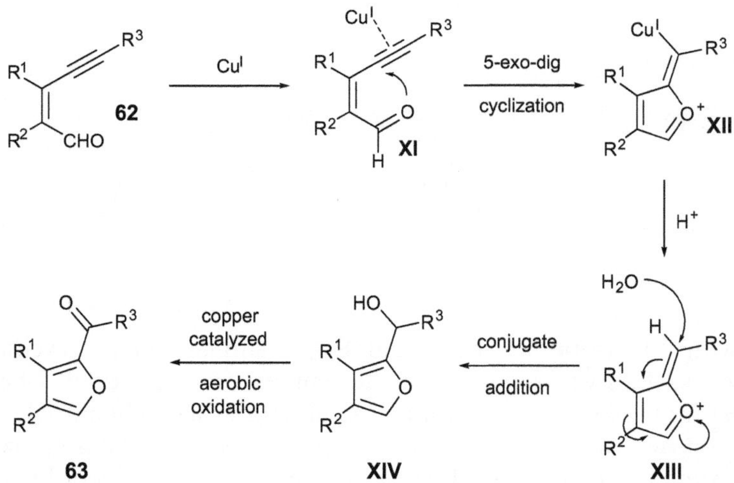

Scheme 7.16 Proposed mechanism of tandem cyclization–oxidation.

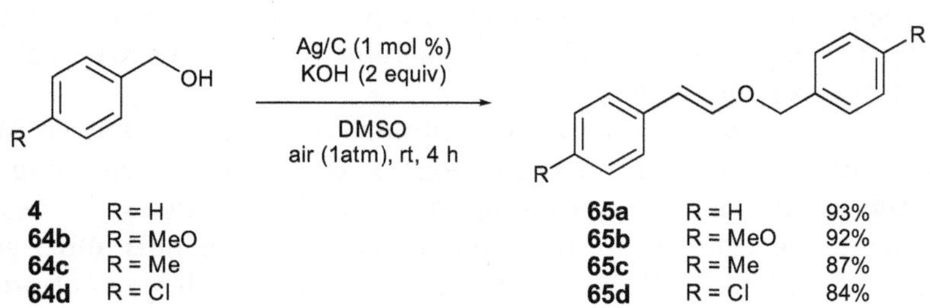

(7.14)

The reaction featured a single carbon transfer and C=C and C–O bond formation. Control experiments established DMSO as the carbon source

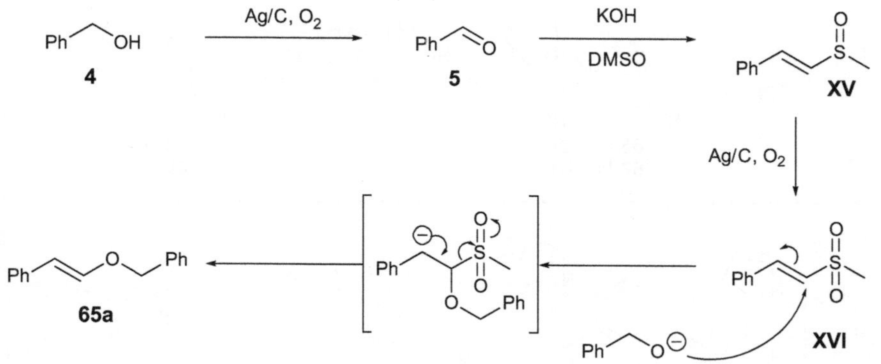

Scheme 7.17 Proposed mechanism of the formation of styryl ethers from benzyl alcohols.

and styryl sulfoxide as the key intermediate. The authors proposed the mechanism shown in Scheme 7.17. Silver nanoparticles first catalyzed the aerobic oxidation of benzyl alcohol **4** to benzaldehyde **5**. The addition of the anion of DMSO followed by elimination of water produced the styryl sulfoxide intermediate **XV**. The latter was oxidized by the silver nanoparticles to afford the corresponding styryl sulfone **XVI**, which underwent an addition–elimination with benzyl alkoxide to produce the observed styryl benzyl ether **65a**.

Iron also mediated cascade redox reactions for the synthesis of *N*-heterocyclic structures possessing interesting biological activities. Pereira and Thiéry reported the synthesis of pyrrolo[1,2-*a*]quinoxalines **68** and pyrrolo[1,2-*a*]pyrazines **69** from alcohols and 2-nitrophenylpyrrole **66** and 3-nitro-2-pyrrolopyridine **67**, respectively, using a stoichiometric mixture of iron powder and hydrochloric acid (Scheme 7.18).[41]

In this reaction, iron played a dual redox role: it not only promoted the aerobic oxidation of the alcohol moiety, but also served to reduce the nitro group, forming the ammonium intermediate **XVII** under strongly acidic conditions (Scheme 7.19). Upon formation of the iminium **XVIII** from the aldehyde and the ammonium components, intramolecular cyclization took place, followed by oxidation (probably mediated by iron species under an air atmosphere), leading to the *N*-heterocyclic product **68** or **69**.

Exploiting the modular combination of transition metal catalysis and organocatalysis, Kudo and co-workers designed an oxidation–asymmetric aldol multicatalytic reaction (Equation 7.15).[42] In a model reaction, 2-nitrobenzyl alcohol **70** was subjected to aerobic oxidation in presence of poly(ethylene glycol)-polystyrene–TEMPO resin (PEG–PS–TEMPO) along with a CuCl–bpy complex. Without purification, the aldehyde intermediate underwent a PEG–PS–proline-catalyzed asymmetric aldol reaction with acetone to form the β-hydroxy ketone **71** in 73% yield with good enantioselectivity (85% *ee*). Importantly, the introduction of triglycerine between the

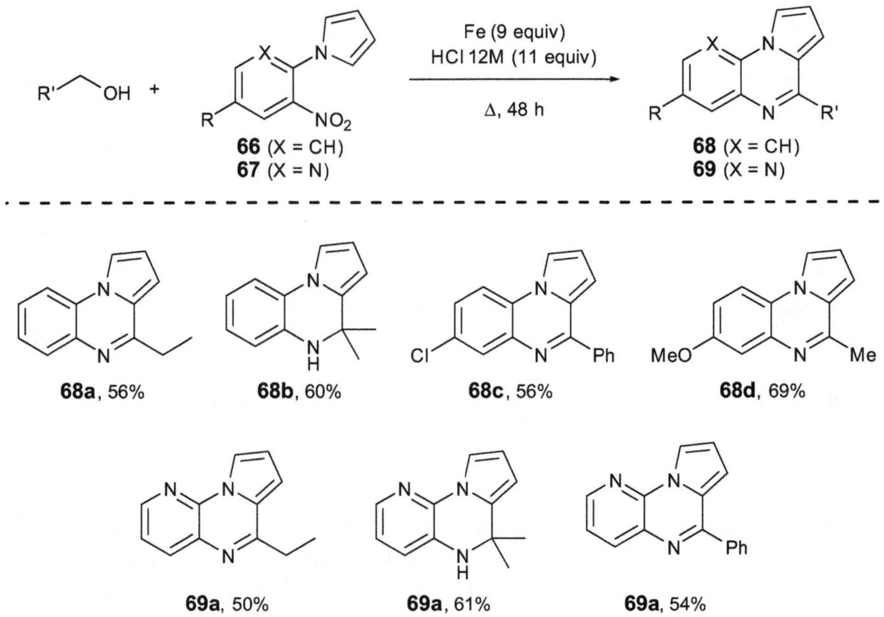

Scheme 7.18 Iron-mediated multistep synthesis of *N*-heterocyclic structures.

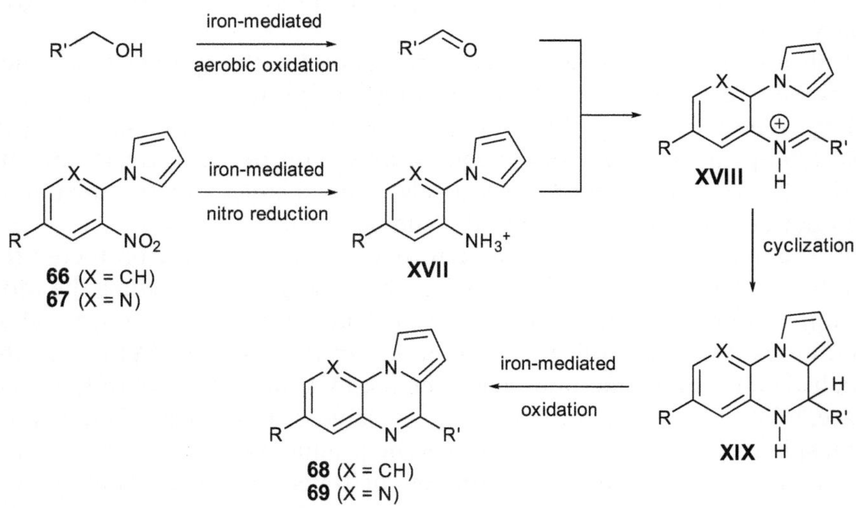

Scheme 7.19 Proposed mechanism of the multistep synthesis of *N*-heterocyclic structures.

PEG–PS resin and TEMPO and also the preadsorption of the CuCl–bpy complex were essential for an efficient aerobic oxidation step. The two PEG–PS catalysts were also easily separable and reusable.

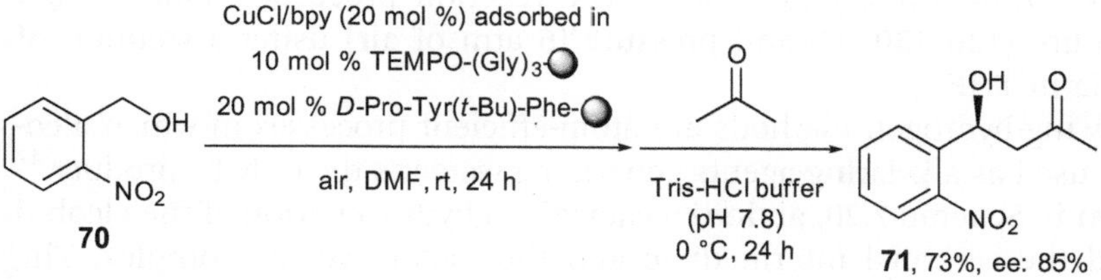

(7.15)

The transformation of an alcohol C–O single bond to a nitrile C≡N triple bond was accomplished through a sequential multistep dehydrogenation reaction. Huang and co-workers disclosed such a process using a catalytic CuI–bpy–TEMPO system.[43] The alcohol was first oxidized by this catalytic system to form the corresponding aldehyde, which reacted with aqueous ammonia affording an imine intermediate. The aerobic oxidation of the latter with the CuI–bpy–TEMPO system produced the desired nitrile. The robustness of this methodology was demonstrated by using KU-0063794 **72**, a specific inhibitor of the mammalian target of rapamycin,[44] as substrate to produce the corresponding nitrile analog **73** (Equation 7.16).

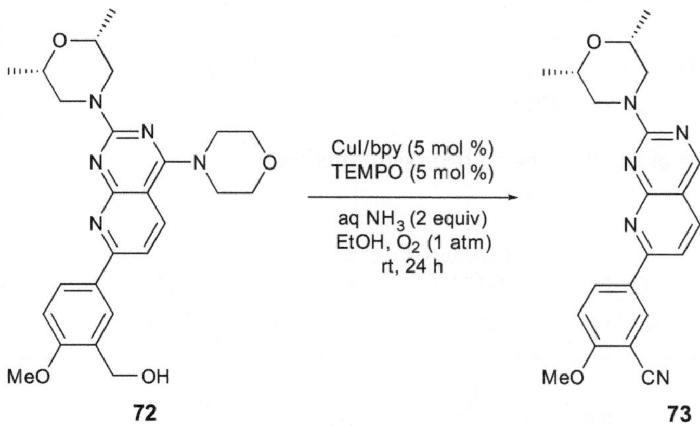

(7.16)

Simultaneously, Cook and co-workers also described similar copper catalysts to perform the oxidation of aldehydes to produce nitriles using aqueous ammonia and air as oxidant at room temperature. In only one example, the authors demonstrated that the reaction proceeded efficiently starting from 1-octanol **74** at 50 °C (Equation 7.17).[45] The corresponding octanenitrile **75** was isolated in quantitative yield.

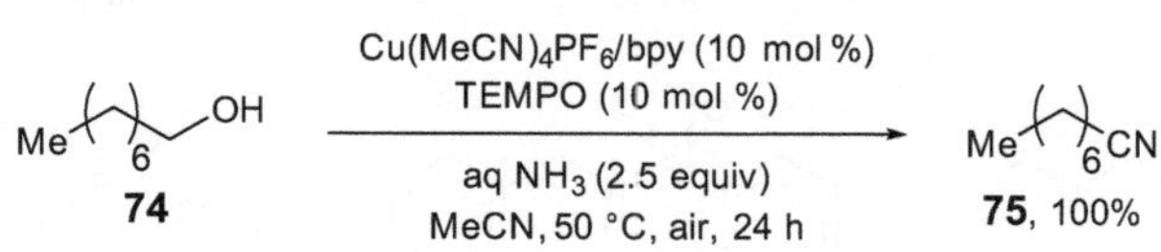

(7.17)

Supported ruthenium hydroxide was also reported by Mizuno and co-workers as a heterogeneous complex to catalyze a similar oxidative synthesis of nitriles from primary alcohols.[46] The reaction proceeded well at high temperature (120–130 °C) and pressure (6 atm of air) using a solution of ammonia in THF.

Borrowing-hydrogen methods are atom-efficient processes in which alcohols are used as alkylating agents generating water as the only by-product.[47] As shown in Scheme 7.20, an initial catalytic dehydrogenation of the alcohol provided the carbonyl intermediate and the metal–hydride complex. The carbonyl intermediate was reacted with amine to form the corresponding imine, which was then reduced by the metal–hydride complex, affording the alkylated amine product and regenerating the initial catalyst.

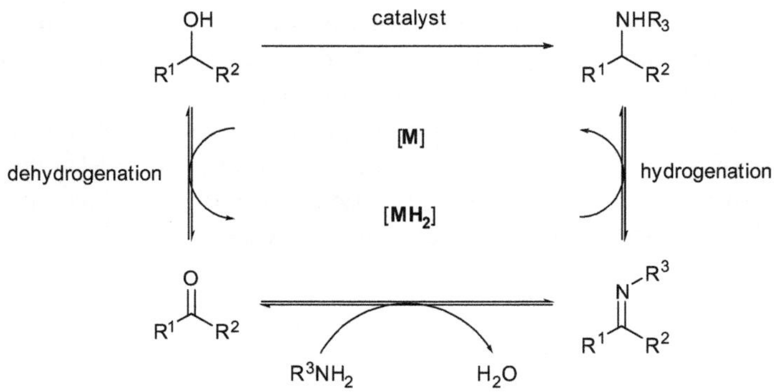

Scheme 7.20 Concept of the 'borrowing-hydrogen' method.

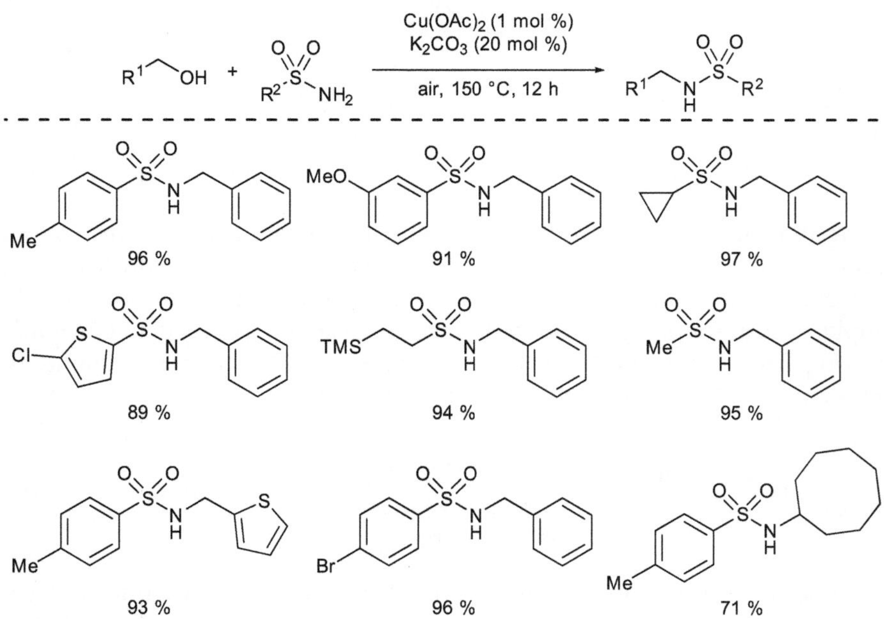

Scheme 7.21 Copper-catalyzed sulfonamide alkylation of alcohols.

Expensive transition metal complexes derived from ruthenium and iridium were the most efficient catalysts for these transformations, which do not require the addition of molecular oxygen as an oxidant. Searching for a more practical catalytic system, Beller and co-workers investigated copper complexes for the alkylation of sulfonamides with alcohols. They found that copper(II) acetate catalyzed such a reaction and produced the alkylated sulfonamide in excellent yields (Scheme 7.21).[48]

Importantly, the presence of an air atmosphere was beneficial whereas an inert or a higher oxygen content atmosphere inhibited the reaction. Based on mechanistic studies, the authors proposed the mechanism shown in Scheme 7.22, similar to that previously described (see Scheme 7.20). HRMS analysis revealed *in situ* formation of bissulfonylated amidine **76**, which played the role of a stabilizing ligand for the copper catalyst (Figure 7.2). The interaction between ligand **76** and copper was observed by UV–visible measurements. Labeling experiments with benzyl-d_7 alcohol showed that the dehydrogenation–imide formation–hydrogenation sequence was a reversible process. Furthermore, the C–H bond cleavage (*i.e.* dehydrogenation step), facilitated by molecular oxygen, is the rate-determining step.

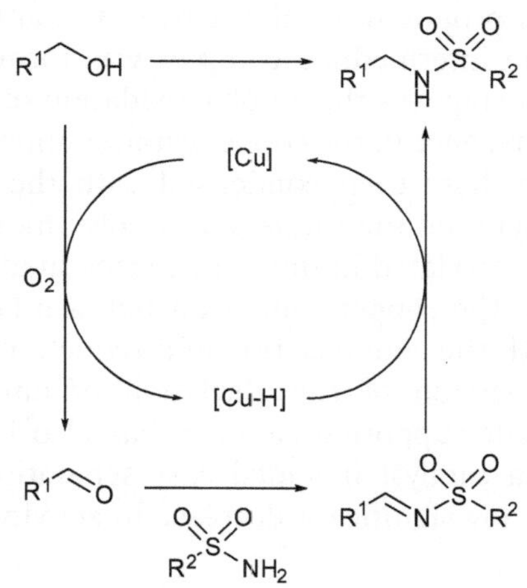

Scheme 7.22 Proposed mechanism of the Cu-catalyzed sulfonamide alkylation of alcohols.

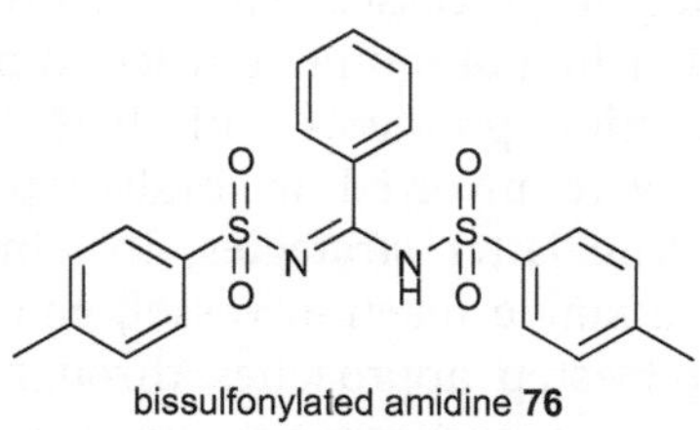

bissulfonylated amidine **76**

Figure 7.2 Copper-stabilizing ligand in catalytic alkylation of sulfonamides.

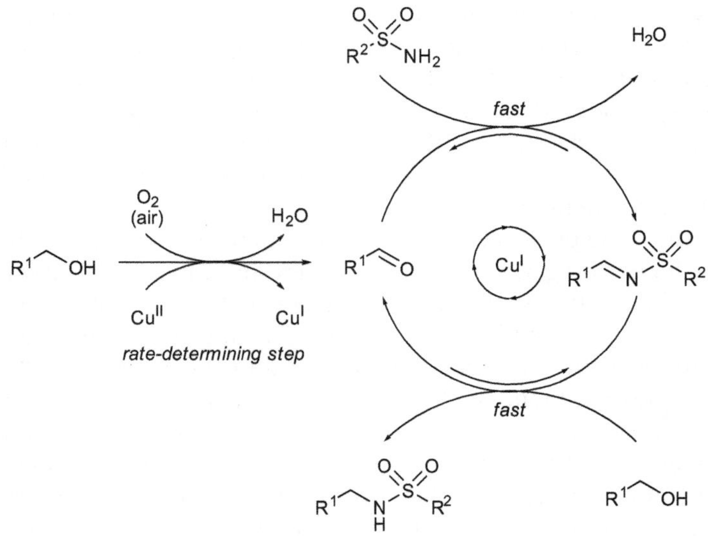

Scheme 7.23 Proposed mechanism of the catalytic *N*-alkylation reaction.

Xu and co-workers proposed an alternative mechanism in which no reactive copper–hydride intermediate complex was formed (Scheme 7.23).[49] The rate-determining step was the aerobic oxidation of the alcohol to form the aldehyde in the presence of the copper catalyst under an air atmosphere. The carbonyl intermediate then condensed with the amine to yield the imine. A reduction between the imine and a novel molecule of the alcohol provided the desired alkylated amine and regenerated the aldehyde intermediate. Presumably, the copper complex acted as a Lewis acid in the fast catalytic cycle for both the condensation and reduction steps.

A heterogeneous version of the alkylation of amines using copper–aluminium hydrotalcite-supported catalysis has also been reported.[50] The advantages of such a catalyst included easy separation and its reuse for several runs without any significant decrease in activity.

7.5 Conclusion

The catalytic aerobic oxidation of alcohols has permitted the development of fascinating multicatalytic processes. Thus, two transition metals could work synergistically either in a biomimetic reaction or within nanoclusters, providing efficient oxidation protocols with high TON and TOF. Other multicatalytic systems were involved in multistep syntheses increasing molecular complexity. The latter strategies, in which multiple chemical bonds were formed in a single reaction vessel, constitute a greener alternative to classical step-by-step approaches through minimization of the amount of solvent, reactants and reagents needed for the syntheses and purifications.

Considering the numerous recent examples of the concept of multicatalysis in the aerobic oxidation of alcohols, it is easy to conceive that these strategies are still in their infancy. Some improvements are expected in the coming years, especially in the field of nanoparticles and new heterogeneous systems. The development of asymmetric multicatalytic reactions also remains a key future objective. With continued advancements, this area will become a synthetically important feature in the organic chemistry community and should find widespread application in academia and industry.

Abbreviations

bpy	2,'2′-bipyridine
DBAD	di-*tert*-butyl azodicarboxylate
DMF	dimethylformamide
ETM	electron-transfer mediator
FMN	flavin mononucleotide
HRMS	high-resolution mass spectrometry
I*i*Pr	bis[1,3-(1,3-diisopropylphenyl)]imidazol-2-ylidene
IMes	bis[1,3-(2,4,6-trimethylphenyl)]imidazol-2-ylidene
i-Pr	2-propyl
MS	molecular sieve
NAD	nicotinamide adenine dinucleotide
NP	nanoparticle
Phen	1,10-phenanthroline
PI-CB	carbon-stabilized polymer-incarcerated
PMAS	poly(2-methoxyaniline-5-sulfonic acid)
t-Bu	*tert*-butyl
TEMPO	2,2,6,6-tetramethylpiperidine-1-oxyl, free radical
THF	tetrahydrofuran
TOF	turnover frequency
TON	turnover number
TPP	5,10,15,20-tetraphenyl-21*H*,23*H*-porphine
Ts	tosyl

References

1. For recent selected reviews, see: (a) L. F. Tietze, G. Brasche and K. Gericke, *Domino Reactions in Organic Synthesis*, Wiley-VCH, Weinheim, 2006; (b) L. F. Tietze, *Domino Reactions*, Wiley-VCH, Weinheim, 2014; (c) C. J. Chapman and C. G. Frost, *Synthesis*, 2007, 1; (d) F.-X. Felpin and E. Fouquet, *ChemSusChem*, 2008, **1**, 718; (e) L. M. Ambrosini and T. H. Lambert, *ChemCatChem*, 2010, **2**, 1373; (f) J. Zhou, *Chem. Asian J.*, 2010, **5**, 422; (g) A. E. Allen and D. W. C. MacMillan, *Chem. Sci.*, 2012, **3**, 633; (h) R. C. Wende and P. R. Schreiner, *Green Chem.*, 2012, **14**, 1821;

(i) C. C. J. Loh and D. Enders, *Chem. Eur. J.*, 2012, **18**, 10212; (j) X. Zeng, *Chem. Rev.*, 2013, **113**, 6864.
2. T. Hudlicky and J. W. Reed, *The Way of Synthesis*, Wiley-VCH, Weinheim, 2007, pp. 101–106.
3. (a) I. E. Markó, P. R. Giles, M. Tsukazaki, A. Gautier, R. Dumeunier, K. Dodo, F. Philippart, I. Chelle-Regnault, J.-L. Mutonkole, S. M. Brown and C. J. Urch, in *Transition Metals for Organic Synthesis*, ed. M. Beller and C. Bolm, Wiley-VCH, Weinheim, 2nd edn, 2004, pp. 437–478; (b) B.-Z. Zhan and A. Thompson, *Tetrahedron*, 2004, **60**, 2917; (c) S. S. Stahl, *Science*, 2005, **309**, 1824; (d) M. J. Schultz and M. S. Sigman, *Tetrahedron*, 2006, **62**, 8227; (e) T. Matsumoto, M. Ueno, N. Wang and S. Kobayashi, *Chem. Asian J.*, 2008, **3**, 196; (f) K. M. Gligorich and M. S. Sigman, *Chem. Commun.*, 2009, 3854; (g) Z. Shi, C. Zhang, C. Tang and N. Jiao, *Chem. Soc. Rev.*, 2012, **41**, 3381; (h) C. Parmeggiani and F. Cardona, *Green Chem.*, 2012, **14**, 547.
4. (a) J. March, in *Advanced Organic Chemistry*, Wiley, New York, 7th edn, 2013, pp. 1437–1496; (b) F. A. Carey and R. J. Sundberg, in *Advanced Organic Chemistry: Part B*, Springer, New York, 5th edn, 2010, pp. 1063–1073.
5. (a) R. A. Sheldon, I. W. C. E. Arends and A. Dijksman, *Catal. Today*, 2000, **57**, 157; (b) R. A. Sheldon, I. W. C. E. Arends, G.-J. ten Brink and A. Dijksman, *Acc. Chem. Res.*, 2002, **35**, 774; (c) R. A. Sheldon, I. Arends and U. Hanefeld, *Green Chemistry and Catalysis*, Wiley-VCH, Weinheim, 2007, pp. 133–221.
6. (a) H. R. Horton, L. A. Moran, R. S. Ochs, J. D. Rawn and K. G. Scrimgeour, *Principles of Biochemistry*, Prentice Hall, Englewood Cliffs, NJ, 3rd edn, 2002; (b) D. L. Nelson and M. M. Cox, *Lehninger Principles of Biochemistry*, Worth Publishers, New York, 2000.
7. J.-E. Bäckvall, A. K. Awasthi and Z. D. Renko, *J. Am. Chem. Soc.*, 1987, **109**, 4750.
8. J. Piera and J.-E. Bäckvall, *Angew. Chem., Int. Ed.*, 2008, **47**, 3506.
9. (a) M. S. Thompson and T. J. Meyer, *J. Am. Chem. Soc.*, 1982, **104**, 4106; (b) S. Shinoda, H. Itagaki and Y. Saito, *J. Chem. Soc., Chem. Commun.*, 1985, 860; (c) L. Roecker and T. J. Meyer, *J. Am. Chem. Soc.*, 1987, **109**, 746.
10. J.-E. Bäckvall, R. L. Chowdhury and U. Karlsson, *J. Chem. Soc., Chem. Commun.*, 1991, 473.
11. M. Abed, I. Goldberg, Z. Stein and Y. Shvo, *Organometallics*, 1988, 7, 2054.
12. (a) G.-Z. Wang, U. Andreasson and J.-E. Bäckvall, *J. Chem. Soc., Chem. Commun.*, 1994, 1037; (b) G. Csjernyik, A. H. Éll, L. Fadini, B. Pugin and J.-E. Bäckvall, *J. Org. Chem.*, 2002, **67**, 1657.
13. Á. Zsigmond, F. Notheisz, G. Csjernyik and J.-E. Bäckvall, *Top. Catal.*, 2002, **19**, 119.
14. E. V. Johnston, E. A. Karlsson, L.-H. Tran, B. Åkermark and J.-E. Bäckvall, *Eur. J. Org. Chem.*, 2010, 1971.

15. Y. Endo and J.-E. Bäckvall, *Chem. Eur. J.*, 2011, **17**, 12596.
16. B. P. Babu, Y. Endo and J.-E. Bäckvall, *Chem. Eur. J.*, 2012, **18**, 11524.
17. D. Saio, T. Amaya and T. Hirao, *Adv. Synth. Catal.*, 2010, **352**, 2177.
18. S.-I. Murahashi, T. Naota and N. Hirai, *J. Org. Chem.*, 1993, **58**, 7318.
19. (a) K. S. Coleman, M. Coppe, C. Thomas and J. A. Osborn, *Tetrahedron Lett.*, 1999, **40**, 3723; (b) J. Muldoon and S. N. Brown, *Org. Lett.*, 2002, **4**, 1043.
20. P. A. Shapley, N. Zhang, J. L. Allen, D. H. Pool and H.-C. Liang, *J. Am. Chem. Soc.*, 2000, **122**, 1079.
21. D. I. Enache, J. K. Edwards, P. Landon, B. Solsona-Espriu, A. F. Carley, A. A. Herzing, M. Watanabe, C. J. Kiely, D. W. Knight and G. J. Hutchings, *Science*, 2006, **311**, 362.
22. A. Villa, G. M. Veith and L. Prati, *Angew. Chem. Int. Ed.*, 2010, **49**, 4499.
23. H. Miyamura, R. Matsubara and S. Kobayashi, *Chem. Commun.*, 2008, 2031.
24. (a) K. Kaizuka, H. Miyamura and S. Kobayashi, *J. Am. Chem. Soc.*, 2010, **132**, 15096; (b) K. Kaizuka, K.-Y. Lee, H. Miyamura and S. Kobayashi, *J. Flow Chem.*, 2012, **2**, 1.
25. W.-J. Yoo, H. Miyamura and S. Kobayashi, *J. Am. Chem. Soc.*, 2011, **133**, 3095.
26. H. Miyamura, G. C. Y. Choo, T. Yasukawa, W.-J. Yoo and S. Kobayashi, *Chem. Commun.*, 2013, **49**, 9917.
27. J.-F. Soulé, H. Miyamura and S. Kobayashi, *Chem. Commun.*, 2013, 355.
28. J.-F. Soulé, H. Miyamura and S. Kobayashi, *J. Am. Chem. Soc.*, 2011, **133**, 18550.
29. J.-F. Soulé, H. Miyamura and S. Kobayashi, *Chem. Asian J.*, 2013, **8**, 2614.
30. For selected examples of multicomponent reactions using one catalytic aerobic oxidation step, see: (a) G. Kim, D. G. Lee and S. Chang, *Bull. Korean Chem. Soc.*, 2001, **22**, 943; (b) J. Brioche, G. Masson and J. Zhu, *Org. Lett.*, 2010, **12**, 1432; (c) F.-Q. Yuan and F.-S. Han, *Org. Lett.*, 2012, **14**, 1218.
31. H. Lebel and V. Paquet, *J. Am. Chem. Soc.*, 2004, **126**, 11152.
32. D. R. Jensen, M. J. Schultz, J. A. Müller and M. S. Sigman, *Angew. Chem. Int. Ed.*, 2003, **42**, 3810.
33. (a) L. Jafarpour, A. C. Hillier and S. P. Nolan, *Organometallics*, 2002, **21**, 442; (b) L. Jafarpour and S. P. Nolan, *J. Organomet. Chem.*, 2001, **617–618**, 17.
34. I. E. Markó, A. Gautier, R. Dumeunier, K. Doda, F. Philippart, S. M. Brown and C. J. Urch, *Angew. Chem. Int. Ed.*, 2004, **43**, 1588.
35. (a) H. Lebel, M. Davi, S. Díez-González and S. P. Nolan, *J. Org. Chem.*, 2007, **72**, 144; (b) H. Lebel, M. Davi and G. T. Stokłosa, *J. Org. Chem.*, 2008, **73**, 6828; (c) H. Lebel and M. Davi, *Adv. Catal. Synth.*, 2008, **350**, 2352.
36. M. Davi and H. Lebel, *Org. Lett.*, 2009, **11**, 41.
37. N. T. Patil, H. Wu and Y. Yamamoto, *J. Org. Chem.*, 2005, **70**, 4531.
38. G. Cheng and Y. Hu, *J. Org. Chem.*, 2008, **73**, 4732.
39. R. Jana, S. Paul, A. Biswas and J. K. Ray, *Tetrahedron Lett.*, 2010, **51**, 273.

40. Q. Zhang, S. Cai, L. Li, Y. Chen, H. Rong, Z. Niu, J. Liu, W. He and Y. Li, *ACS Catal.*, 2013, **3**, 1681.
41. M. F. Pereira and V. Thiéry, *Org. Lett.*, 2012, **14**, 4754.
42. K. Akagawa, S. Takigawa, E. Mano and K. Kudo, *Tetrahedron Lett.*, 2011, **52**, 770.
43. W. Yin, C. Wang and Y. Huang, *Org. Lett.*, 2013, **15**, 1850.
44. J. M. García-Martínez, J. Moran, R. G. Clarke, A. Gray, S. C. Cosulich, C. M. Chresta and D. R. Alessi, *Biochem. J.*, 2009, **421**, 29.
45. L. M. Dornan, Q. Cao, J. C. A. Flanagan, J. J. Crawford, M. J. Cook and M. J. Muldoon, *Chem. Commun.*, 2013, **49**, 6030.
46. T. Oishi, K. Yamagushi and N. Mizuno, *Top. Catal.*, 2010, **53**, 479.
47. (a) M. H. S. A. Hamid, P. A. Slatford and J. M. J. Williams, *Adv. Synth. Catal.*, 2007, **349**, 1555; (b) S. Bähn, S. Imm, L. Neubert, M. Zhang, H. Neumann and M. Beller, *ChemCatChem*, 2011, **3**, 1853.
48. (a) F. Shi, M. K. Tse, X. Cui, D. Gördes, D. Michalik, K. Thurow, Y. Deng and M. Beller, *Angew. Chem. Int. Ed.*, 2009, **48**, 5912; (b) X. Cui, F. Shi, M. K. Tse, D. Gördes, K. Thurow, M. Beller and Y. Deng, *Adv. Synth. Catal.*, 2009, **351**, 2949.
49. Q. Li, S. Fan, Q. Sun, H. Tian, X. Yu and Q. Xu, *Org. Biomol. Chem.*, 2012, **10**, 2966.
50. P. R. Likhar, R. Arundhati, M. L. Kantam and P. S. Prathima, *Eur. J. Org. Chem.*, 2009, 5383.

CHAPTER 8

Gas-Phase Oxidation of Alcohols: Innovation in Industrial Technologies and Recent Developments

JULIANA VELASQUEZ OCHOA AND FABRIZIO CAVANI*

Dipartimento di Chimica Industriale 'Toso Montanari', Università di Bologna, Viale Risorgimento 4, 40136 Bologna, Italy
*Email: fabrizio.cavani@unibo.it

8.1 Introduction

The gas-phase oxidation of alcohols is a reaction of industrial relevance, although nowadays the only commercially important application is the production of formaldehyde, since other routes are preferred for the synthesis of higher aldehydes or acids.[1,2] On the other hand, this class of reactions has been the subject of investigation for many years, especially during the 1980s and 1990s, mainly with regard to the gas-phase oxidation of ethanol catalyzed by either single transition metal oxides or mixed oxides. These studies aimed to discover which catalyst properties were needed in order to address the reaction pathway either to the selective formation of acetaldehyde, *via* oxidative dehydrogenation, or to the consecutive conversion into acetic acid. Furthermore, aliphatic and aromatic alcohols show a wide range of possible transformations, including (i) dehydration and condensation to give olefins and dialkyl ethers, respectively, both catalyzed

RSC Green Chemistry No. 28
Transition Metal Catalysis in Aerobic Alcohol Oxidation
Edited by Francesca Cardona and Camilla Parmeggiani

Published by the Royal Society of Chemistry, www.rsc.org

by acid sites, (ii) anaerobic dehydrogenation to aldehydes or ketones, a reaction catalyzed by basic sites, and (iii) oxidative dehydrogenation. Aldehydes can also easily undergo consecutive reactions leading to various products, such as dimerization or condensation (*via* the Tishchenko reaction or aldol condensation), both reactions also being catalyzed by basic sites, hemiacetalization and acetalization by reaction with the starting alcohol and oxidation to carboxylic acids. Lastly, alcohols may act as reducing molecules towards carbonyls and even C=C double bonds, *via* the Meerwein–Ponndorf–Verley–Oppenauer reaction, which again can be catalyzed by basic sites. Therefore, alcohols are excellent probe molecules for the characterization of surface acid–base and redox properties of solid catalysts, being more versatile than conventional methods that make use of more specific and site-selective probe molecules. This peculiarity led researchers to use alcohols for investigating both catalyst multifunctional properties and complex reaction patterns leading to by-products during oxidation reactions.

During the last decade, however, interest in the gas-phase oxidation of alcohols, especially ethanol, has grown considerably again, because of the concomitance of two factors which have greatly stimulated research into applied catalysis, *i.e.*, the use of bioalcohols for the production of fuels, fuel additives and chemicals and the use of supported metal nanoparticles (NPs), especially Au and Au-based alloys, as catalysts for both liquid- and gas-phase reduction and oxidation reactions. Indeed, most scientific papers dealing with the gas-phase oxidation of alcohols published in recent years report the use of supported Au NPs as catalysts.

In this chapter, following a brief overview of the main industrial technologies currently used for gas-phase alcohol oxidation, we describe the most recent improvements in this area, also referring to papers published earlier on this topic, in order to provide an updated overview of the main catalyst features needed for this class of reaction. The focus is mainly on the selective oxidation of bioalcohols into various chemicals.

8.2 Gas-Phase Oxidation of Methanol

8.2.1 Industrial Process of Methanol Oxidative Dehydrogenation to Formaldehyde

Formaldehyde can be produced by either *oxidehydrogenation* of methanol (the Perstorp–Reichhold Formox process, recently acquired by Johnson Matthey):

$$CH_3OH + \tfrac{1}{2}O_2 \rightarrow CH_2O + H_2O \qquad \Delta H^\circ = -37\ \text{kcal mol}^{-1} \tag{8.1}$$

or *dehydrogenation of methanol* coupled with oxidation of the produced molecular hydrogen to water (the BASF process):

$$CH_3OH \rightarrow CH_2O + H_2 \qquad \Delta H^\circ = 20\ \text{kcal mol}^{-1} \tag{8.2}$$

$$H_2 + 0.5O_2 \rightarrow H_2O \qquad \Delta H^\circ = -57\ \text{kcal mol}^{-1} \tag{8.3}$$

These processes differ both in the operating conditions and in the catalyst used.

The oxidehydrogenation of methanol is an exothermal process and is carried out with an excess of oxygen (air) compared with the stoichiometric requirement. The catalyst is an Fe–Mo mixed oxide, made of $Fe_2(MoO_4)_3$ and MoO_3.

The process of methanol oxidehydrogenation is performed at temperatures around 300 °C (the inlet temperature may be between 260 and 280 °C), slightly above atmospheric pressure, with air, in a multitubular reactor. A 98–99% conversion of methanol is obtained, with a yield of formaldehyde of around 92%; a 50–60% formaldehyde solution is produced, containing less than 1% methanol. In many cases, a small adiabatic reactor is added downstream, to improve the methanol conversion by up to 99.8%. At the highest methanol conversion, the yield of formaldehyde is around 94–95%. Older processes operate at 6 vol.% methanol in air (just below the lower limit of flammability, the upper limit is close to 60%). The drawbacks of this process are relatively low productivity, lack of purity of formaldehyde (as it contains relatively high amounts of formic acid), limited lifetime of the catalyst and all the costs associated with the presence of large amounts of inert gas, *i.e.* large equipment, costs for compression and large volumes of exhaust gas.

In recent years, the productivity of many plants has been streamlined by increasing the concentration of methanol and, at the same time, decreasing the concentration of oxygen, achieved through recycling part of the spent gas. Thus a methanol concentration of up to 8.2% can be reached, with a maximum inlet oxygen concentration of 9.0%. Further decreases in oxygen concentration or increases in methanol concentration cannot be obtained, because the catalyst becomes over-reduced and its activity declines. One disadvantage of operating at higher concentrations of methanol is an increase in the hot spot temperature, which accelerates catalyst deactivation. This phenomenon has been partly solved by diluting the catalyst along the bed, in order to distribute the heat of the reaction better. Newer generations of catalysts include formulations that differ along the catalytic bed.

Better control of the bed temperature also improves the selectivity to formaldehyde, while decreasing undesired decomposition or combustion. In addition, it is possible to increase the methanol concentration (while remaining outside the flammability bell) and thus increase throughput and reduce vent emissions.

Perstorp (now Johnson Matthey) catalyst: Since the implementation of the first Formox plant in 1959, the core of the process has been the iron–molybdenum oxide catalyst. The traditional composition comprises an Mo to Fe ratio that varies from 1 to 5, but is usually higher than 1.5. The Mo excess is needed to prevent the formation of Fe-rich phases on the catalyst surface during the reaction, due to the loss of MoO_3 (by sublimation); however, the stoichiometric iron molybdate $Fe_2(MoO_4)_3$ is considered to be the active phase.[3] Changes in composition may further include small amounts of additional oxides such as V_2O_5, CuO, Cr_2O_3, CoO and P_2O_5, which may help to increase the activity or stability of the catalyst.

At present, plants using the Formox process are operating with the CAP (Catalyst Activity Profile) concept. This new concept implies a non-homogeneous distribution of the catalyst composition along the bed; instead, each part of the catalytic bed is loaded according to the heat release needed to maintain a more evenly distributed heat of reaction. The average lifetime of an industrial iron molybdate catalyst ranges from 6 to 12 months.

Clariant catalyst: The specialty chemical company Clariant also provides catalysts for formaldehyde production from methanol. Under the FAMAX label they have developed materials that are able to catalyze this reaction. Their formulation is based on Fe–Mo oxides, but they are provided in various forms and are often diluted with inert materials to optimize operation. During a recent conference, it was claimed that this catalyst series had been improved in terms of new shape, new formulation and a low surface area, thus making it possible to achieve a lower pressure drop of around 10–15%, higher selectivity (at high methanol conversion), less CO formation and a 0.5% improvement in molar yield.[4]

The dehydrogenation of methanol with oxidation of hydrogen, on the other hand, is carried out under conditions of oxygen deficiency compared with the stoichiometric requirement. Indeed, in this case, methanol dehydrogenation accounts for approximately half the formaldehyde produced in the reaction, while the remainder is produced by the oxidehydrogenation of methanol. This process is autothermal since the heat developed by the oxidation of molecular hydrogen to water is used for the endothermal dehydrogenation of methanol. The conversion of methanol is a function of both the reaction temperature and the composition of the feed. High temperatures (>500 °C) are necessary in order to obtain a good conversion, since the reaction of methanol dehydrogenation is thermodynamically limited (for instance, conversion at 400 °C is around 50%). The catalyst is made of unsupported crystalline silver.

Methanol dehydrogenation operates either at temperatures higher than 600 °C (typically between 680 and 720 °C; in this case there is no need to recycle unconverted methanol, since its amount is less than 2%; selectivity is around 98–99%, thus higher than that of the methanol-lean process) or at 510–550 °C, for processes which involve excess methanol and for which the per pass conversion is relatively low (60–80%). In the latter case, unconverted methanol is recycled. When temperatures higher than 600 °C are involved, sometimes the exhaust gas is recycled (and steam is also added) in order to dilute oxygen (the typical inlet concentration is 10–12 vol.%) and to increase methanol concentration (>40%), to remain above the upper limit of flammability and increase productivity. The concentration of formaldehyde obtained can range from 40 to 55%, depending on the technology used. At such high temperatures, very short residence times are necessary in order to avoid the decomposition of formaldehyde.

BASF catalyst: Formaldehyde plants using the silver catalyst (BASF process) remain the more popular of the two technologies available for manufacturing formaldehyde, with the ratio being about four silver plants to every one mixed metal oxide system (Formox). BASF's advantages are relatively low investment, high yield and stable production. The catalyst for this kind of process consists mainly of electrolytic silver grains from 1 to 5 mm in size sandwiched between silver gauze. Both the morphology and chemical composition of the silver catalyst play an important role in the final reaction rate and selectivity, especially at the high temperatures used in this process, since the kinetics of the reactions involved are significantly higher than at low temperature.[5]

The following side products may form in both processes: dimethyl ether (catalyzed by the acid centers of the catalyst), methyl formate (by the dimerization of formaldehyde), CO_2 (by the combustion of both methanol and formaldehyde), CO (by the dehydrogenation of formaldehyde) and formic acid (by the oxidation of formaldehyde). The dehydrogenation of formaldehyde to CO mainly occurs homogeneously at the high temperature used in the methanol dehydrogenation process and it is therefore very important to quench the gaseous stream rapidly after reaction. The best catalyst must not possess either strong acid sites or basic sites. Large surface areas are also undesirable, in order to decrease the contribution of consecutive reactions.

8.2.2 Recent Improvements to Industrial Catalysts and Technology for Methanol Oxidehydrogenation

The key points concerning improvements on current technologies and catalysts for formaldehyde production include:

- Increased capacity, achieved either by increasing the inlet concentration of methanol (while limiting the hot spots in the catalytic bed, the main reason for catalyst deactivation) or by increasing the space velocity, which is typically limited by the pressure drop in the reactor.
- Operation with other ballast (preferably steam), to improve the heat transfer properties of the system, to lower the hot spot temperature and increase the catalyst lifetime.[6]
- The *in situ* production of the catalyst by mixing the precursor oxides in the reactor under conventional conversion conditions. This new approach involves more common and less expensive reagents, does not require water and reduces the formation of pollutants, such as nitrogen oxides (NO_x), ammonia (NH_3) and hydrochloric acid (HCl), which are formed during conventional synthesis. In this case, the catalytically active $Fe_2(MoO_4)_3$–MoO_3 component is formed in the reactor during the process.[7]

- Fluidized bed reactors as an option to avoid the well-known hot spot temperatures. However, owing to their low mechanical resistance, unsupported iron molybdates are not suitable for this configuration. Currently, research is focused on finding a suitable catalyst. Until now, low surface area silica seemed to be the best support, but the activity of this supported iron molybdate is much lower than that of the unsupported material.[3] The replacement of part of the Mo by other elements (such as V, Sb, P, Sn and Mn) could probably lead to more active supported catalysts. Nevertheless, there is a need for more research in this area.

8.2.3 New Classes of Catalysts for Methanol Oxidation

Vanadium oxides have redox properties that make them very interesting catalysts for oxidation reactions, especially in the highest oxidation states (V_2O_5). For instance, when using bare powdered V_2O_5 at 50% methanol conversion the main product is formaldehyde, whereas at higher conversion methylformate is the main product.[8] Different kinds of vanadates have been extensively investigated by Wachs *et al.*;[9–11] more specifically, in a patent[12] they claimed that bulk metal vanadates possess higher activity than the traditional MoO_3–$Fe_2(MoO_4)_3$ catalysts for methanol selective oxidation, thus making it possible to use them at lower temperatures while maintaining selectivity and also avoiding the problems related to catalyst deactivation by volatilization. Among the catalysts tested, those that performed better (at methanol conversion higher than 96%) were $CrVO_4$, $NbVO_5$ and $Ni_3(VO_4)_2$ with 90.4, 90.0 and 94.0% selectivity to formaldehyde, respectively (at 300 °C and a methanol–oxygen–helium feed gas composition of 6 : 13 : 81 mol%).

In another investigation by Wachs and co-workers[13] on methanol oxidation over bulk $FeVO_4$ and supported 4% V_2O_5–α-Fe_2O_3, it was found that both surfaces are similar and consist of VO_x species which are the catalytically active sites. The study was performed with temperature-programmed IR spectroscopy comparing the individual oxides, supports and the bulk catalyst while methanol was being adsorbed. They found that in all cases the species CH_3OH^* and CH_3O^* are present on the catalyst surface with the exception of the acidic α-Fe_2O_3, yielding dimethyl ether (DME). Instead, for V_2O_5, $FeVO_4$ and supported 4% V_2O_5–α-Fe_2O_3, both surface intermediates gave rise mainly to HCHO (see Scheme 8.1). The results were confirmed with HR-TEM images showing an amorphous VO_x-enriched layer of ~1 nm at the outer surface of the bulk $FeVO_4$ catalysts and also by the increased V : Fe atomic ratio in the surface of the $FeVO_4$ particles found in the elemental analysis by electron energy loss spectroscopy (EELS).

Recently, a catalyst formulation containing V and Sb supported on silica was patented by Wacker Chemie,[14] claiming a high yield for methanol oxidation to formaldehyde (92,1%), which is of great importance in moving towards fluidized bed technology.

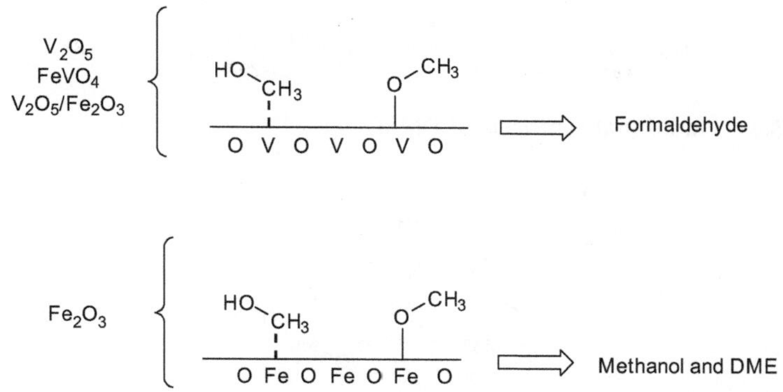

Scheme 8.1 Surface intermediate species detected on different catalysts when adsorbing methanol at 100 °C and their respective products released at higher temperature.[13]

8.2.4 New Reactions and Applications of Methanol Gas-Phase Oxidation

8.2.4.1 Dimethoxymethane Synthesis

Dimethoxymethane (DMM) is generally synthesized after the oxidation of methanol to formaldehyde by using an acidic catalyst and additional methanol to perform acetalization (Scheme 8.2). Among the most commonly used catalysts are ion-exchange resins (which may degrade and cause clogging of the reactor) and inorganic acids (such as sulfuric acid, with all its related complications such as corrosion, environmental pollution, the need for neutralization of waste water and irreversible loss of the acid).

Recently, DMM has gained importance as a green solvent, fuel additive and precursor of the so-called POMM [poly(oxymethylene dimethyl ether)], which is a good candidate to replace methanol in fuel cells, with lower toxicity and higher efficiency.[15] However, to make it attractive from the economical point of view, research in progress is aiming to produce it in a single step directly from methanol.

Among the catalysts used in the attempt to synthesize DMM directly, the most successful ones include bulk and supported Keggin structures[16] [$H_4PVMo_{11}O_{40}$–SiO_2 (DMM selectivity of 58.1% at a methanol conversion of 68.2%)], RuO_2–Al_2O_3[17] (DMM selectivity of 66.8% at a methanol conversion of only 20%), Re oxide and Re-based mixed oxides supported on TiO_2, V_2O_5, ZrO_2, Fe_2O_3 and Al_2O_3[18] (maximum DMM selectivity between 60 and 93% for a methanol conversion of 15–60%) and also $(SO_4)^{2-}$–V_2O_5–TiO_2[19] (with a DMM selectivity of 90% for 64% methanol conversion). However, the catalytic system with the best productivity for this reaction appears to be industrial Fe–Mo ($\sim$4.6 kg_{DMM} h^{-1} $kg_{cat}{}^{-1}$ at 280 °C at a methanol conversion of 56% and a DMM selectivity of 90%) as reported recently by Dumeignil *et al.*[20,21] What really makes the difference and directs the reaction towards DMM production is mainly the use of a feed rich in

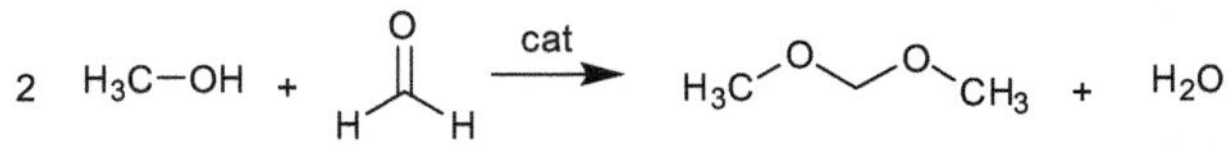

Scheme 8.2 Traditional synthesis of DMM.

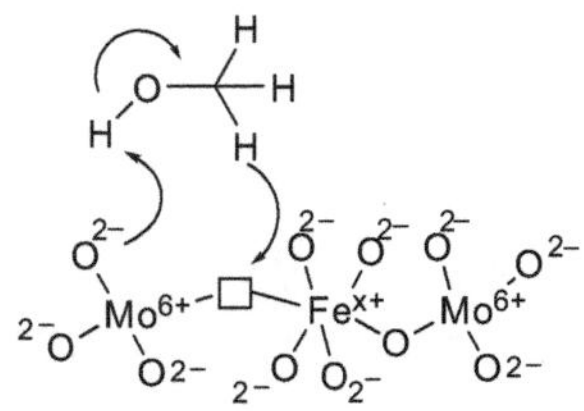

Scheme 8.3 Proposed active sites for methanol transformation over Fe–Mo–O catalyst. □, Anionic vacancy Fe^{x+}: Fe^{3+} or Fe^{2+}.
Source: adapted from Thavornprasert *et al.*[20]

methanol (CH_3OH:O_2:He = 40:13:47 mol%). This possibility was not explored before, mainly due to the increase in hot spot temperatures, but nowadays there is better control of the catalytic bed temperature and less pressure drop, thus making the production of DMM very attractive since the overall throughput is increased (more methanol, less carrier, less vent emissions). In the same study,[20] a series of Fe–Mo–O catalysts were characterized by advanced techniques (low-energy ion scattering, X-ray photoelectron spectroscopy, electron paramagnetic resonance and temperature-programmed desorption), finding that under these reaction conditions partial reduction is taking place intensively on iron whereas the Mo coordinatively unsaturated sites (CUS) generate Lewis acidity which are the sites for methanol adsorption, but at the same time there is a synergy of the iron species to these CUS which increases when the Fe becomes partially reduced. Therefore, the rate-determining step, related to the rupture of the C–H bond, corresponds to a heterolytic break with the subtraction of a hydride species from methanol, which then would be located in the anionic vacancy close to the Fe and Mo cations (see Scheme 8.3).

8.3 Gas-Phase Oxidation of Ethanol

The gas-phase oxidation of ethanol, a reaction belonging to the class of oxidative dehydrogenations, is one of the oldest routes for the preparation of acetaldehyde, but is still used industrially although nowadays the majority of this compound is produced by means of the Wacker process of ethylene oxidative hydration.[1,2] However, the oxidation of ethanol may be a preferable choice when small capacities are needed. This process is exothermal (ΔH° −43 kcal mol^{-1}) and sometimes is also referred to as autothermal dehydrogenation, when the mechanism involves the combustion of molecular hydrogen formed by ethanol dehydrogenation, rather than a true abstraction of hydrogen from the reactant with water formation;

the latter mechanism involves a cycle of electron exchange, first between the substrate and the metal cation and then between the latter and O_2.

The nature of the mechanism is a function of the catalyst type and reaction conditions used; for instance, in the case of the Veba-Chemie process, which makes use of an Ag catalyst in the form of wire gauze or crystals, the mechanism is a dehydrogenation combined with hydrogen combustion. Depending on the reaction temperature (which can range from 500 to 650 °C, depending on the alcohol : air inlet ratio used and the type of reactant used, either air or oxygen), the per pass conversion can be between 30 and 70%, with a selectivity between 90 and 99%. The main by-product is acetic acid, but minor amounts of formic acid, ethyl acetate, CO and CO_2 may also form. Other catalysts claimed are based on Cu oxide.

In recent years, because of the increasing interest in the development of a more sustainable chemical industry, the transformation of bioethanol into chemicals has been attracting considerable interest.[22,23] Indeed, it has to be recognized that the value-adding chain based on biomass-based ethanol is not new, since many processes for the transformation of this platform molecule into chemicals were developed and extensively used since the mid-twentieth century, but were later abandoned because of the more favorable costs when the same chemicals started to be produced from oil. Even nowadays, the economics for the production of base chemicals from bioethanol are fairly uncertain and, in most cases, an economic assessment would discourage the use of this molecule for the synthesis of chemicals, at least in Europe. The case is different in regions such as South America, India or even North America, where the margins of profitability are greater and plants have already been started up to produce ethylene through ethanol dehydration, the olefin being then used for polyethylene or PVC production. Recently, Abengoa claimed a new process for the Guerbet condensation of ethanol into 1-butanol, a fuel additive.[24,25]

Nevertheless, over the last 15 years, the scientific community has shown increasing interest in reactions aimed at the valorization of ethanol.[26–31] Here, some important achievements in this field are outlined, with a specific focus on gas-phase oxidations, such as oxidation to acetaldehyde (and acetic acid), direct oxidation to ethene oxide, oxychlorination to 1,2-dichloroethane and ammoxidation to acetonitrile.

8.3.1 Oxidation of Ethanol to Acetaldehyde and Acetic Acid: Metallic Catalysts

Catalysts for the gas-phase oxidation of ethanol to acetaldehyde and acetic acid include noble metals, which usually catalyze the oxidation to acetic acid (with the exception of Au, the selectivity of which is, however, remarkably affected by the reaction conditions used) and other transition metal oxides (or mixed metal oxides), which are more selective towards acetaldehyde, such as supported V oxides, Mo oxides and polyoxometallates.

There is no doubt that in recent years gold-based catalysts have attracted the attention of researchers as catalysts for both liquid- and gas-phase oxidations.[26,32–42] Also in the case of alcohol oxidation, when used either in the form of unsupported nanoporous Au or as supported nanoparticles (NPs), these catalysts show peculiar reactivity properties. In general, high selectivity to acetaldehyde is shown, with remarkably low selectivity to CO_2, which is unexpected because of the generation of highly electrophilic O species on the Au surface when the NPs are supported over conventional materials, such as silica or titania. However, depending on the reaction conditions and catalyst type, the prevailing product can be either acetic acid (in this case, however, there is typically also greater selectivity to CO_2, because of the higher temperatures needed) or ethyl acetate (formed by consecutive dimerization of acetaldehyde). The behavior is different from that shown in liquid-phase oxidation with Au-based catalysts, which may provide fairly high selectivity to acetic acid and ethyl acetate (by esterification of acetic acid with ethanol) at moderate temperature and pressure (Scheme 8.4).[43,44]

For example, nanoporous Au, prepared by the dealloying of Au–Ag alloys, has a stable structure that is active without any support. It catalyzes the selective oxidative coupling of methanol to methyl formate with selectivity above 97% and high turnover frequency at temperatures below 80 °C,

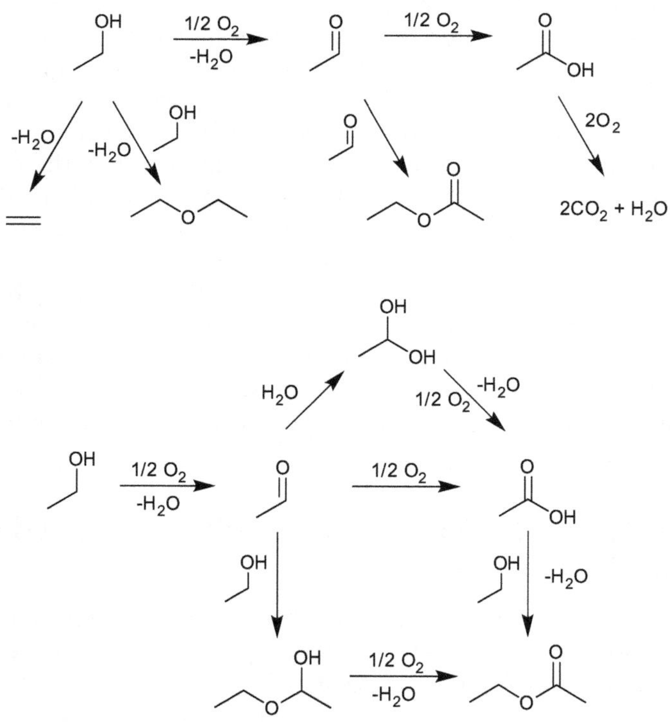

Scheme 8.4 Main reactions occurring during the gas-phase (top) and liquid-phase (bottom) oxidation of ethanol.

because of (i) its ability to activate O_2 at low temperature, (ii) the comparatively weak adsorption of the partial oxidation products shown, compared with Pt and Pd, which saves the product from consecutive overoxidation reactions,[45] and (iii) its low tendency to activate C–H bonds.[46] The same material (but with residual Ag in the dealloyed material) catalyzed the oxidation of ethanol to acetaldehyde or ethyl acetate (at 160 °C, with a selectivity of 48% and 32% ethanol conversion) or into acetic acid (at 240 °C, with selectivity 68% and conversion over 70%), as the prevailing compound, depending on the reaction temperature, O_2 : ethanol molar feed ratio and amount of residual Ag in the dealloyed material.[47] On the other hand, the role of Ag has also been demonstrated by Ding *et al.*,[48] who reported 82% ethanol conversion with 94% selectivity to acetaldehyde, using nanoporous Ag at 250 °C. Surprisingly, the catalyst showed high selectivity to aldehydes from various primary alcohols, except methanol, which gave methylformate as the prevailing compound. This was attributed to the fact that the β-H elimination of alkoxy species in methanol is more difficult than with higher alcohols; therefore, the methoxy species accumulated on the catalyst surface further react with the absorbed formaldehyde to yield the dimeric compound.

The nature of the support may have a profound influence on the reactivity of NPs; the support itself may also show catalytic properties, as in the case of systems described by Takei *et al.*[49,50] For example, a very high yield of acetaldehyde was shown using Au NPs supported over acidic MoO_3 (94%, with 5% ethylene at 240 °C, catalyst with 1% Au loading) or weakly basic La_2O_3. Oxidation to acetic acid took place principally over Au NPs deposited on n-type metal oxides such as ZnO (44% yield of acetaldehyde, 46% of acetic acid, 6% of ethyl acetate, at 220 °C) and V_2O_5, whereas complete oxidation was preferentially obtained over p-type semiconducting metal oxides such as MnO_2 and Co_3O_4. Even though many of these metal oxides proved active in ethanol oxidation, the presence of Au enhanced the activity, and the selectivity was also greatly affected by the type of metal oxide support. In the mechanism proposed by the authors, following the generation of the ethoxy species over either basic or acidic metal oxide or n-type semiconducting metal oxide supports, the former reacts with O^- – generated at the Au NP surface or at the interface between Au and the support – to produce acetaldehyde. The presence of an excessive concentration of activated O species on the surface of the p-type metal oxide support was the reason for the high catalytic activity shown and the combustion of the C_2 species, with prevailing formation of carbon oxides.

Evidence for the role of electrophilic, highly reactive O species were also reported by Sobolevet *et al.*,[51–53] who showed that TiO_2-supported Au NPs, approximately 2 nm in size, gave rise to 'double-peak' catalytic activity during the gas-phase oxidation of ethanol, which was not observed for NPs supported over either Al_2O_3 or SiO_2. The low-temperature peak, at 120 °C, was attributed to the formation of a specific active oxygen form on the Au/TiO_2 surface, generated under mild reaction conditions. The prevailing products were acetaldehyde and ethyl acetate, with ethanol conversion in the range 40–60% and a feed inlet with ethanol:oxygen composition 2 : 18 (mol%). This O species was postulated to be an O^- radical, similar to the α-O generated in Fe-ZSM5,[54–56] or an O_2^- superoxide ion; the latter were indeed generated on some thermally activated gold catalysts by oxygen adsorption at room temperature.[57,58] For example, Tsunoyama *et al.*[59] hypothesized the stabilization of a superoxo-like molecular oxygen species on small gold clusters, which facilitates hydrogen abstraction. Also, H_2 was hypothesized to be a co-factor in the generation of these active species, since ethanol dehydrogenation may occur at low temperatures on Au/TiO_2.[60,61] Furthermore, the TiO_2 matrix ensured the thermostability of the generated oxygen species over a broader range of temperatures compared with alumina or silica. At more elevated reaction temperatures, therefore, this O species did not contribute to catalytic behavior and it is possible that, indeed, under these conditions the mechanism involved ethanol dehydrogenation with hydrogen combustion. It is worth noting that even in the case of the gas-phase epoxidation of propylene with O_2–H_2 mixtures, the best catalysts are based on Au/TiO_2;[62] in this case, the hypothesized active species is an AuOOH hydroperoxo species, generated *in situ* by the reaction of oxygen and hydrogen on the Au NP surface. With a ceria support, Au catalyzed the selective oxidative dehydrogenation of ethanol to acetaldehyde at low temperature (200–300 °C), whereas at higher temperatures the selectivity switched towards acetone and finally methane.[63–67] The same group also carried out extensive research on the interaction of ethanol in an oxygen atmosphere with ceria-supported Pt, Pd, Rh and bimetallic alloys. It was found that the addition of any of these metals to CeO_2 suppressed the oxidation of ethanol to acetates, but facilitated ethanol dehydrogenation to acetaldehyde, which was the predominant product on non-reduced surfaces. However, Rh promoted the breaking of the C–C bond in ethanol and the formation of CO.[63,65]

The chemistry of the interaction of alcohols with coinage metals was studied by Mullins and co-workers.[68–73] Based on temperature-programmed desorption and molecular beam reactive scattering on an atomic oxygen-precovered Au(111) surface, the authors suggested that the abstraction of hydrogen is the initial step in the surface decomposition of methanol, with the bound methoxy group as the primary reaction intermediate in methanol oxidation. With higher alcohols, such as 1- and 2-propanol or 2-butanol, the surface-bound alkoxy species is selectively transformed into the corresponding aldehyde or ketone at temperatures lower than 80 °C, at low oxygen coverage. Oxygen atoms and the hydroxyls formed by preadsorption were found to be responsible for the distinct catalytic behavior of Au, participating in reactions as either a Brønsted base or a nucleophilic base.

Recently, using Au nanoparticles supported over $MgCuCr_2O_4$ spinel, a 95% yield of acetaldehyde was achieved at total ethanol conversion. The catalyst showed a stable performance over 500 h. This outstanding behavior was attributed to the synergy between metallic Au NPs and the surface Cu^+

species, which act as sites for O_2 activation. Oxygen clearly suppressed the formation of coke. In the mechanism proposed, the electrophilic O species facilitated O–H bond cleavage and metal–alcoholate formation, whereas Au NPs in close proximity acted as sites for C–H cleavage.[74]

The role of the size of NPs on catalytic behavior has also been addressed. Zheng and Stucky[75] showed that silica-supported Au NPs with a nominal size of 6 nm are more active than 3.5 and 8.2 nm particles in the gas-phase oxidation of ethanol. Similar results were also presented by Rossi and co-workers, at the World Congress on Oxidation Catalysis held in Berlin in 2001;[76] they showed that a high selectivity to the corresponding aldehyde could be obtained in the gas-phase oxidation of various aliphatic alcohols,[77] and the highest activity was observed with Au particles larger than 10 nm. Guan and Hensen[60] found a strong influence of the gold particle size on the non-oxidative dehydrogenation of ethanol, with nanoparticles of about 6 nm showing much higher activity than smaller or larger particles. This optimal catalytic activity was attributed to the existence of a maximum density of surface steps with a suitable geometry for the removal of β-H atoms from adsorbed ethoxide. In the presence of oxygen, the generation of adsorbed oxygen species accelerated the reaction rate and decreased selectivity because of the contribution of combustion. Notably, in this case the highest intrinsic activity was shown with particles larger than about 7 nm: an effect which was attributed to a higher density of strongly adsorbed oxygen adatoms, which are much scarcer on smaller gold particles.

Few papers have reported the reactivity of bimetallic particles. Guan and Hensen[78] reported a system of Au–Ir metallic NPs of size 2–3 nm (where Au retards the bulk oxidation and sintering of Ir in an oxidizing atmosphere at temperatures higher than 350 °C). The NPs displayed enhanced activity in ethanol oxidation to acetaldehyde, outperforming their monometallic counterparts, but with slightly lower selectivity to C_2 compounds (acetaldehyde with traces of acetic acid) compared with Au NPs (92% *versus* 99% for the latter). The model proposed involves intimate contact between Au sites for dissociative ethanol adsorption and Ir sites, covered by O adatoms, which catalyze C–H bond cleavage to yield acetaldehyde.

The catalytic and electrocatalytic activity of Pd alloyed with transition metals such as Cu for ethanol oxidation was enhanced compared with Pd alone.[79] In the gas-phase reaction, the nanoalloy Pd–Cu catalyst was shown to suppress the formation of acetic acid; the effect was due both to a surface enrichment of Pd in the alloy and to the oxophilicity of the base metal Cu alloyed with Pd, as demonstrated by a significant positive shift of the reduction potential of the oxygenated Pd species on the surface.

Table 8.1 lists some catalyst types based on Au and Ag which have been reported in the literature for ethanol gas-phase oxidation; in most cases, acetaldehyde was the main product of oxidation, but in some cases acetic acid was also produced in remarkable yields. In these cases, formation of the acid was shown at temperatures higher than 250 °C with catalysts which started to be active in ethanol conversion in the range 150–200 °C.[52]

Table 8.1 Summary of catalytic results for the gas-phase oxidation of ethanol over noble metal-based catalysts.

Catalyst	Feed composition, ethanol : O_2 (mol%)	Best acetaldehyde yield (%), T (°C)	Best acetic acid yield (%), T (°C)	Ref.
Ag–Fe/Si_3N_4	2 : 18	95 (300)	–	80, 81
5% Au/TiO_2	2 : 18	65 (130)	45 (290)	52
2% Au/Al_2O_3	2 : 18	62 (225)	43 (295)	52
6.7% Au/MCM41	1 : N.R.	~70 (350)	–	60
Au/$MgCuCr_2O_4$	1.5 : 4.5	95 (250)	–	74
$AuIr_3/SiO_2$	1 : 1	52 (180)	–	78
Au(Ag) nanoporous	29 : 29	40 (240), ethyl acetate 26 (240)	–	47
1% Au/MoO_3	0.7 : 2.3	94 (240)		50
11.3% Au/ZnO	0.7 : 2.3	44 (220)	46 (220)	50
Ag nanoporous	10 : 5	78 (250)	–	48
Au/CeO_2	7 : 11	56 (200)	Acetone 37 (500)	65

However, the nature of the support used and the type of interaction between the latter and ethanol was one of the main factors affecting the distribution of products and finally the acetic acid yield.[50]

8.3.2 Oxidation of Ethanol to Acetaldehyde and Acetic Acid: Metal Oxide and Mixed Oxide Catalysts

An overview of mixed metal oxides and supported metal oxides for the gas-phase oxidation of ethanol was given by Haruta and co-workers.[49] Here, we summarize the main features of the reaction and those catalyst characteristics that are generally required to obtain high selectivity to either acetaldehyde or acetic acid.

Several classes of metal oxides catalyze the oxidation of ethanol into acetaldehyde or acetic acid, such as V_2O_5–TiO_2, Mo–V–Nb–Te–O, P–Mo–V–O polyoxometallate, CeO_x–TiO_2, CuO–Fe_2O_3 and Co_3O_4–γ-Al_2O_3–cordierite, to list just a few published more recently.[82–89] An unconventional oxidic catalyst is based on graphite nanofibers, where oxygen groups terminating the prismatic edge sites of graphene planes were supposed to be responsible for the catalytic activity in oxidative dehydrogenation.[90]

A systematic study of several metal oxides, *i.e.*, CaO, SiO_2, TiO_2, Fe_2O_3 and Fe_3O_4, in the oxidative dehydrogenation of ethanol was carried out by Idriss and Seebauer.[64] The main reaction product in all cases was acetaldehyde, with acetone and ethyl acetate as secondary products; the ketone was considered to form *via* acetate ketonization, while ethyl acetate (formed with selectivity as high as 35% at low ethanol conversion over Fe_2O_3), was produced *via* the Tishchenko reaction. The number and strength of basic sites were found to be the catalyst features that most affected the catalytic behavior.

Outstanding behavior was reported for catalysts based on Mo–V mixed oxides, where the incorporation of either Nb or Te or both (to produce the so-called M1 phase, active and selective in propane oxidation and ammoxidation to acrylic acid and acrylonitrile, respectively), has a profound effect on catalytic performance. As reported by Sobolev and Koltunov,[85,91] the Mo–V–Nb–Te four-component catalyst proved to be very efficient in the synthesis of acetic acid, with no need for co-fed water; an acetic acid yield higher than 95% was observed at 250 °C, whereas at lower temperatures and partial ethanol conversion the prevailing product was acetaldehyde. The ethanol content in the feed could be increased up to 20% and the productivity of acetic acid and ethyl acetate remained above 3 g g_{cat}^{-1} h^{-1}. The ratio between acetic acid and ethyl acetate yields was greatly affected by the spatial velocity used; under certain conditions, at 200 °C, similar yields to ethyl acetate and acetic acid (~42% each) were found. The three-component mixed oxide, containing Mo, V and Nb, prepared by precipitation in the presence of colloidal TiO_2, was studied by Li and Iglesia;[84] a yield of acetic acid as high as 95% was achieved, where titania favored the dispersion of the active component without negatively affecting the catalytic behavior. In this case, water was shown to increase the selectivity by inhibiting acetaldehyde synthesis more than its conversion to acetic acid, thus minimizing the undesired combustion of acetaldehyde.

Other catalysts based on molybdenum oxide were studied (less recently) by Appel and colleagues.[92–96] In SnO_2-supported MoO_3 catalysts, the active site was the four-coordinate isolated molybdate species, formed by a reaction with basic hydroxyl species and probably also with Lewis acid sites. The adsorption of ethanol to the molybdenyl unit generated an alkoxide species, which was then selectively oxidized to acetic acid. Under the best reaction conditions, an acetic acid yield close to 45% was achieved at 275 °C; at lower temperatures, acetaldehyde was the prevailing product. The addition of water to the feed gas decreased the ethanol conversion and significantly increased the selectivity to acetic acid; in fact, water promoted the desorption of ethanol and carboxylates, present as bridging and monodentate species, because of active site blocking by hydroxyl groups. Doping with Ce increased both activity and selectivity to acetic acid and acetaldehyde.

Another class of catalysts showing excellent performance in the direct oxidation of ethanol to acetic acid is based on titania-supported V_2O_5.[82,83,91] For example, using 50 wt% aqueous ethanol, acetaldehyde was the prevailing product at 175–200 °C and 2.7 bar pressure, with yields above 90%. In contrast, acetic acid was the predominant product at low space velocity; a selectivity above 80%, even at temperatures as low as 165 °C, could be reached.[83] In general, however, for fixed reaction conditions, acetaldehyde was the preferred compound at low temperatures, whereas acetic acid formed more readily at high temperatures; a yield of the latter compound as high as 75%, at 200 °C, was achieved even in the absence of co-fed water.[91]

In conclusion, the results reported in the literature highlight that catalysts based either on molybdenum or vanadium oxide show the greatest selectivity to acetic acid, in the presence of co-fed steam, especially if temperatures exceed 200–250 °C. Conversely, lower temperatures lead to the preferential formation of acetaldehyde. Acidic features of the catalyst or of the support over which the active phase is deposited are a requisite to achieve high selectivity to the acid, since desorption into the gas phase is facilitated, thus saving the product from consecutive oxidative degradations.

8.3.3 Direct Ammoxidation, Oxychlorination and Epoxidation of Ethanol

Ethanol has also been used as a reactant for the synthesis of chemicals which are typically produced starting from other molecules, such as (a) acetonitrile, by means of ethanol ammoxidation, whereas nitrile is typically obtained as a by-product of propylene ammoxidation, (b) 1,2-dichloroethane, by means of ethanol oxychlorination, whereas it is typically produced through ethylene oxychlorination, and (c) ethylene oxide, typically being produced by olefin epoxidation. In this perspective, the mechanism involved in each of the above reactions can be considered sequential, since two in-series reactions occur over a single, but bifunctional, catalyst.

Ethanol ammoxidation can be carried out with the same reactor and process by which propylene is ammoxidized to acrylonitrile;[97,98] in fact, acetonitrile is a by-product of the acrylonitrile process and conditions aimed at the ammoxidation of ethanol are not much different from those necessary for propylene ammoxidation. However, some authors have investigated the ammoxidation of primary alcohols to nitriles using dedicated catalysts and conditions. Table 8.2 summarizes the results reported in the literature. Outstanding results were obtained with crystalline silicon aluminium phosphate (SAPO) and vanadium aluminium phosphate (VAPO).

In the aerobic ammoxidation of alcohols to nitriles, the activation of the strong N–H bond in ammonia (107 kcal mol^{-1}) require severe reaction conditions. One key reaction intermediate, as reported in the literature, is the hemiaminal compound [1-aminoethanol, $CH_3CH(OH)NH_2$, in the case of ethanol ammoxidation], obtained by addition of ammonia to the carbonyl moiety in acetaldehyde; the hemiaminal may be the precursor of either acetaldehyde imine (ethanimine, *via* dehydration) or acetamide (*via* dehydrogenation or oxidative dehydrogenation). In this regard, it is claimed

Table 8.2 Summary of catalysts and conditions for the gas-phase ammoxidation of ethanol to acetonitrile.

Catalyst	*T* (°C)	Ethanol : H_2O : NH_3 : O_2	Ethanol conversion, acetonitrile yield (%)	By-products	Ref.
V/P/Sb/O–Al_2O_3	400	Ethanol : water 1 : 2; NH_3 : air 2.1 : 1	84, 82^a	Acetaldehyde	99
SAPO	350	1 : 1 : 5 : air	100, 99^b	–	100
VAPO	350	1 : 1 : 5 : air	100, 96.5^c	Acetaldehyde	101

^{a}The authors hypothesize a mechanism *via* acetic acid and acetamide.
^{b}The authors hypothesize a mechanism *via* either aldehyde and imine or ethylamine.
^{c}The authors hypothesize a mechanism *via* aldehyde and imine.

that the presence of an acid functionality in the catalyst may be essential to accelerate the dehydration of 1-aminoethanol during the non-reductive amination of ethanol.[102]

The conversion of ethanol to 1,2-dichloroethane, which is an intermediate for the synthesis of vinyl chloride monomer, formally consists of two steps: (i) ethanol dehydration to ethylene and (ii) oxychlorination of the olefin to the dichlorinated compound. Once again, in principle, the two reactions can be carried out in sequence over a single bifunctional catalyst, with a catalyst showing both acid features and redox properties.[103,104] On the other hand, a two-step reaction with two catalytic beds in sequence, each one aimed at fostering one specific reaction, can also be envisaged. This approach has been described, for instance, in the case of glycerol oxidehydration to acrylic acid (see below). This type of process configuration, which can be referred to the intensification rules as listed in the green chemistry principles, is possible only when the co-product and by-products generated in the first section do not interfere with the catalyst and reactions occurring in the second catalytic bed. This avoids expensive separation procedures aimed at the purification of the product which should serve as the reactant for the second step carried out over the downstream catalytic section. On the other hand, if all the reactants needed for reactions occurring over both catalytic beds are co-fed all together at the top of the first bed, reactants for the second section should not interfere with reactions occurring in the first section. If instead this is the case, a differentiated feed, with reactants separation at the inlet of the two beds, is alternatively recommended (Scheme 8.5).

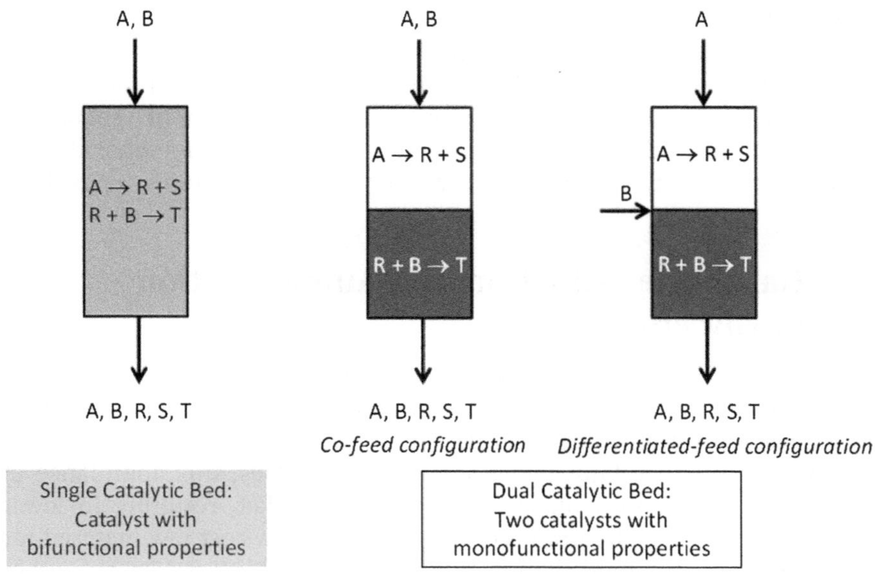

Scheme 8.5 Possible reactor configuration for sequential reactions, including two steps: (a) A → R + S; and (b) R + B → T.

Recently, a two-bed configuration was suggested for ethanol conversion to 1,2-dichloroethane: the first bed consists of a ZSM5 zeolite and the second bed of a conventional oxychlorination catalyst made of Na-doped $CuCl_2$–γ-Al_2O_3 catalyst.[105] Under the optimized reaction conditions, at 300 °C and with a feed containing 8% ethanol, 18% HCl and 31% air, an 82% yield of dichloroethane was obtained. However, the mechanism proposed by the authors was not one that would be readily envisaged; in fact, it was demonstrated that the main product generated in the zeolitic bed was ethyl chloride, produced by ethanol hydrochlorination, followed by the disproportionation of two ethyl chloride molecules into ethylene and dichloroethane. Conversely, ethanol dehydration followed by ethylene oxychlorination played a minor role in the temperature range studied.

The direct transformation of ethanol into ethylene oxide was studied by Lippits and Nieuwenhuys, using Al_2O_3-supported catalysts based on metallic Au, Ag or Cu doped with Li_2O.[106,107] Lithium affected the selectivity by suppressing the formation of diethyl ether and ethylene, two compounds formed mainly over the acidic support. A selectivity to ethylene oxide as high as 50%, with ethanol conversion close to 80% at 300 °C, was obtained with the Au–Li_2O/Al_2O_3 catalyst in the absence of oxygen; the product, however, was observed only during the first heating cycle, because the deposition of coke poisoned the active sites responsible for ethylene oxide formation. In the presence of co-fed oxygen, with an equimolar content in the feed of ethanol and O_2, the same catalyst showed an almost 90% selectivity to ethylene oxide at 200 °C, with 80% ethanol conversion. Higher temperatures favored the formation of ethylene, diethyl ether and CO_2; oxygen also prevented the formation of coke on the catalysts. Similar behavior was shown by Li_2O-doped Ag NPs, supported over alumina and also by supported Cu NPs; in this case, however, the presence of Li did not affect the catalytic behavior. The authors did not discuss the possible mechanism for ethylene oxide formation; however, the results shown were not in favor of a sequential mechanism with ethylene as the reaction intermediate for epoxide formation, but rather the direct transformation of ethanol into the epoxide.

8.4 Gas-Phase Oxidation and Ammoxidation of Glycerol

The growing trend of biodiesel production shown over recent decades in Europe has led to the availability of large volumes of glycerol, produced *via* the transesterification of triglycerides, in addition to the traditional direct hydrolysis to produce fatty acids. Since the late 1990s, this abundance of glycerol has significantly affected the glycerin market, resulting in lower prices, while making glycerol become a desirable, low-cost raw material for conversion into different chemicals and biofuel additives. On the other hand, from 2007, the glycerin market has faced very volatile conditions, recent price increases and more recently a supply shortage of refined glycerol. This makes any accurate economic evaluation of chemical processes designed for its transformation rather difficult. Nevertheless, the transformation of glycerol into more added-value products still remains a key issue for developing a new bio-based chemical industry.[108–111]

Of the several options focused on glycerol conversion, most involve liquid-phase transformations, but some gas-phase oxidative approaches have also been attracting the interest of academic researchers and industry, *i.e.*, oxidation to acrylic acid[112–121] and ammoxidation to acrylonitrile.[122–124] Concerning both reactions, processes proposed include either the one-pot approach (with a single but bifunctional catalyst) or two in-series catalytic beds. In fact, both reactions include a first step of acid-catalyzed dehydration of glycerol to acrolein and a second step in which acrolein is either oxidized to acrylic acid or ammoxidised into acrylonitrile. Another option recently considered is the ammoxidation of an alternative alcohol also derived from glycerol, allyl alcohol.[124]

As regards the one-pot oxidehydration of acrolein to acrylic acid, the catalysts with the best performance are those based on hexagonal-structure tungsten oxide bronzes (HTBs) doped with V and Nb (Table 8.3). The key property of the catalyst is that of fostering a fast consecutive oxidation step, in order to avoid the formation of undesired heavy compounds generated by both the ketal formation and the oligomerization reactions that take place from the intermediately formed unsaturated aldehyde. The W–V–Nb mixed oxide is characterized by a relatively large surface area and high concentration of stronger acid sites: two properties which are also important for an efficient dehydration of glycerol to acrolein. Compared with the bicomponent W–V HTBs, the presence of Nb leads to a remarkable increase in surface acidity, and also to the development of a mixed oxide characterized by a larger and more stable surface area.[112] On the other hand, the presence of V confers on the catalyst the redox properties needed for the efficient oxidation of acrolein to acrylic acid.

Excellent performance was also obtained with a two-bed reactor configuration, where the first catalyst showed acidic features (for example, zeolite ZSM5) and the second catalyst oxidizing properties for acrolein transformation, such as a V–Mo mixed oxide.[119,120]

A detailed study of the reaction mechanism occurring in the oxidation of glycerol to acrylic acid was carried out by Ueda and co-workers[116] (Scheme 8.6). The first loss of water from glycerol leads to

Table 8.3 Summary of catalytic behavior for the oxidehydration of glycerol to acrylic acid.

Catalyst	*T* (°C)	Inlet feed: glycerol: oxygen : water (mol%)	Yield of acrolein, acrylic acid (%)	Ref.
W–V–O	318	2 : 4 : 54	29, 18	114
W–V–Nb–O	265	6 : 12 : 40	3.5, 50.5	113
Mo–V–Nb–Te–O	N.R.	3 : 6 : 19	1.7, 28.4	116
Fe–V–O/Fe–O	300	1.5 : 1.7 : 30.3	28, 14	117

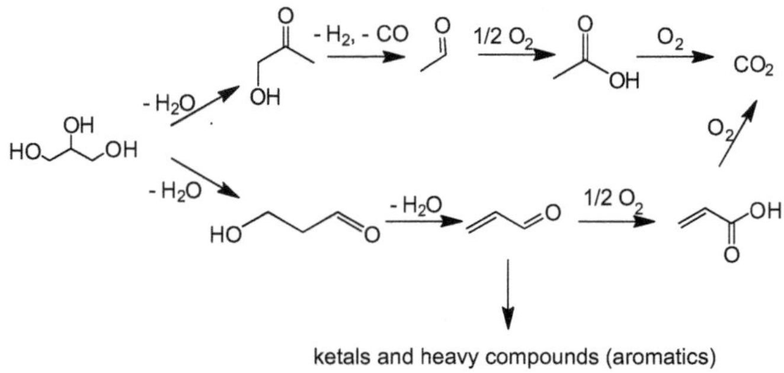

Scheme 8.6 Main reactions occurring in glycerol oxidehydration to acrylic acid.

3-hydroxypropanal and hydroxyacetone; the latter can be further converted into acetaldehyde and oxidized to acetic acid, whereas the former dehydrates to acrolein. However, through a retroaldol reaction, 3-hydroxypropanal can lead to acetaldehyde and formaldehyde. Furthermore, aldehydes can react with unconverted glycerol to form cyclic ether products. In fact, one key issue in the aim of limiting the formation of by-products is to operate under conditions at which no residual glycerol originates along the catalytic bed.[113]

Regarding gas-phase glycerol ammoxidation, Guerrero-Pérez and Bañares[122] reported that, by using a V–Nb–Sb mixed oxide with the rutile structure, it was possible to convert glycerol directly into acrylonitrile, with 58.3% selectivity at 82.6% conversion. An alternative two-bed approach was proposed by Dumeignil and colleagues;[109,123] here, glycerol is first dehydrated to acrolein on to a WO_3/TiO_2 catalyst, while in the subsequent bed, acrolein is ammoxidized with a V–Fe–Sb–O catalyst. The same group also proposed an innovative process from allyl alcohol, which can also be obtained from glycerol, by using a rutile/iron antimonate catalyst; under optimized reaction conditions, a yield of acrylonitrile as high as 83% was obtained.[124]

8.5 Gas-Phase Photocatalytic Oxidation of Alcohols

In recent years, gas-phase photocatalytic oxidation (PCO) has gained importance as a green technology in obtaining partial oxidation products of alcohols under mild conditions. This kind of process uses semiconductor materials as photocatalysts, the most important and widely applied being titanium dioxide (TiO_2) owing to its high photoactivity, low cost, low toxicity and good chemical and thermal stability. It is, however, a UV light absorber, but recently important advances have been achieved, making it possible to activate it with visible light.[125]

The use of photocatalysis for the abatement of pollutants (including alcohols) by complete oxidation has been the subject of many research studies and reviews. For instance, the study of the complete oxidation of ethanol using SiO_2–TiO_2 composites,[126] $UO_2(NO_3)_2$–TiO_2,[127] CdS/TiO_2[128] and

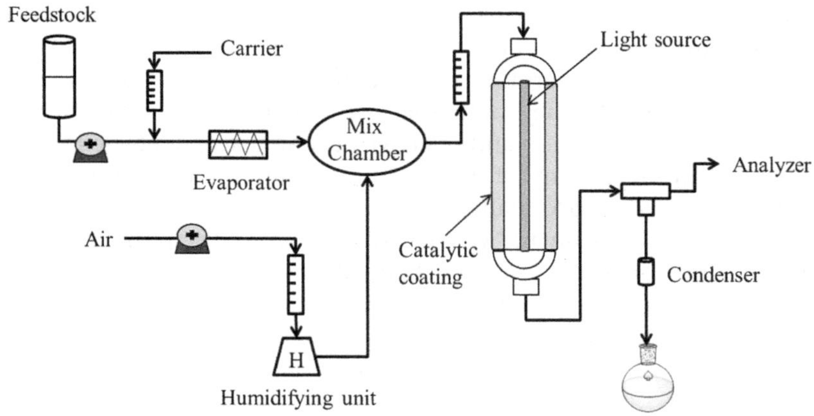

Scheme 8.7 Typical reaction system for gas-phase photooxidation of alcohols.

so on has been described. However, the selective photo-oxidation of alcohols would be much more interesting from the environmental and economic points of view.[129] This kind of process has been reported in the liquid phase; however, to perform it in the gas phase could bring several advantages, such as easier recovery of the product, lower levels of light energy required (photons absorbed by air are low), higher decomposition rates (better diffusion of reagents and products) and also the fact that neither OH radical scavengers present in the aqueous phase (such as chlorides) nor alkalinity interfere.[130] In this regard, very few studies can be found in the literature. For instance, Pillai and Sahle-Demessie[131] reported the gas-phase oxidation of higher alcohols on a TiO_2 surface (1-pentanol, 2-methylbutan-1-ol, 3-methylbutan-1-ol, 1-hexanol, cyclopentanol, 3-methyl-3-buten-1-ol, cyclohexanol, phenylmethanol, 1-phenylethanol and 2-phenylethanol). This study concluded that this process may be suitable for oxidizing primary and secondary aliphatic alcohols to their corresponding carbonyl products. Conversely, aromatic alcohols showed a preference to form secondary reaction products (such as styrene for 1-phenylethanol and benzaldehyde for 2-phenylethanol). Another example is the study by Lopez-Tenllado *et al.*,[132] who used several catalysts for the selective photo-oxidation of 2-propanol to acetone. After testing 25 different solids ($M–TiO_2$ where M is a transition metal) they concluded that Fe, Pd and Zn have a negative effect on the activity whereas Zr and especially Au improve the results compared with pure titania, while even some of those gold-containing systems displayed better activity than Degussa P25 (a commercial catalyst). An example of this kind of reaction system is shown in Scheme 8.7.

8.5.1 Photocatalytic Synthesis of Acetaldehyde from Ethanol

In the attempt to produce acetaldehyde in a 'greener way', some studies concerning the aerobic photocatalytic oxidation of ethanol have been reported.

Antoniadou *et al.*[133] studied the photocatalytic oxidation of ethanol over Ru-doped TiO_2 in three different ways: oxidation in the gas phase (100 °C, ethanol:oxygen = 1:2), reforming in the aqueous phase and as a fuel in a photoactivated fuel cell. They found that in the first case (which is the one of interest here), the conversion and selectivity towards acetaldehyde increased with increase in the amount of dopant but only up to a certain point, after which it started to decrease again (reaching a maximum of 97% selectivity at 55% of conversion over RuO_2/TiO_2 loaded at 0.3 wt%). From this trend, they concluded that the overall photoactivity depends on the availability of adsorption sites on the catalyst, *i.e.* the TiO_2 not covered by the RuO_x, while the Ru species play an essential role in addressing the selectivity to acetaldehyde, thus making it necessary to find an adequate balance in composition. In other studies performed by the same research team, different catalysts were tested including Au/TiO_2[134] (82% conversion and 95% selectivity at 60 °C), Pt/TiO_2[135] (85% conversion and 90% selectivity at 60 °C) and VO_x/TiO_2[136] (66% conversion and 99% selectivity at 60 °C) in the presence of blue phosphors, but, up to now, the best results they obtained were with RuO_x–VO_x/TiO_2[137] (100% conversion and selectivity at 100 °C).

8.6 Conclusion

The gas-phase oxidation of alcohols – a reaction known and industrially applied for many years – is still the subject of intense investigation, with the aim of finding more active, selective and stable catalysts both for industrial methanol oxidative dehydrogenation and higher alcohol oxidation, especially in the case of ethanol, which is now the subject of renewed interest as a bioplatform molecule. The literature on this topic is mainly focused on the catalyst properties needed to obtain high selectivity either to acetaldehyde or to acetic acid, with interest also in the synthesis of dimerization molecules such as ethyl acetate or condensation compounds. Also in this field, as in many others, the majority of studies reported in the literature have been focused on catalysts based on supported Au nanoparticles. New products based on biomolecules, new technology for old reactions and improved catalysts for already established processes are all fields with scientific, technical and economic opportunities included in the vision of a more sustainable chemical industry based on biomass as feedstock.

References

1. P. Arpentinier, F. Cavani and F. Trifirò, *The Technology of Catalytic Oxidations*, Editions Technip, Paris, 2001.
2. G. Centi, F. Cavani and F. Trifirò, *Selective Oxidation by Heterogeneous Catalysis*, Kluwer Academic/Plenum, New York, 2001.
3. A. P. V. Soares, M. F. Portela and A. Kiennemann, *Catal. Rev.*, 2005, 47, 125.

4. U. Duerr, Famax catalysts for formaldehyde synthesis, presented at the Defining the Future VI Conference, Munich, 2013.
5. A. Nagy and G. Mestl, *Appl. Catal. A: Gen.*, 1999, **188**, 337.
6. M. Qian, M. Liauw and G. Emig, *Appl. Catal. A: Gen.*, 2003, **238**, 211.
7. I. E. Wachs and L. E. Briand, Lehigh University, *US Pat.*, 6 624 332, 2003.
8. A. Alshehri, *Methanol Oxidation on Transition Element Oxides*, PhD Thesis, Cardiff University, 2013.
9. K. Routray, W. Zhou, C. J. Kiely and I. E. Wachs, *ACS Catal.*, 2011, **1**, 54.
10. X. Wang and I. E. Wachs, *Catal. Today*, 2004, **96**, 211.
11. L. Burcham, G. Deo, X. Gao and I. Wachs, *Top. Catal.*, 2000, **12**, 85.
12. I. Wachs and L. Briand, *US Pat.*, 7 193 117, 2007; assignee: Lehigh University.
13. K. Routray, W. Zhou, C. J. Kiely and I. E. Wachs, *ACS Catal.*, 2011, **1**, 54.
14. A. Zipp, A. Drochner, H.-J. Eberle, M. Schumann and H. Vogel, Wacker Chemie, *Int. Pat.*, WO 2013041501 A1.
15. J. Gornay, X. Secordel, M. Capron, G. Tesquet, P. Fongarland, E. Payen, J. L. Dubois and F. Dumeignil, *Oil Gas Sci. Technol. – Rev. IFP Energies Nouv.*, 2010, **65**, 751.
16. H. Liu and E. Iglesia, *J. Phys. Chem. B*, 2003, **107**, 10840.
17. H. Liu and E. Iglesia, *J. Phys. Chem. B*, 2005, **109**, 2155.
18. Y. Iwasawa, T. Shido, N. Tamura and Y. Yuan, *Int. Pat.*, WO 2001021566 A1, 2001; assignee: Japan Science and Technology Corporation.
19. X. Lu, Z. Qin, M. Dong, H. Zhu, G. Wang, Y. Zhao, W. Fan and J. Wang, *Fuel*, 2011, **90**, 1335.
20. K. Thavornprasert, M. Capron, L. Jalowiecki-Duhamel, O. Gardoll, M. Trentesaux, A.-S. Mamede, G. Fang, J. Faye, N. Touati, H. Vezin, J.-L. Dubois, J.-L. Couturier and F. Dumeignil, *Appl. Catal. B: Environ.*, 2014, **145**, 126.
21. J. Gornay, X. Sécordel, G. Tesquet, B. de Ménorval, S. Cristol, P. Fongarland, M. Capron, L. Duhamel, E. Payen, J.-L. Dubois and F. Dumeignil, *Green Chem.*, 2010, **12**, 1722.
22. J. A. Posada, A. D. Patel, A. Roes, K. Blok, A. P. C. Faaij and M. K. Patel, *Bioresour. Technol.*, 2013, **135**, 490.
23. J. Rass-Hansen, H. Falsig, B. Jørgensen and C. H. Christensen, *J. Chem. Technol. Biotechnol.*, 2007, **82**, 329.
24. R. Arjona Antolin, J. L. Sanz Yague, A. Corma Canos and M. E. Domine, Abengoa, Bioenergia Nuevas Tecnologias, *Eur. Pat.*, 2 679 303 A1, 2013.
25. R. Arjona Antolin, J. L. Sanz Yague, A. Corma Canos, M. E. Domine, F. Vidal Barrero and F. A. Ladron de Guevara Vidal, Abengoa, Bioenergia Nuevas Tecnologias, *Eur. Pat.*, 2 679 304 A1, 2013.
26. A. Corma and H. Garcia, *Chem. Soc. Rev.*, 2008, **37**, 2096.
27. S. E. Davis, M. S. Ide and R. J. Davis, *Green Chem.*, 2013, **15**, 17.
28. Y. Mikami, A. Dhakshinamoorthy, M. Alvar and H. Garcia, *Catal. Sci. Technol.*, 2013, **3**, 58.

29. C. Parmeggiani and F. Cardona, *Green Chem.*, 2012, **14**, 547.
30. T. Punniyamurthy, S. Velusamy and J. Iqbal, *Chem. Rev.*, 2005, **105**, 2329.
31. C. Vinod, K. Wilson and A. Lee, *J. Chem. Technol. Biotechnol.*, 2011, **86**, 161.
32. M. Stratakis and H. Garcia, *Chem. Rev.*, 2012, **112**, 4469.
33. T. Ishida and M. Haruta, *Angew. Chem. Int. Ed. Engl.*, 2007, **46**, 7154.
34. N. Dimitratos, J. A. Lopez-Sanchez and G. J. Hutchings, *Chem. Sci.*, 2012, **3**, 20.
35. A. S. K. Hashmi and G. J. Hutchings, *Catal. Sci. Technol.*, 2013, **3**, 2861.
36. G. J. Hutchings and J. K. Edwards, *Front. Nanosci.*, 2012, **3**, 249–293.
37. G. J. Hutchings and C. J. Kiely, *Acc. Chem. Res.*, 2013, **46**, 1759.
38. C. Della Pina, E. Falletta and M. Rossi, *Chem. Soc. Rev.*, 2012, **41**, 350.
39. C. Della Pina, E. Falletta and M. Rossi, in *Liquid Phase Oxidation via Heterogeneous Catalysis: Organic Synthesis and Industrial Applications*, ed. M. G. Clerici and O. Kholdeeva, Wiley-VCH, Weinheim, 2013, pp. 221–262.
40. C. Della Pina, E. Falletta, L. Prati and M. Rossi, *Chem. Soc. Rev.*, 2008, **37**, 2077.
41. C. Della Pina, E. Falletta and M. Rossi, in *Modern Gold Catalyzed Synthesis*, ed. S. K. Hashmi and F. D. Toste, Wiley-VCH, Wenheim, 2012, pp. 309–329.
42. C. Della Pina, E. Falletta and M. Rossi, in *Nanoparticles and Catalysis*, ed. D. Astruc, Wiley-VCH, Wenheim, 2008, pp. 427–455.
43. C. H. Christensen, B. Jørgensen, J. Rass-Hansen, K. Egeblad, R. Madsen, S. K. Klitgaard, S. M. Hansen, M. R. Hansen, H. C. Andersen and A. Riisager, *Angew. Chem. Int. Ed. Engl.*, 2006, **45**, 4648.
44. B. Jørgensen, S. E. Christiansen, M. L. D. Thomsen and C. H. Christensen, *J. Catal.*, 2007, **251**, 332.
45. A. Ciftci, D. A. J. M. Ligthart, P. Pastorino and E. J. M. Hensen, *Appl. Catal. B: Environ.*, 2013, **130–131**, 325.
46. A. Wittstock, V. Zielasek, J. Biener, C. M. Friend and M. Bäumer, *Science*, 2010, **327**, 319.
47. X.-M. Kong and L.-L. Shen, *Catal. Commun.*, 2012, **24**, 34.
48. Z. Li, J. Xu, X. Gu, K. Wang, W. Wang, X. Zhang, Z. Zhang and Y. Ding, *ChemCatChem*, 2013, **5**, 1705.
49. T. Takei, N. Iguchi and M. Haruta, *Catal. Surv. Asia*, 2011, **15**, 80.
50. T. Takei, N. Iguchi and M. Haruta, *New J. Chem.*, 2011, **35**, 2227.
51. V. I. Sobolev, O. A. Simakova and K. Y. Koltunov, *ChemCatChem*, 2011, **3**, 1422.
52. V. I. Sobolev, K. Y. Koltunov, O. A. Simakova, A.-R. Leino and D. Y. Murzin, *Appl. Catal. A: Gen.*, 2012, **433–434**, 88.
53. O. A. Simakova, V. I. Sobolev, K. Y. Koltunov, B. Campo, A.-R. Leino, K. Kordás and D. Y. Murzin, *ChemCatChem*, 2010, **2**, 1535.
54. G. I. Pannov, V. I. Sobolev and A. S. Kharitonov, *J. Mol. Catal.*, 1990, **61**, 85.

55. G. I. Panov, A. K. Uriarte, M. A. Rodkin and V. I. Sobolev, *Catal. Today*, 1998, **41**, 365.
56. M. V. Parfenov, E. V. Starokon, S. V. Semikolenov and G. I. Panov, *J. Catal.*, 2009, **263**, 173.
57. H. Liu, A. I. Kozlov, A. P. Kozlova, T. Shido, K. Asakura and Y. Iwasawa, *J. Catal.*, 1999, **185**, 252.
58. Z. Hao, L. Fen, G. Lu, J. Liu, L. An and H. Wang, *Appl. Catal. A: Gen.*, 2001, **213**, 173.
59. H. Tsunoyama, H. Sakurai, Y. Negishi and T. Tsukuda, *J. Am. Chem. Soc.*, 2005, **127**, 9374.
60. Y. Guan and E. J. M. Hensen, *Appl. Catal. A: Gen.*, 2009, **361**, 49.
61. W. Fang, Q. Zhang, J. Chen, W. Deng and Y. Wang, *Chem. Commun.*, 2010, **46**, 1547.
62. F. Cavani and A. M. Gaffney, in *Sustainable Industrial Chemistry*, ed. F. Cavani, G. Centi, S. Perathoner and F. Trifirò, Wiley-VCH, Weinheim, 2009, pp. 319–366.
63. P.-Y. Sheng, G. A. Bowmaker and H. Idriss, *Appl. Catal. A: Gen.*, 2004, **261**, 171.
64. H. Idriss and E. G. Seebauer, *J. Mol. Catal. A: Chem.*, 2000, **152**, 201.
65. H. Idriss, *Platinum Met. Rev.*, 2004, **48**, 105.
66. A. Yee, S. J. Morrison and H. Idriss, *J. Catal.*, 1999, **186**, 279.
67. A. Yee, S. J. Morrison and H. Idriss, *Catal. Today*, 2000, **63**, 327.
68. J. Gong, D. W. Flaherty, R. A. Ojifinni, J. M. White and C. B. Mullins, *J. Phys. Chem. C*, 2008, **112**, 5501.
69. J. Gong, D. W. Flaherty, T. Yan and C. B. Mullins, *ChemPhysChem*, 2008, **9**, 2461.
70. J. Gong and C. B. Mullins, *Acc. Chem. Res.*, 2009, **42**, 1063.
71. J. Gong, T. Yan and C. B. Mullins, *Chem. Commun.*, 2009, 761.
72. R. a Ojifinni, N. S. Froemming, J. Gong, M. Pan, T. S. Kim, J. M. White, G. Henkelman and C. B. Mullins, *J. Am. Chem. Soc.*, 2008, **130**, 6801.
73. T. Yan, J. Gong and C. B. Mullins, *J. Am. Chem. Soc.*, 2009, **131**, 16189.
74. P. Liu and E. J. M. Hensen, *J. Am. Chem. Soc.*, 2013, **135**, 14032.
75. N. Zheng and G. D. Stucky, *J. Am. Chem. Soc.*, 2006, **128**, 14278.
76. S. Biella, L. Prati and M. Rossi, in *IV World Congress on Oxidation Catalysis*, Berlin, 2001, p. 371.
77. S. Biella and M. Rossi, *Chem. Commun.*, 2003, **9**, 378.
78. Y. Guan and E. J. M. Hensen, *J. Catal.*, 2013, **305**, 135.
79. J. Yin, S. Shan, M. S. Ng, L. Yang, D. Mott, W. Fang, N. Kang, J. Luo and C.-J. Zhong, *Langmuir*, 2013, **29**, 9249.
80. O. V. Vodyankina, A. S. Blokhina, I. A. Kurzina, V. I. Sobolev, K. Y. Koltunov, L. N. Chukhlomina and E. S. Dvilis, *Catal. Today*, 2013, **203**, 127.
81. A. S. Blokhina, I. A. Kurzina, V. I. Sobolev, K. Y. Koltunov, G. V. Mamontov and O. V. Vodyankina, *Kinet. Catal.*, 2012, **53**, 477.
82. M. Gubelmann-Bonneau, *US Pat.*, 5 840 971, 1998; assignee: Rhodia.
83. B. Jørgensen, S. B. Kristensen, A. J. Kunov-Kruse, R. Fehrmann, C. H. Christensen and A. Riisager, *Top. Catal.*, 2009, **52**, 253.

84. X. Li and E. Iglesia, *Chem. Eur. J.*, 2007, **13**, 9324.
85. V. I. Sobolev and K. Y. Koltunov, *ChemCatChem*, 2011, **3**, 1143.
86. A. A. Spojakina, N. G. Kostova, B. Sow, M. W. Stamenova and K. Jiratova, *Catal. Today*, 2001, **65**, 315.
87. G. Litt and C. Almquist, *Appl. Catal. B: Environ.*, 2009, **90**, 10.
88. Y. Eguchi, D. Abe and H. Yoshitake, *Micropor. Mesopor. Mater.*, 2008, **116**, 44.
89. C. P. Rodrigues, V. T. da Silva and M. Schmal, *Appl. Catal. B: Environ.*, 2010, **96**, 1.
90. R. D. Weinstein, A. R. Ferens, R. J. Orange and P. Lemaire, *Carbon N. Y.*, 2011, **49**, 701.
91. K. Y. Koltunov and V. I. Sobolev, *Catal. Ind.*, 2012, **4**, 247.
92. F. Gonçalves, P. R. S. Medeiros, J. G. Eon and L. G. Appel, *Appl. Catal. A: Gen.*, 2000, **193**, 195.
93. M. Daturi and L. G. Appel, *J. Catal.*, 2002, **209**, 427.
94. F. M. Gonçalves, P. R. S. Medeiros and L. G. Appel, *Appl. Catal. A: Gen.*, 2001, **208**, 265.
95. P. R. S. Medeiros and L. G. Appel, *Appl. Catal. A: Gen.*, 2002, **231**, 125.
96. P. R. S. Medeiros, J. G. Eon and L. G. Appel, *Catal. Lett.*, 2000, **69**, 79.
97. S. P. Godbole, M. J. Seely and D. Suresh, *US Pat.*, 6 204 407, 2001; assignee: Standard Oil.
98. J. L. Callahan and J. A. Stamm, *US Pat.*, 3 106 574, 1963; assignee: Standard Oil.
99. B. M. Reddy and B. Manohar, *J. Chem. Soc., Chem. Commun.*, 1993, 234.
100. S. J. Kulkarni, R. Ramachandra Rao, M. Subrahmanyam and A. V. Rama Rao, *J. Chem. Soc., Chem. Commun.*, 1994, 273.
101. R. Ramachandra Rao, N. Srinivas, S. Kulkarni, M. Subrahmanyam and K. V. Raghavan, *Indian J. Chem.*, 1997, **36A**, 708.
102. Y. Hu, J. Cao, J. Deng, B. Cui, M. Tan, J. Li and H. Zhang, *React. Kinet. Catal. Lett.*, 2012, **106**, 127.
103. R. W. Diesen, *US Pat.*, 4 910 176, 1990; assignee: Dow Chemical.
104. R. W. Diesen, *US Pat.*, 4 814 527, 1989; assignee: Dow Chemical.
105. R. Al-Hajri and D. Chadwick, *Appl. Catal. A: Gen.*, 2010, **388**, 96.
106. M. J. Lippits and B. E. Nieuwenhuys, *Catal. Today*, 2010, **154**, 127.
107. M. J. Lippits and B. E. Nieuwenhuys, *J. Catal.*, 2010, **274**, 142.
108. J. A. Kenar, *Lipid Technol.*, 2007, **19**, 249.
109. B. Katryniok, H. Kimura, E. Skrzyńska, J.-S. Girardon, P. Fongarland, M. Capron, R. Ducoulombier, N. Mimura, S. Paul and F. Dumeignil, *Green Chem.*, 2011, **13**, 1960.
110. M. Pagliaro, R. Ciriminna, H. Kimura, M. Rossi and C. Della Pina, *Angew. Chem. Int. Ed. Engl.*, 2007, **46**, 4434.
111. M. Ayoub and A. Z. Abdullah, *Renew. Sustain. Energy Rev.*, 2012, **16**, 2671.
112. A. Chieregato, F. Basile, P. Concepción, S. Guidetti, G. Liosi, M. D. Soriano, C. Trevisanut, F. Cavani and J. M. L. Nieto, *Catal. Today*, 2012, **197**, 58.

113. A. Chieregato, M. D. Soriano, F. Basile, G. Liosi, S. Zamora, P. Concepción, F. Cavani and J. M. López Nieto, *Appl. Catal. B: Environ.*, 2014, **150–151**, 37.
114. M. D. Soriano, P. Concepción, J. M. L. Nieto, F. Cavani, S. Guidetti and C. Trevisanut, *Green Chem.*, 2011, **13**, 2954.
115. F. Wang, J.-L. Dubois and W. Ueda, *J. Catal.*, 2009, **268**, 260.
116. J. Deleplanque, J.-L. Dubois, J.-F. Devaux and W. Ueda, *Catal. Today*, 2010, **157**, 351.
117. F. Wang, J. Xu, J.-L. Dubois and W. Ueda, *ChemSusChem*, 2010, **3**, 1383.
118. F. Wang, J.-L. Dubois and W. Ueda, *Appl. Catal. A: Gen.*, 2010, **376**, 25.
119. J.-L. Dubois, C. Duquenne and W. Holderich, *US Pat.*, 7 910 771, 2006; assignee: Arkema France.
120. A. Witsuthammakul and T. Sooknoi, *Appl. Catal. A: Gen.*, 2012, **413–414**, 109.
121. C. F. M. Pestana, A. C. Guerra, G. Ferreira, C. Turci and J. Mota, *J. Braz. Chem. Soc.*, 2013, **24**, 100.
122. M. O. Guerrero-Pérez and M. A. Bañares, *ChemSusChem*, 2008, **1**, 511.
123. C. Liebig, S. Paul, B. Katryniok, C. Guillon, J.-L. Couturier, J.-L. Dubois, F. Dumeignil and W. F. Hoelderich, *Appl. Catal. B Environ.*, 2013, **132–133**, 170.
124. C. Guillon, C. Liebig, S. Paul, A.-S. Mamede, W. F. Hölderich, F. Dumeignil and B. Katryniok, *Green Chem.*, 2013, **15**, 3015.
125. M. Pelaez, N. T. Nolan, S. C. Pillai, M. K. Seery, P. Falaras, A. G. Kontos, P. S. M. Dunlop, J. W. J. Hamilton, J. A. Byrne, K. O'Shea, M. H. Entezari and D. D. Dionysiou, *Appl. Catal. B: Environ.*, 2012, **125**, 331.
126. J. L. Coutts, L. H. Levine, J. T. Richards and D. W. Mazyck, *J. Photochem. Photobiol. A: Chem.*, 2011, **225**, 58.
127. P. a. Kolinko, T. N. Filippov, D. V. Kozlov and V. N. Parmon, *J. Photochem. Photobiol. A: Chem.*, 2012, **250**, 72.
128. E. a. Kozlova, N. S. Kozhevnikova, S. V. Cherepanova, T. P. Lyubina, E. Yu. Gerasimov, V. V. Kaichev, A. V. Vorontsov, S. V. Tsybulya, A. A. Rempel and V. N. Parmon, *J. Photochem. Photobiol. A: Chem.*, 2012, **250**, 103.
129. G. Palmisano, E. García-López, G. Marcì, V. Loddo, S. Yurdakal, V. Augugliaro and L. Palmisano, *Chem. Commun.*, 2010, **46**, 7074.
130. E. Piera, J. Ayllón, X. Doménech and J. Peral, *Catal. Today*, 2002, **76**, 259.
131. U. R. Pillai and E. Sahle-Demessie, *J. Catal.*, 2002, **211**, 434.
132. F. J. López-Tenllado, a. Marinas, F. J. Urbano, J. C. Colmenares, M. C. Hidalgo, J. M. Marinas and J. M. Moreno, *Appl. Catal. B: Environ.*, 2012, **128**, 150.
133. M. Antoniadou, V. Vaiano, D. Sannino and P. Lianos, *Chem. Eng. J.*, 2013, **224**, 144.
134. D. Sannino, V. Vaiano, P. Ciambelli, M. C. Hidalgo, J. J. Murcia and J. A. Navío, *J. Adv. Oxid. Technol.*, 2012, **15**, 284.

135. J. J. Murcia, M. C. Hidalgo, J. A. Navío, V. Vaiano, P. Ciambelli and D. Sannino, *Int. J. Photoenergy*, 2012, **2012**, 687262.
136. D. Sannino, V. Vaiano, P. Ciambelli, G. Carotenuto, M. Di Serio and E. Santacesaria, *Catal. Today*, 2013, **209**, 159.
137. D. Sannino, V. Vaiano and P. Ciambelli, *Chem. Eng. Trans.*, 2013, **32**, 625.

CHAPTER 9

Asymmetric Oxidation of Alcohols and Phenol Derivatives with Air as Oxidant

TAKUYA OGUMA[a,b] AND TSUTOMU KATSUKI*[b,c,d]

[a] Department of Chemistry, Graduate School of Science, Kyushu University, Hakozaki, Higashi-ku, Fukuoka, 812-8581, Japan; [b] International Institute for Carbon-Neutral Energy Research (I2CNER), Kyushu University, Hakozaki, Higashi-ku, Fukuoka, 812-8581, Japan; [c] Institute for Advanced Study, Kyushu University, Hakozaki, Higashi-ku, Fukuoka, 812-8581, Japan; [d] ACT-C, JST, Hakozaki, Higashi-ku, Fukuoka 812-8581, Japan

*Email: katsuscc@chem.kyushu-univ.jp

9.1 Introduction

Alcohol, phenol, and their derivatives are substructures found in many natural products and pharmaceuticals and also serve as versatile building blocks. Thus, the development of efficient methods for their synthesis and transformation has been an important objective in organic synthesis. Most alcohols and phenols are readily available and undergo oxidation to give various functional groups, and therefore, their asymmetric oxidation reactions are valuable tools for the enantioenriched synthesis of functionalized molecules. The efficiency of oxidation reactions largely depends on the oxidant used. Accordingly, many oxidants have been introduced to increase the efficiency of the oxidative conversions. Of the available oxidants, dioxygen is ideal because of its high atom efficiency and low environmental burden.[1] Moreover, dioxygen is ubiquitous and abundant in air, comprising

RSC Green Chemistry No. 28
Transition Metal Catalysis in Aerobic Alcohol Oxidation
Edited by Francesca Cardona and Camilla Parmeggiani

Published by the Royal Society of Chemistry, www.rsc.org

20.9% by volume, and the use of air as oxidant provides other advantages in cost and safety. Hence the asymmetric catalysis of oxidation reactions using air as the oxidant has been actively investigated. Generally, oxygen activation requires electron transfer and redox-active transition metal complexes can be efficient catalysts for aerobic oxidation, and various efficient transition metal catalysts have recently been developed for asymmetric aerobic oxidation.[2] In this chapter, asymmetric oxidation of both alcohol and phenol derivatives using air as oxidant is described with a focus on ruthenium-, iridium- and iron-based catalysts.

9.2 Asymmetric Oxidation of Alcohols with Air

Asymmetric oxidation of alcohols is roughly classified into two types, (i) oxidative kinetic resolution (OKR) of racemic secondary alcohols and (ii) oxidative desymmetrization of *meso*-diols (Scheme 9.1).

Optically active secondary alcohols are versatile chiral building blocks and their synthesis continues to be of growing importance. Although many methods for the asymmetric synthesis of enantioenriched secondary alcohols have been reported,[3] few methods are available for obtaining optically pure or almost optically pure secondary alcohols, especially aliphatic alcohols, except for enzymatic processes that show (almost) complete enantioselectivity, albeit generally with narrow substrate scopes.[4] OKR of racemic secondary alcohols, most of which are readily available in bulk, is a route to enantiopure alcohols,[5] although at least half of the racemic alcohols are consumed.[6]

Asymmetric oxidative desymmetrization of *meso*-diols is a potent tool for obtaining optically active hydroxy ketones, lactols and lactones, which are useful synthetic intermediates for various natural product syntheses.

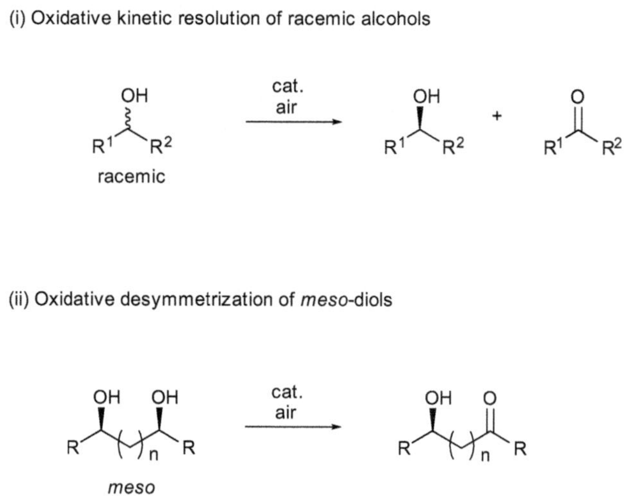

Scheme 9.1 Examples of typical asymmetric oxidation of alcohols.

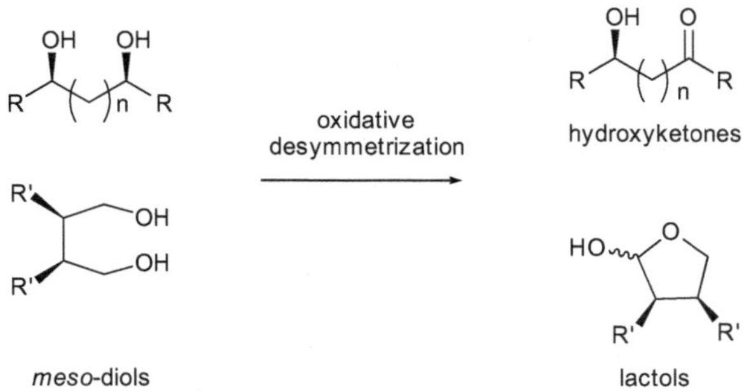

Scheme 9.2 Representative examples of desymmetrization of *meso*-diols.

Which product is obtained by the reaction depends on the type of diols used and reaction selectivity. For example, acyclic 1,3-diols give chiral β-hydroxy ketones, whereas cyclic and acyclic *meso*-1,4-diols affords γ-hydroxy ketones and lactols, respectively (Scheme 9.2). Further, an important advantage of desymmetrization is that the theoretical yield of product is 100% at maximum.

9.2.1 Oxidative Kinetic Resolution of Racemic Alcohols

OKR is a useful strategy for obtaining extremely enantioenriched alcohols. In particular, aerobic OKR has many advantages with respect to atom economy, cost, environmental harmony, *etc.* Hence, since the seminal report of aerobic OKR of alcohols by Beckett and Homer in 1986,[7] much effort has been directed towards developing the catalysis of enantiomer-differentiating oxidation using dioxygen as the oxidant. In 2000, Katsuki and co-workers reported that (ON)ruthenium(salen) complex **1a** catalyzes OKR of racemic secondary alcohols using air as oxidant at room temperature under photo-irradiation (Scheme 9.3).[8a] This method can be applied to both activated and non-activated secondary alcohols. The oxidation has been inferred to proceed *via* a single electron transfer (SET) step (discussed in Section 9.2.2). Recently, (aqua)ruthenium(salen) complex **1b** was found to catalyze OKR of secondary alcohols using air at room temperature without irradiation in a highly enantioselective manner (Scheme 9.3).[8c] This result indicates that irradiation is essential not for SET from an Ru^{III} intermediate to dioxygen but for dissociation of the ON group from **1a** (*cf.* Scheme 9.14).

While studying asymmetric desymmetrization of *meso*-1,3-diols using complex **1a** (described in Section 9.2.2), they found an interesting phenomenon that the addition of bidentate ligand **2** enhanced the relative reaction rate (k_{rel}) for the OKR (Scheme 9.4).[8b]

In 2001, Sigman and co-workers[9a] and Ferreira and Stoltz[9b] independently discovered that a Pd–sparteine complex catalyzes aerobic OKR. The initially reported OKR reactions were carried out using dioxygen as oxidant at 60–80 °C. Bagdanoff and Stoltz, however, reported in 2004 that the OKR

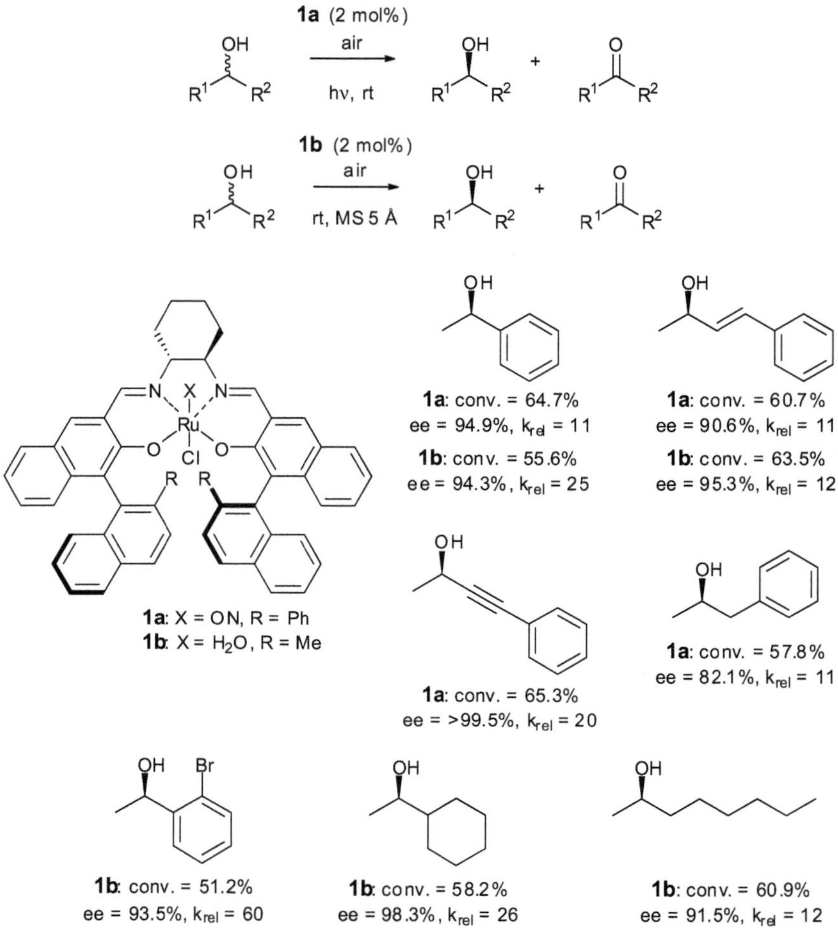

Scheme 9.3 Ru(salen)-catalyzed aerobic oxidative kinetic resolution of racemic secondary alcohols.

could be performed with air at room temperature by using Cs_2CO_3–chloroform-modified conditions (Scheme 9.5).[10] These palladium-based aerobic oxidations of alcohols is particularly described in Chapter 4.

Chen and co-workers[11] and Toste and co-workers[12] independently reported that vanadium–Schiff base complexes serve as efficient catalysts for the aerobic OKR of α-hydroxy esters and related compounds, albeit under a dioxygen atmosphere. On the other hand, focusing on group 9 metals, Ikariya and co-workers developed efficient chiral iridium and rhodium complexes **3** for asymmetric transfer hydrogenation reactions in 1999 (Scheme 9.6).[13] Later, Heiden and Rauchfuss[14] and Ikariya and co-workers[15] independently discovered that these Cp*metal–amine complexes **3c** are oxidized by molecular oxygen to give Cp*iridium–amide complexes **4** (Scheme 9.7).

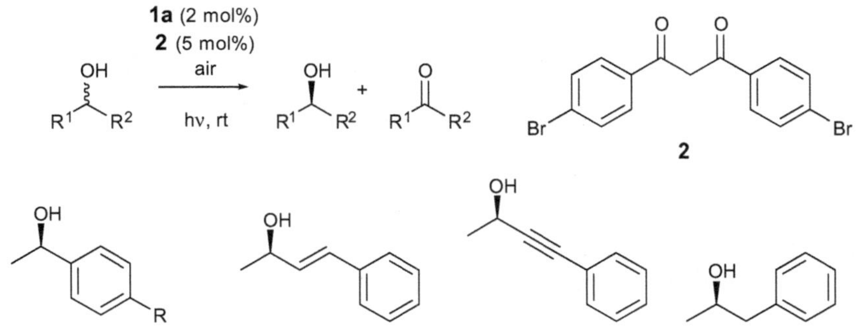

R = H conv. = 61%
ee = 98%, k_{rel} = 19 (±1)
R = Cl conv. = 57%
ee = 95%, k_{rel} = 23 (±1)
R = Br conv. = 54%
ee = 90%, k_{rel} = 22 (±1)

conv. = 54%
ee = 82%, k_{rel} = 14 (±1)

conv. = 58%
ee = 99%, k_{rel} = 30 (±2)

conv. = 41%
ee = 55%, k_{rel} = 15 (±1)

Scheme 9.4 Kinetic resolution of racemic secondary alcohols with bidentate ligand-supported ruthenium–salen complex **1a**.

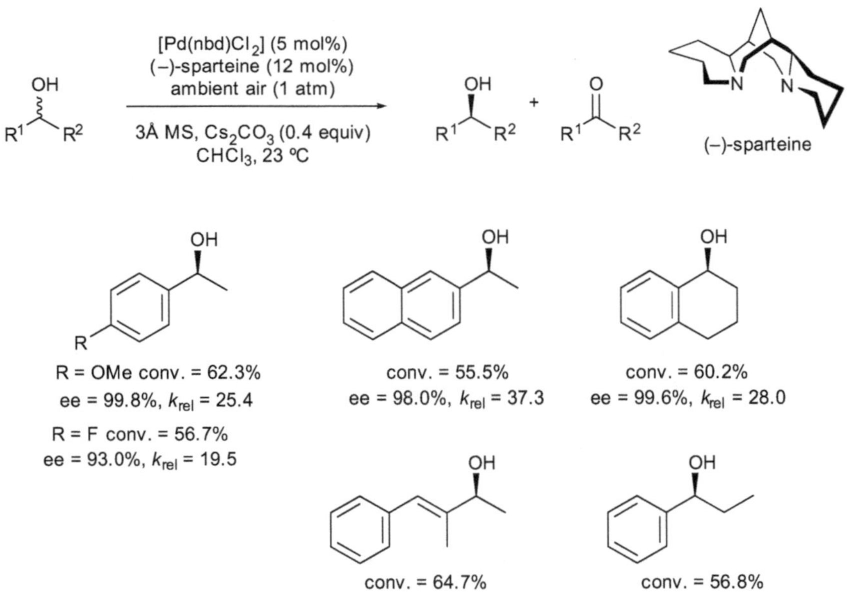

R = OMe conv. = 62.3%
ee = 99.8%, k_{rel} = 25.4
R = F conv. = 56.7%
ee = 93.0%, k_{rel} = 19.5

conv. = 55.5%
ee = 98.0%, k_{rel} = 37.3

conv. = 60.2%
ee = 99.6%, k_{rel} = 28.0

conv. = 64.7%
ee = 98.9%, k_{rel} = 15.7

conv. = 56.8%
ee = 94.9%, k_{rel} = 21.7

Scheme 9.5 Pd–sparteine complex-catalyzed oxidative kinetic resolution of secondary alcohols under ambient air.

With the notion that the Cp*iridium–amide complex should show a reverse transfer dehydrogenation catalysis, Ikariya and co-workers examined the OKR of secondary alcohols with Cp*iridium–amide complex **4a** bearing an *N*-menthanesulfonylated diamine ligand and developed the highly

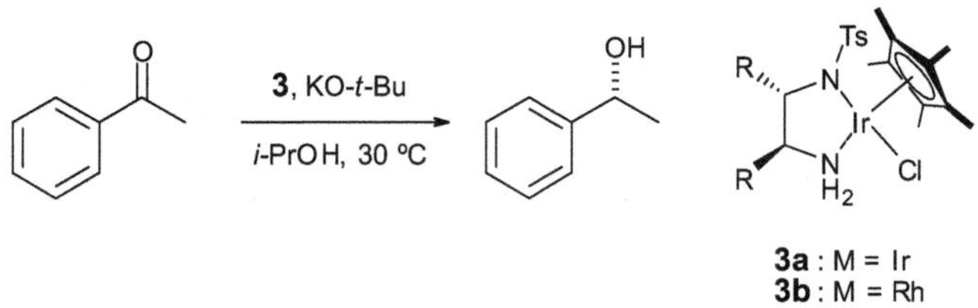

Scheme 9.6 Asymmetric transfer hydrogenation using iridium- and rhodium-based catalysts.

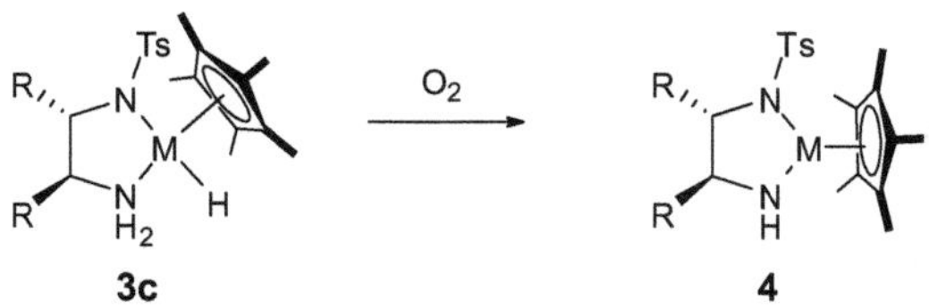

Scheme 9.7 Oxidation of Cp*metal–amine complexes to the metal amide complexes by O_2.

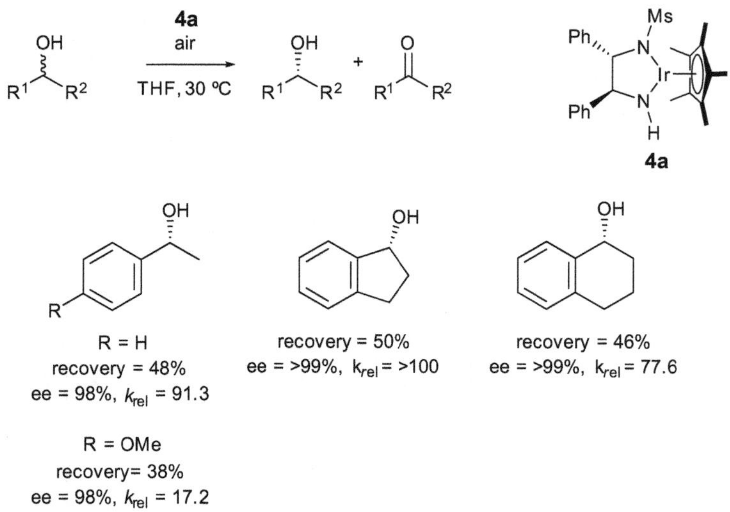

Scheme 9.8 Cp*iridium–chiral diamine catalyzed OKR of secondary alcohols.

enantioselective OKR of racemic secondary aromatic alcohols such as 1-phenylethanols, hydroxyindane and 1-hydroxytetralin in air under mild conditions at 30 °C (Scheme 9.8). In particular, the resolution of the alcohols proceeds with significant enantioselectivities at lower substrate concentration (0.1–0.2 M).

The catalysis of OKR described above requires noble transition metals such as ruthenium, palladium or iridium. In 2009, Sekar and co-workers reported for the first time that an iron complex with chiral Schiff base ligand

5 catalyzes OKR of racemic benzoins using dioxygen in the presence of a catalytic amount of 2,2,6,6-tetramethylpiperidine-1-oxyl (TEMPO) at 60 °C (Scheme 9.9).[16] OKR of alcohols other than benzoins has not been reported.

In the same year, Egami and Katsuki reported iron(salan)-catalyzed oxidative coupling of 2-naphthols (Scheme 9.10; described in detail in Section

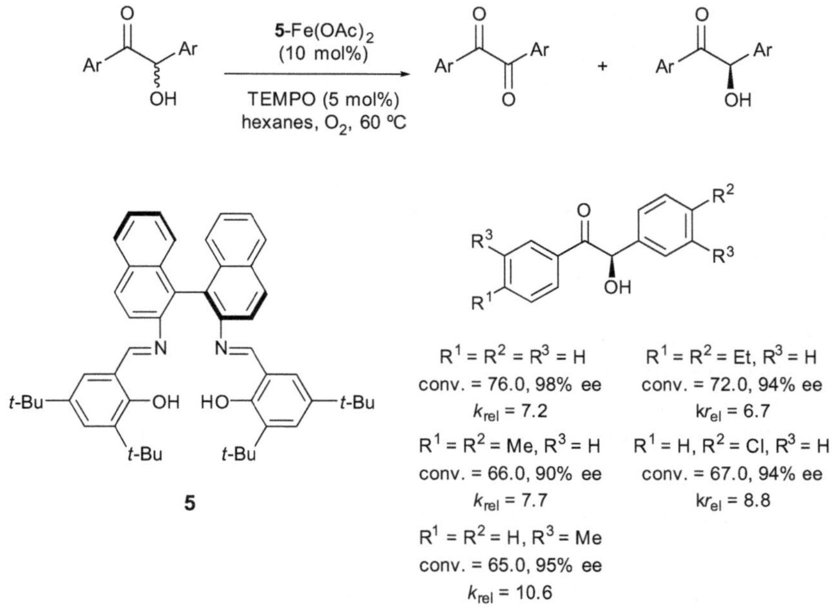

Scheme 9.9 Iron-catalyzed OKR of benzoins with a chiral Schiff base.

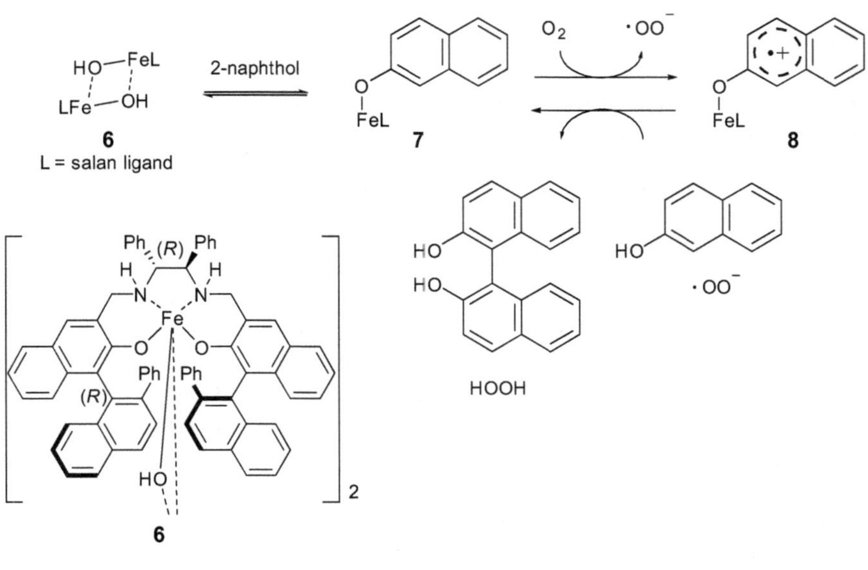

Scheme 9.10 Iron(salan)-catalyzed oxidative coupling of 2-naphthols with dioxygen.

9.3.1).[17a] The mechanistic study of this coupling reaction revealed that the dimeric iron(salan) complex **6** is dissociated in the presence of 2-naphthol to the monomeric iron(salan)(2-naphthoxo) complex 7, which transfers an electron to dioxygen and affords radical cation species **8**.[17b,c]

The radical cation in **8** should be distributed over the naphthoxy oxygen and a similar radical cation structure has been proposed as the active site for alcohol oxidation by galactose oxidase.[18] The participation of a similar radical cation intermediate has also been proposed for alcohol oxidation using a ruthenium–salen complex as the catalyst (Section 9.2.2., Scheme 9.14).[19] Hence it was expected that the iron(salan) complex **6** would also catalyze asymmetric alcohol oxidation in the presence of a naphthol. However, if the additive is 2-naphthol, the coupling of 2-naphthol should compete with alcohol oxidation. Therefore, several naphthols were screened and 1-naphthol was found to be the additive of choice. Indeed, complex **6** catalyzed OKR of secondary alcohols using air as oxidant in the presence of catalytic amount of 1-naphthol at 50–60 °C.[20] This method was successfully applied not only to activated secondary alcohols such as benzylic and allylic alcohols, irrespective of the presence or absence of either electron-donating or -withdrawing groups, but also to aliphatic alcohols such as 4-phenylbutan-2-ol and 1-cyclopentylethan-1-ol (Scheme 9.11).

9.2.2 Asymmetric Oxidative Desymmetrization of *meso*-Diols

Some catalysts for OKR described in Section 9.2.1 have been applied to the aerobic oxidative desymmetrization of *meso*-diols (Scheme 9.2). For example, the desymmetrization of *meso*-diols with the palladium–sparteine system, which is an efficient catalyst system for OKR, successfully provides chiral acyclic and cyclic hydroxy ketones under O_2.[21] Katsuki and co-workers reported the first example of desymmetrization using ambient air as oxidant.[19,23] They discovered that achiral ruthenium(salen) complex **9** bearing quaternary carbons at its ethylenediamine moiety catalyzes the selective oxidation of a primary alcohol in the presence of a secondary alcohol (Scheme 9.12) and, moreover, the oxidation of 1,*n*-diols to give lactols or *n*-hydroxyaldehydes.[22] Based on these results, they further developed asymmetric oxidative desymmetrization of *meso*-diols using ruthenium(salen) complexes as catalyst in air at room temperature under photoirradiation (Scheme 9.13).[19,23]

Complexes **10** that bear chiral quaternary carbons at the diamine moiety or a bulky aryl substituent at C2″ oxidize *meso*-diols enantioselectively to give the optically active lactols, the enantiomeric excesses (*ee*) of which were determined after pyridinium dichromate (PDC) oxidation to the corresponding lactones. During this study, it was found that the enantioselectivity of the reaction is affected by the nature of the apical ligand of the catalyst. Hence the scope of this desymmetrization can be broadened by an appropriate combination of the apical ligand and the C2″-aryl group.

Investigations of reaction mechanisms were conducted through kinetic, kinetic isotope effect (KIE) and spectroscopic studies and it was found that

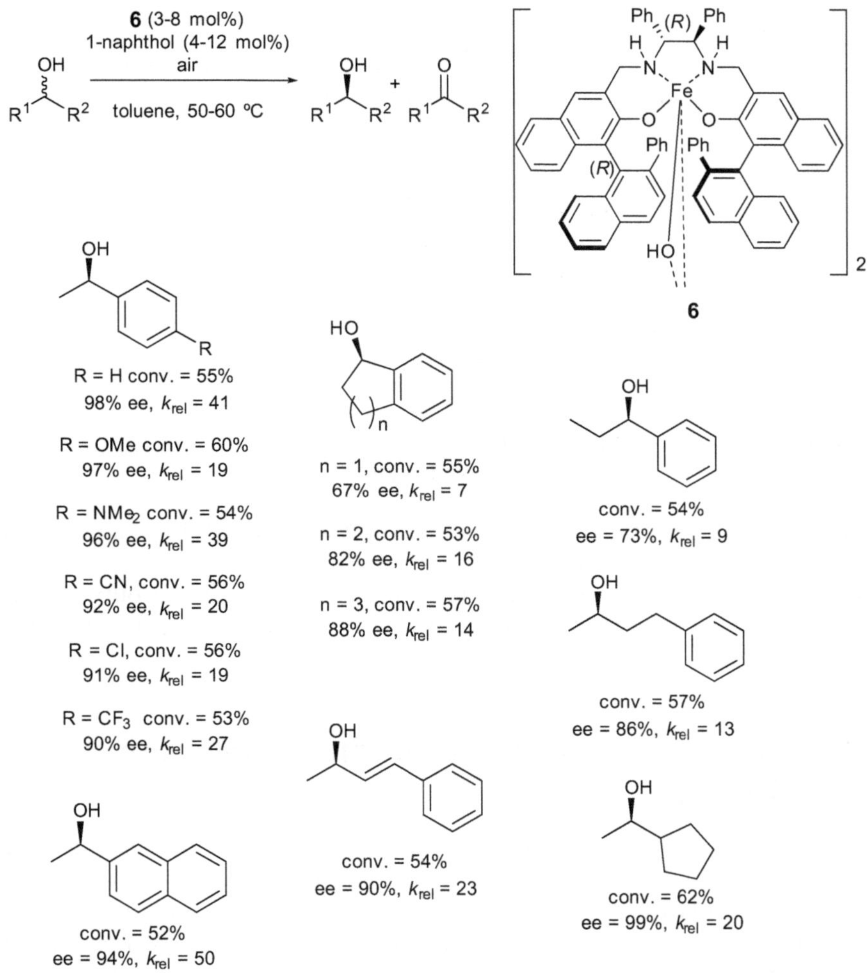

Scheme 9.11 Iron(salan)-catalyzed OKR of aromatic and aliphatic secondary alcohols under air.

ruthenium(NO)(salen) complexes are precatalysts and NO ligand dissociation by irradiation *in situ* generates a genuine ruthenium(salen) catalyst, which catalyzes alcohol oxidation *via* SET from alcohol-bound ruthenium complexes to dioxygen and subsequent hydrogen atom transfer (HAT) by the resulting oxygen radical cation. It is noteworthy that the rate-determining step (RDS) depends on the nature of apical ligand (X); when the apical ligand is Cl, the RDS is the SET step, and when the apical ligand is OH, the RDS is the HAT step (Scheme 9.14). However, the KIE for diol oxidation indicated that a radical cation is generated not on the naphthol oxygen but on the alcohol oxygen, due to the stabilization of the radical cation by hydrogen-bond formation,[24] and the subsequent HAT from the radical cation is brought about by a superoxide in an intermolecular manner (Scheme 9.14,

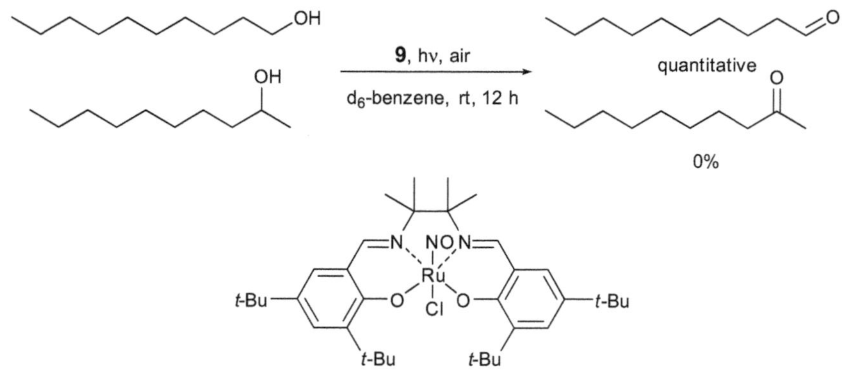

Scheme 9.12 Aerobic primary alcohol oxidation with achiral Ru(salen) catalyst in the presence of secondary alcohols.

right-hand cycle). This explains why the complexes bearing chiral quaternary carbons at the diamine moiety are generally favorable catalysts for the desymmetrization of *meso*-diols. The quaternary methyl group at the diamine moiety has been considered to play an important role in regulating the conformation of *meso*-diols.

9.3 Asymmetric Oxidation of Phenol Derivatives with Air

Phenol derivatives are abundant in fossil resources and plants. They are electron rich and undergo various oxidation reactions. Therefore, their asymmetric oxidative transformation using chiral transition metal complexes as catalysts is a topic of growing interest. Especially two classes of asymmetric oxidative reactions of phenol derivatives using ambient air have been actively investigated: (i) asymmetric oxidative coupling (AOC) and (ii) asymmetric oxidative dearomatization. As is well known, the complexes that catalyze the oxidation of alcohols not infrequently also catalyze the oxidation of phenol and its derivatives. In this section, these two types of oxidation of phenol derivatives are described.

9.3.1 Asymmetric Oxidative Coupling of Phenol Derivatives

Optically active biphenols or binaphthols (BINOLs) and their derivatives have been utilized not only as versatile ligands in asymmetric synthesis but also as intermediates for the synthesis of natural products bearing axial chirality.[25] Oxidative coupling of phenol derivatives is a direct and atom-efficient tactic since it does not require any prefunctional group such as a halogen or metal. Since the first report of the catalytic AOC of 2-naphthol using dioxygen in the presence of a copper–diamine catalyst in 1995,[26]

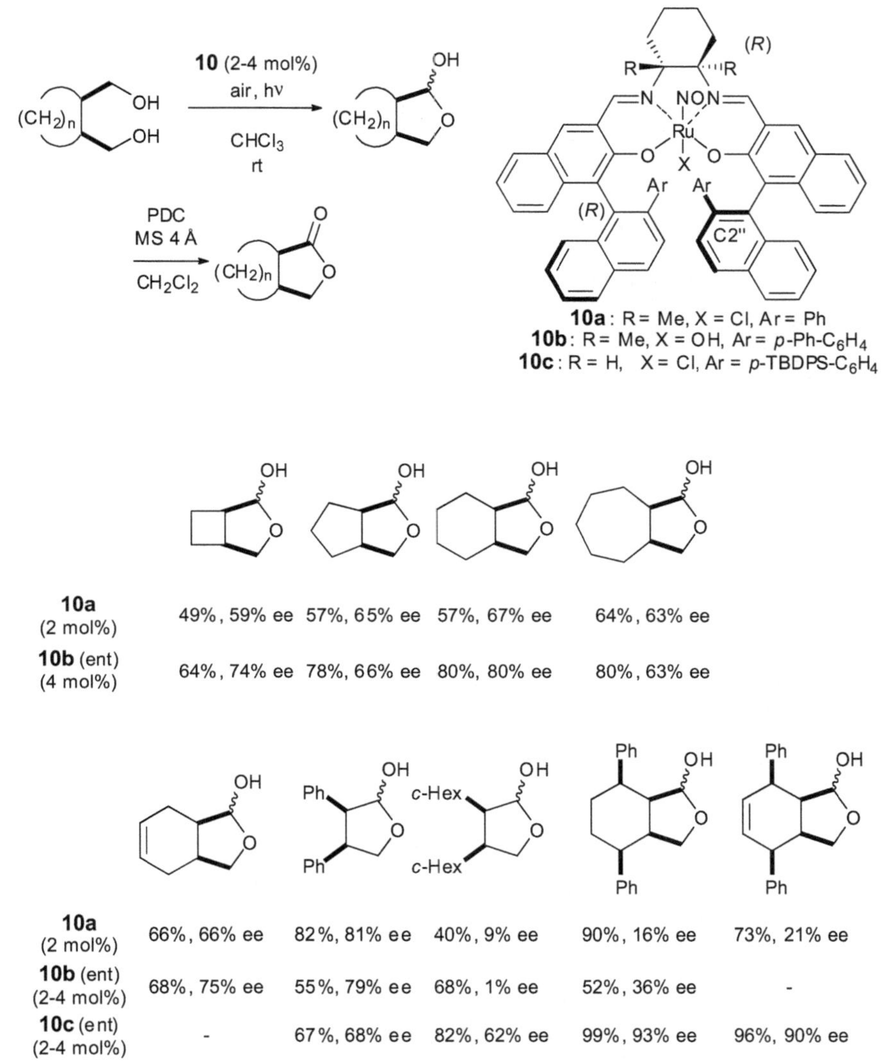

Scheme 9.13 Ru(salen)-catalyzed aerobic oxidative desymmetrization of *meso*-diols under air.

numerous studies on transition metal-catalyzed asymmetric aerobic oxidative coupling of 2-naphthols have been performed[27] but most of them use pure dioxygen as oxidant, owing to the low partial pressure of oxygen in air. Ruthenium–salen complex **1**, which is an efficient catalyst for OKR of secondary alcohols using air as oxidant, was found to catalyze the AOC of 2-naphthols in air under irradiation, albeit with moderate selectivity (Scheme 9.15).[28a] A similar Ru(salen) complex bearing diphenylethylenediamine moiety **11** also catalyzed the oxidative intramolecular coupling of

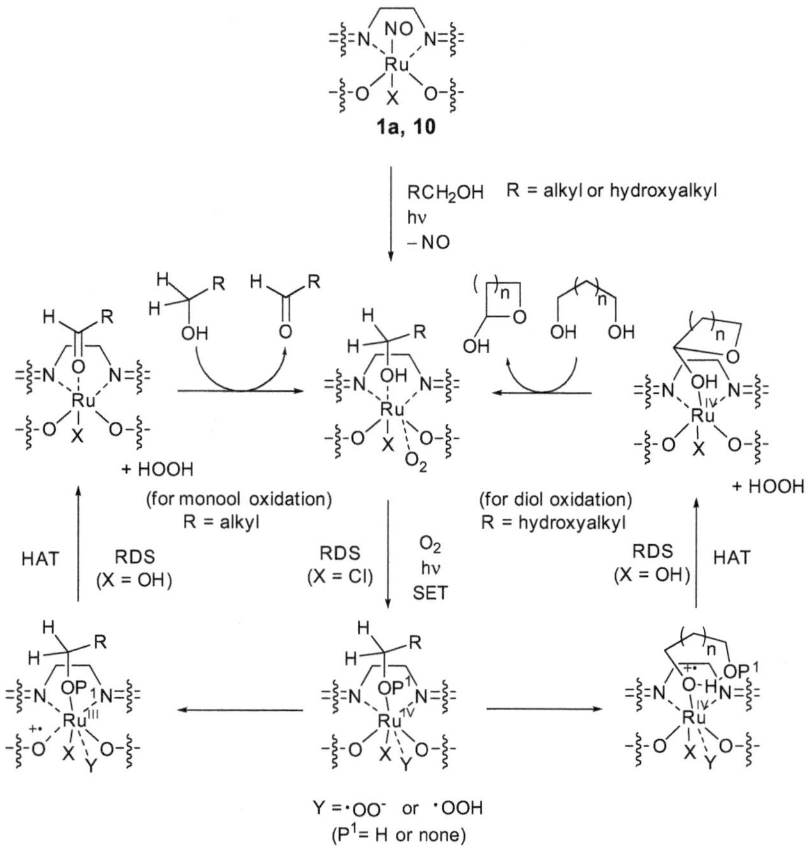

Scheme 9.14 Proposed mechanisms for Ru(salen)-catalyzed aerobic alcohol oxidation.

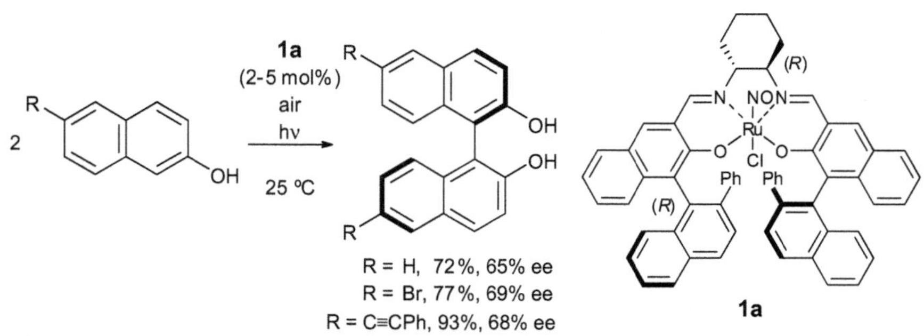

Scheme 9.15 Ru(salen)-catalyzed oxidative coupling of 2-naphthols under photoirradiation.

2,2′-dihydroxystilbene derivatives with moderate to good enantioselectivity to give chiral 2,3-dihydrobenzofurans (Scheme 9.16).[28b] Recently, highly enantioselective catalysis of aerobic oxidative coupling of 2-naphthols by vanadium complexes (**12** and **13**) have been reported (Scheme 9.17).[29–31]

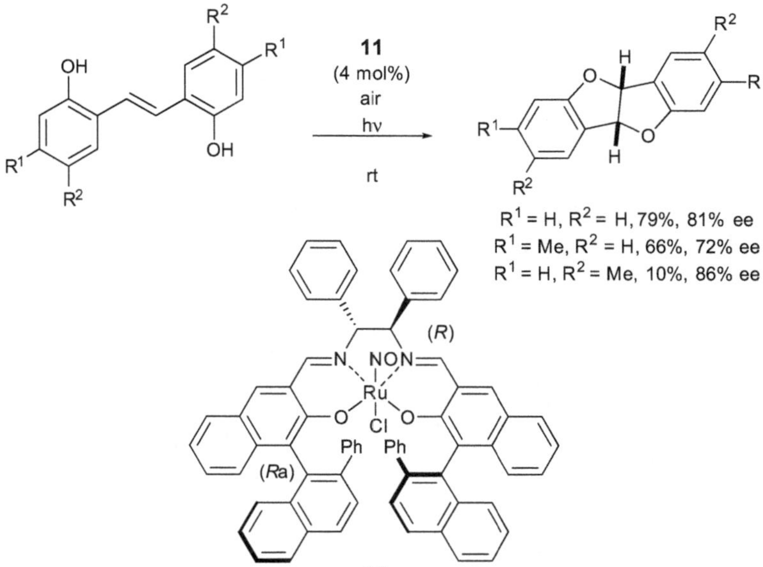

11

Scheme 9.16 Enantioselective cyclization catalyzed by an Ru(salen) complex.

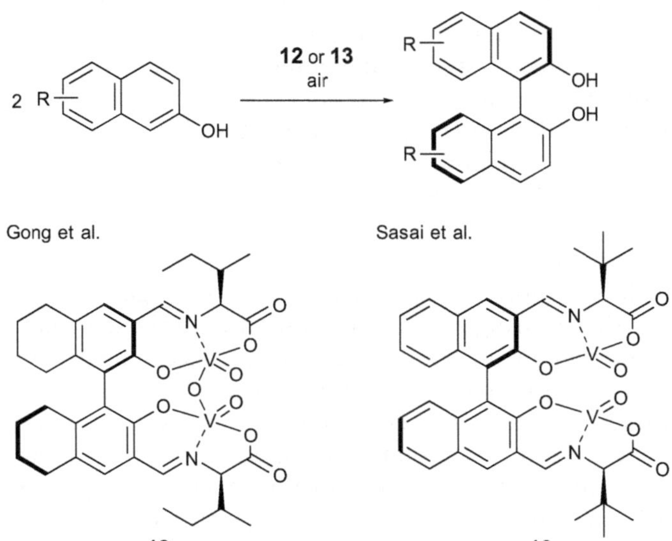

12 **13**

Scheme 9.17 Binuclear vanadium complex-catalyzed aerobic oxidative coupling of 2-naphthols.

In 2009, Egami and Katsuki reported the iron-catalyzed AOC of 2-naphthols with air (Scheme 9.18).[17a] Iron(salan) complex **6** catalyzes the highly enantioselective oxidative coupling of 2-naphthol derivatives that

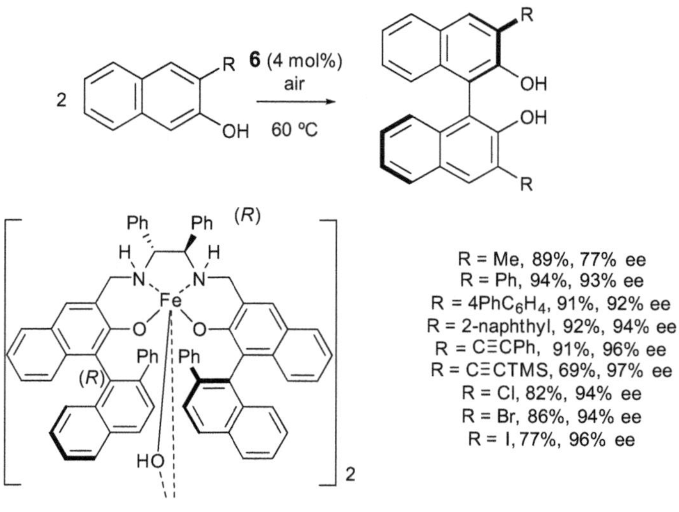

Scheme 9.18 Iron(salan)-catalyzed AOC of C3-substituted 2-naphthols.

possess a non-coordinating group at their C3 position, whereas the corresponding iron(salen) complex does not catalyze the coupling, mainly owing to its high oxidation potential and inflexibility.[17b,32]

Based on mechanistic studies including kinetics, correlation between *ee* values of the catalyst and product and X-ray analysis, it was proposed that the dimeric complex **6** dissociates into monomeric species **7** in the presence of 2-naphthol, which is oxidized to give a radical cation species **8** and undergoes coupling *via* a radical anion mechanism (Scheme 9.19).[17b,c]

The SET oxidation is the rate-determining step and the monomeric species bearing a more electron-rich naphthoxo ligand is oxidized in preference to that bearing a less electron-rich naphthoxo group. In agreement with this mechanism, the monomeric species can be isolated at lower temperature and participates in the asymmetric catalysis of 2-naphthol coupling under the coupling conditions, whereas the complexes that cannot dissociate into a monomeric species do not show catalysis of oxidative coupling. Such a catalytically inactive complex **14** gives a μ-naphthoxo-μ-oxo–iron(salan) complex **15** upon treatment with 3-bromo-2-naphthol **16** (Scheme 9.20).[17b]

Understanding of the mechanism of the coupling gave a clue to the highly enantio- and cross-selective coupling of electronically different 2-naphthols. Previously reported cross-coupling reactions gave a mixture of two homo- and one cross-coupling products. According to the proposed mechanism, it is anticipated that a 2-naphthol derivative bearing a strong electron-withdrawing group such as an ester or ketone cannot be oxidized and the homo-coupling of the derivative does not proceed. However,

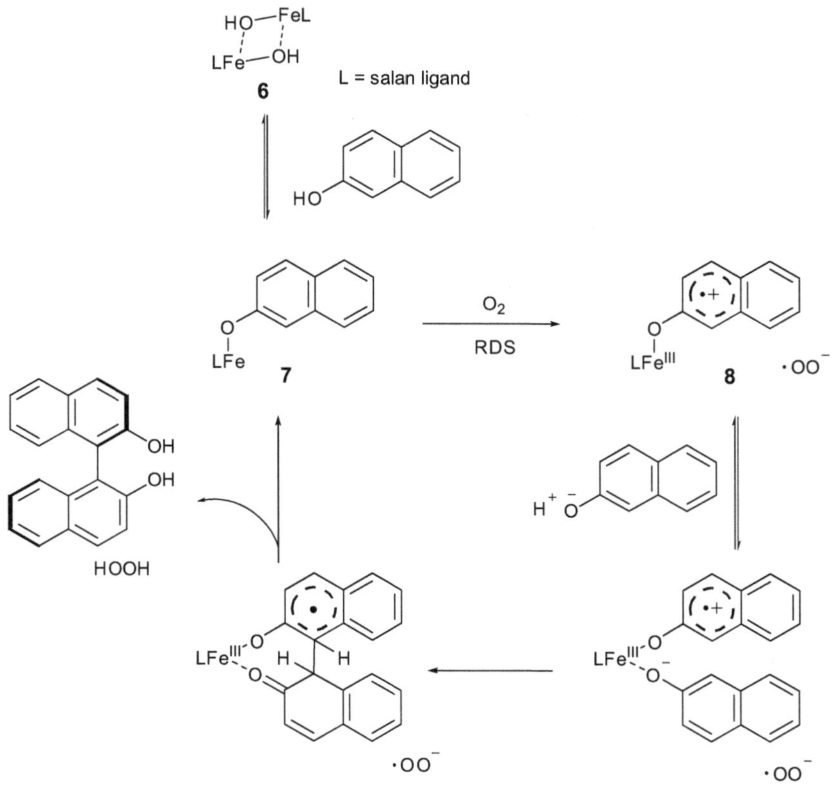

Scheme 9.19 Radical anion mechanism proposed for iron(salan)-catalyzed oxidative coupling of 2-naphthols.

the presence of an electron-withdrawing group renders the 2-naphthol derivative more acidic and it readily dissociates to generate a 2-naphthoxide anion, a counterpart of the cross-coupling reaction. Hence the coupling reaction between electron-rich and -poor substrates gives the cross-coupling product **18** as the major product and one homo-coupling product of an electron-rich 2-naphthol derivative **17** as the minor product (Table 9.1).

9.3.2 Asymmetric Oxidative Dearomatization of Phenol Derivatives

Asymmetric oxidative dearomatization (AOD) of phenol derivatives gives optically active cyclic enones or their dimers through two-electron oxidation and nucleophilic addition.[33] Various kinds of AOD have been developed using a series of stoichiometric oxidants such as high-valent transition metals, hypervalent iodine and organic peracids. An example of AOD using dioxygen was reported by Porco and co-workers, in which the stoichiometric

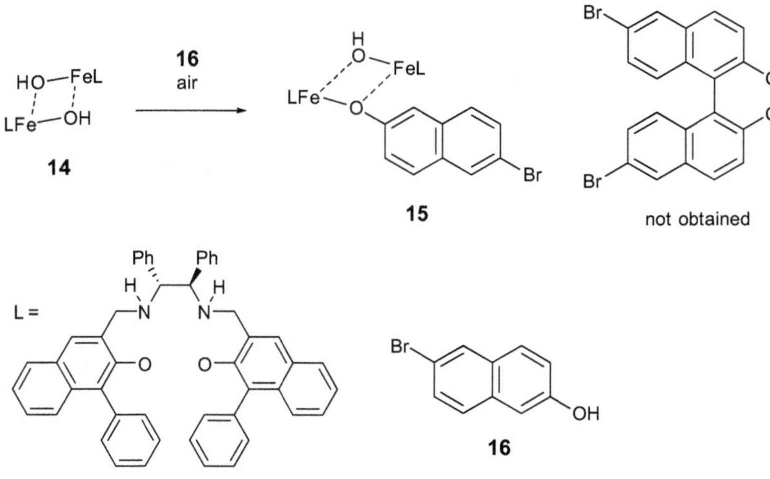

Scheme 9.20 Formation of a μ-naphthoxo-μ-oxo–iron(salan) species by treatment of catalytically inactive iron(salan) complex **14** with 6-bromo-2-naphthol (**16**).

Table 9.1 Iron(salan)-catalyzed asymmetric oxidative cross-coupling of 2-naphthols under air.

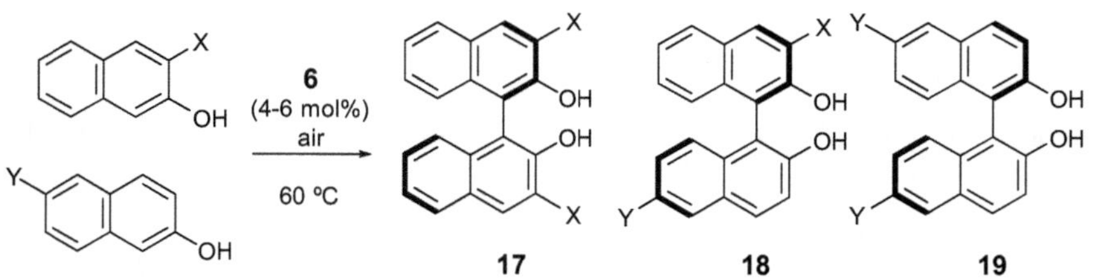

Entry	X	Y	Cross-coupling product **18** Yield (%)	*ee* (%)	**17** : **18** : **19**
1	2-Naphthyl	Br	52	89	2 : 5 : 2
2	PhCC	Br	44	88	4 : 9 : 3
3	Ph	CO_2Me	52	90	2 : 5 : 0
4[a]	Ph	CO_2Me	70	90	2 : 11 : 0
5[a]	Ph	Ac	65	88	2 : 15 : 0
6[a]	Ph	CHO	60	88	1 : 6 : 0
7[a]	Ph	CN	68	87	1 : 10 : trace
8[a]	TMSCC	Br	53	93	2 : 21 : 0
9[a]	Br	CO_2Me	54	93	1 : 15 : 0
10[a]	I	CO_2Me	50	95	1 : 20 : 0

[a]2 equivalents of the electron-deficient 2-naphthols were used and the electron-rich 2-naphthols were added in three batches.

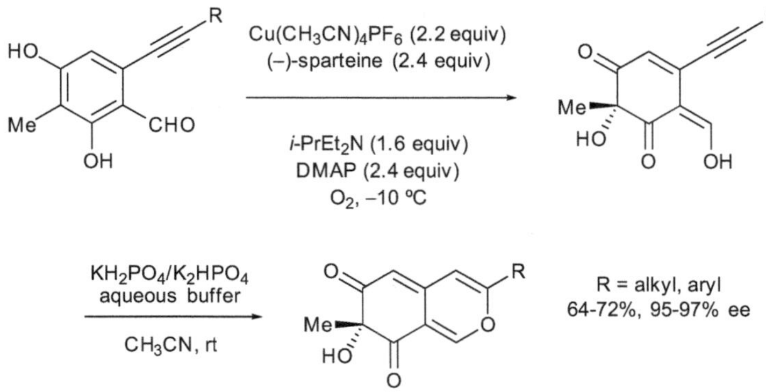

Scheme 9.21 Enantioselective oxidative dearomatization of phenol derivative with O_2.

oxidant is a copper(II) salt generated *in situ* by the oxidation of a Cu(I) precursor with dioxygen. The resulting di-μ-oxo–copper complex also acts as an oxygen nucleophile (Scheme 9.21).[34]

The unique catalysis of binaphthol coupling by iron(salan) complex **6** also opened an efficient route for asymmetric oxidative dearomatization using air as the oxidant.[35] Oguma and Katsuki discovered that complex **6** catalyzes the dearomatization of 1,3-disubstituted 2-naphthols in the presence of nitroalkanes, the pK_a of which are almost equal to that of 2-naphthol, as the nucleophile to give the optically active naphthalene-2(1*H*)-ones bearing an all-carbon quaternary stereocenter at the C1 position (Scheme 9.22).[35] The C3 substituent included alkyl, allyl, halogen, phenyl and propargyl groups and the C1 substituent included methyl, *n*-alkyl and cyclohexyl groups. The use of nitroethane or 1-nitropropane as the nucleophile instead of nitromethane afforded the dearomatized products with good diastereoselectivity of 9:1 to 10:1 together with high enantio-selectivity. This is the first asymmetric oxidative dearomatization with air and also the first asymmetric intermolecular C–C bond-forming oxidative dearomatization. It is noteworthy that 2-naphthol bearing a bulky C1 substituent such as cyclohexyl reacts with nitromethane to give the desired product.

The proposed mechanism for this reaction, based on a coupling reaction, (i) SET from **7** to dioxygen to generate a radical cation intermediate **8**, (ii) coordination of the *in situ*-generated anion of nitroalkane to the Fe(III) ion and (iii) nucleophilic attack of the anion on the C1 position of the intermediate in an intramolecular fashion, explains these results (Scheme 9.23).

While further studying the dearomatization of 2,4,6-trimethylphenol, the formation of 2,6-dimethyl-4-(2-nitroethyl)phenol was observed. This result suggests that dehydrogenation of a radical cation species probably by superoxide competes with the nucleophilic attack of the nitroalkane-derived anion to the species and that the resulting *p*-quinone methide (*p*-QM)

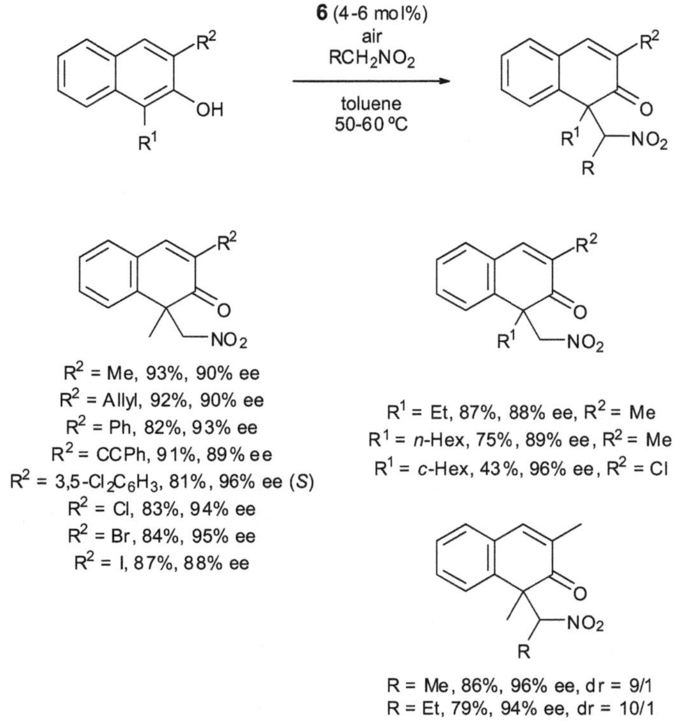

Scheme 9.22 Iron-catalyzed asymmetric aerobic dearomatization of naphthols with nitroalkanes.

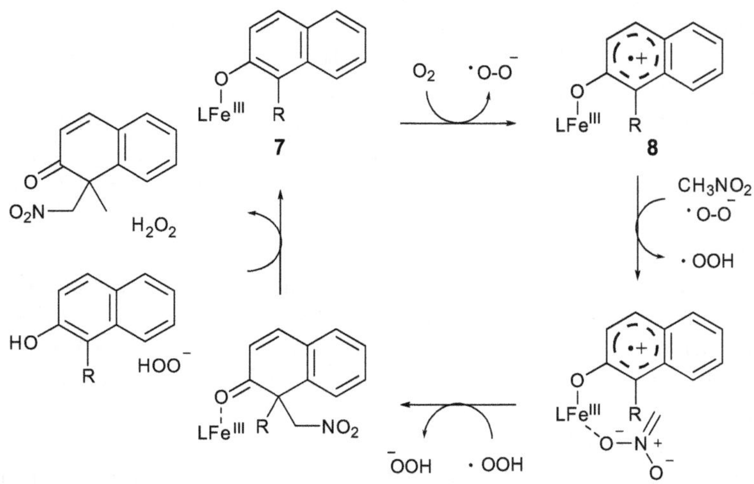

Scheme 9.23 Proposed mechanism of iron-catalyzed oxidative dearomatization of 2-naphthols.

undergoes a Michael addition reaction. Hence we expected that the radical cation species **8** would preferentially undergo dehydrogenation to generate *in situ* a synthetically useful *o*-quinone methide (*o*-QM)[36] in the presence of less nucleophilic anion than nitroalkane-derived anion. It is well known that *o*-QMs serve as Michael acceptors. This feature of *o*-QMs has been sophisticatedly harnessed to construct spirocyclic compounds in the presence of an arenol that is almost equally acidic to nitromethane and gives the *in situ*-generated arene oxide anion upon dissociation, which is a less nucleophilic ambident anion. Thus, in the presence of phenol, dehydrogenation of intermediate **7** is expected to occur preferentially under the oxidation conditions and the resulting *o*-QM intermediate to undergo Michael addition with the phenoxide to produce a 1,1′-methylene(phenol)(naphthoxo) intermediate. The aerobic dearomatization of the intermediate gives spirocyclic compound **20**, in which phenoxide anion serves as an oxygen nucleophile. It is noteworthy that this three-step tandem reaction proceeds without isolating any intermediate (Scheme 9.24).[37]

As expected, the reactions of 1-methyl-3-alkyl-2-naphthols and phenols substituted with various groups such as bromo, ester, nitro, cyano, acetal and *tert*-butyl proceeded using complex **6** or **21** as the catalyst at 90 °C under air to give the spirocycles with high enantioselectivity and acceptable yields (Scheme 9.25). Higher enantioselectivity was obtained with complex **21**. The reaction of 3,5-dimethylphenols showed lower enantioselectivity but a single recrystallization gave an almost optically pure product.

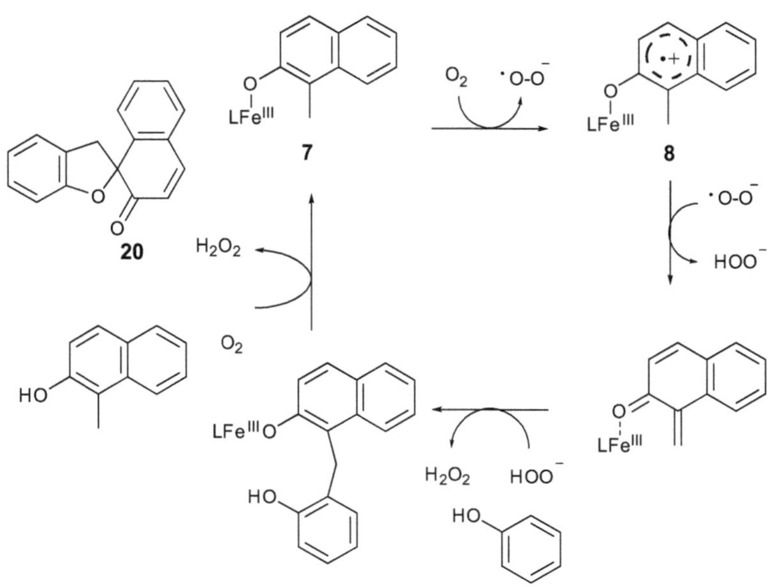

Scheme 9.24 An envisioned tandem approach for the construction of a spirocyclic framework *via* an *o*-QM intermediate.

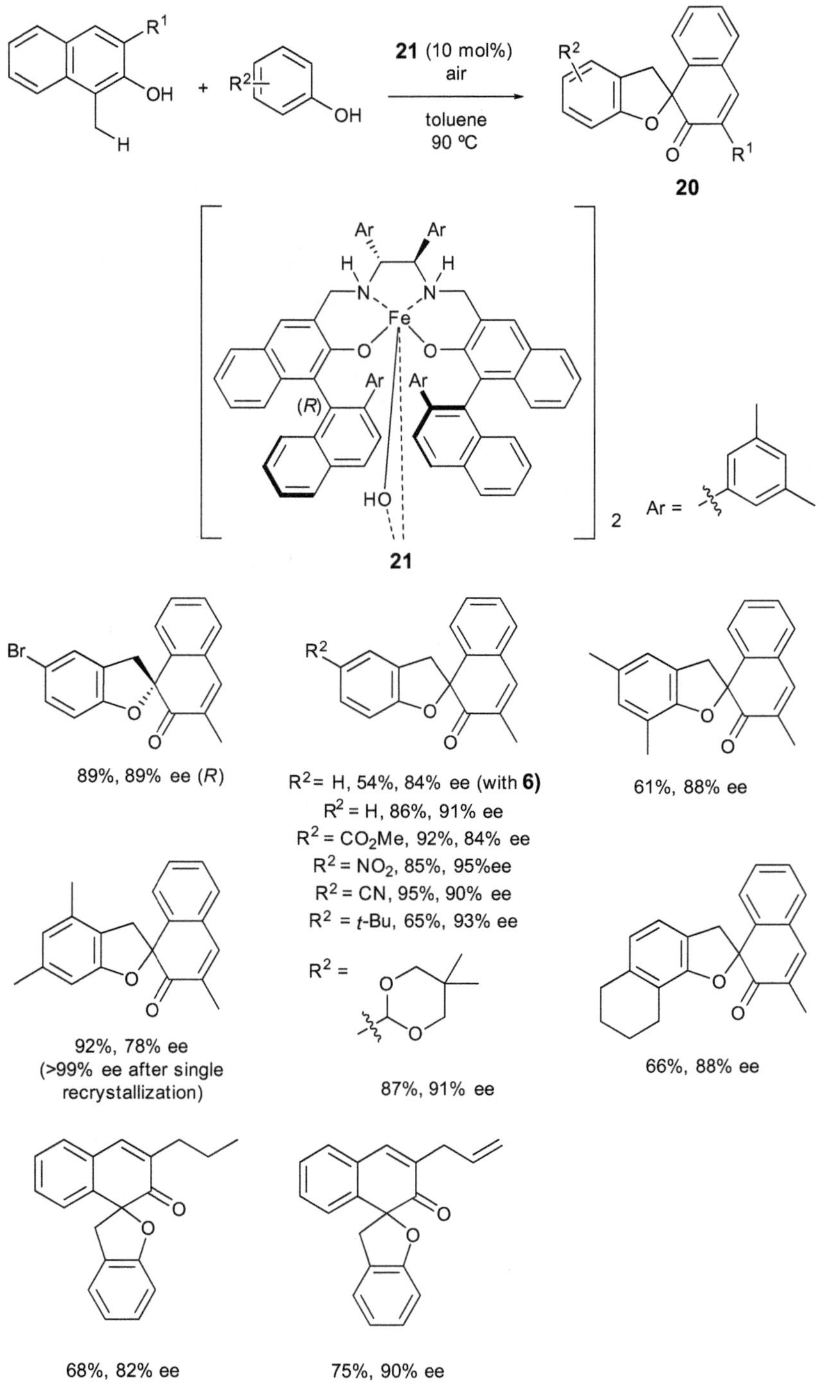

Scheme 9.25 Iron-catalyzed asymmetric tandem construction of spirocyclic compounds from 1-methyl-2-naphthols and phenols using air as oxidant.

9.4 Conclusion

Dioxygen is ubiquitous in air (the oxygen content is 20.9%) and oxidation with air has several advantages over that with pure dioxygen in terms of cost, transport energy, safety, and so on. In this chapter, asymmetric oxidations of alcohols and phenol derivatives with air have been described. Oxidation of alcohols and phenols is a dehydrogenation reaction and the metal-catalyzed oxidation with dioxygen falls formally into two categories based on whether hydrogen is transferred as a hydrogen atom [hydrogen atom transfer (HAT)] or as a hydrogen anion (hydride migration). In the first category of reactions, dioxygen serves as a single electron acceptor and converts a catalyst to the corresponding radical (or radical cationic) intermediate, which undergoes an intramolecular HAT or coupling reaction. Some copper- (Chapter 2), ruthenium- (Section 9.2.1),[8,17,19] and iron-based complexes (Section 9.2.1)[20] catalyze this type of alcohol oxidation. Galactose oxidase, which has a copper ion at the active site, a typical biological alcohol oxidase, catalyzes this class of oxidation, in which proton transfer (PT), SET and HAT occur in an intramolecular fashion and proceed under mild conditions.[18] The second category of alcohol oxidation includes some palladium-catalyzed oxidation reactions[9,10,21] that proceed *via* β-hydride elimination (Chapter 4 and Section 9.2.1) and ruthenium- or iridium-catalyzed oxidation reactions[15] that proceed *via* a hydrogen transfer process. Metal hydrido species or lower valent metal species *via* reductive elimination of the metal hydrido species are generated in these oxidation reactions. The role of dioxygen in these reactions is to reoxidize the metal hydrido or lower valent species and to complete the catalytic cycle.

The mechanism of aerobic oxidation *via* the SET process, which is a crucial step for reactions of the first category, has been extensively studied. The mechanistic studies of ruthenium–salen-[19] and iron–salan-catalyzed[17b,c] aerobic oxidations by Katsuki and co-workers revealed that the reactions proceed *via* a radical cation species generated *in situ* by the SET process. Thus, ruthenium–salen and iron–salan complexes catalyze enantioselective alcohol oxidation[8,19,20,23] and naphthol coupling[17,28] using air as the hydrogen acceptor. Moreover, the radical cationic species generated in the iron–salan-catalyzed oxidation of arenols have cationic properties and react with an appropriate carbon and heteroatom nucleophile. By taking advantage of the radical cationic species, highly enantioselective dearomatization[35] and spirocyclization[37] have been achieved. These examples demonstrate the high synthetic potential of such a radical cationic species.

On the other hand, Ikariya and co-workers discovered that a Cp*iridium–amine complex is dehydrogenated by dioxygen to give the Cp*iridium–amide complex, which undergoes highly enantioselective oxidation of secondary alcohols *via* a hydrogen transfer process, *i.e.* oxidation of the second category.[15]

Although asymmetric alcohol and arenol oxidations with air as oxidant are still limited to a few examples, high enantioselectivity has been obtained under mild conditions, which demonstrates the high potential of oxidation with air. These advances have been enhanced by the mechanistic understanding of the reactions. However, the mechanisms of most aerobic oxidation reactions are still unclear and their elucidation is essential for the rational design of new catalysts, which in turn forms the basis for future developments in aerobic alcohol and arenol oxidation.

Acknowledgements

The support from Advanced Catalytic Transformation Program for Carbon Utilization (ACT-C), JST, Nissan Chemical Industries, Ltd., the International Institute for Carbon-Neutral Energy Research (WPI-I2CNER) and MEXT, Japan, is gratefully acknowledged. T.O. is grateful for a JSPS Research Fellowship for Young Scientists.

References

1. (a) T. Mukaiyama and T. Yamada, *Bull. Chem. Soc. Jpn.*, 1995, **68**, 17; (b) *Advances in Catalytic Activation of Dioxygen by Metal Complexes*, ed. L. L. Simandi, Kluwer Academic, Dordrecht, The Netherlands, 2003; (c) J. Piera and J.-E. Bäckvall, *Angew. Chem. Int. Ed.*, 2008, **47**, 3506; (d) E. Roduner, W. Kaim, B. Sarkar, V. B. Urlacher, J. Pleiss, R. Gläser, W.-D. Einicke, G. A. Sprenger, U. Beifuß, E. Klemm, C. Liebner, H. Hieronymus, S.-F. Hsu, B. Plietker and S. Laschat, *ChemCatChem*, 2013, **5**, 82.
2. (a) R. Irie and T. Katsuki, *Chem. Record.*, 2004, **4**, 96; (b) C. Parmeggiani and F. Cardona, *Green Chem.*, 2012, **14**, 547.
3. (a) *Asymmetric Synthesis with Chemical and Biological Methods*, ed. D. Enders and K.-E. Jaeger, Wiley-VCH, Weinheim, 2007; (b) *Catalytic Asymmetric Synthesis*, ed. I. Ojima, John Wiley, Hoboken, NJ, 3rd edn, 2010; (c) *Privileged Chiral Ligands and Catalysts*, ed. Q.-L. Zhou, Wiley-VCH, Weinheim, 2011.
4. (a) H. Gröger, W. Hummel, S. Borchert and M. Krausser, in *Enzyme Catalysis in Organic Synthesis*, ed. K. Drauz, H. Gröger and O. May, Wiley-VCH, Weinheim, 3rd edn, 2012, vol. 2, pp. 1001–1033; (b) J. Hasegawa, H. Nanba and Y. Yasohara, in *Asymmetric Catalysis on Industrial Scale: Challenges, Application and Solution*, ed. H.-U. Blaser and H.-J. Jürgen, Wiley-VCH, Weinheim, 3rd edn, 2011, pp. 171–185.
5. (a) E. Vedejs and M. Jure, *Angew. Chem. Int. Ed.*, 2005, **44**, 3974; (b) M. J. Schultz and M. S. Sigman, *Tetrahedron*, 2006, **62**, 8227; (c) M. Wills, *Angew. Chem. Int. Ed.*, 2008, **47**, 4264.
6. (a) V. S. Martin, S. S. Woodard, T. Katsuki, Y. Yamada, M. Ikeda and K. B. Sharpless, *J. Am. Chem. Soc.*, 1981, **103**, 6237; (b) J. C. Fiaud and H. B. Kagan, *Top. Stereochem.*, 1988, **18**, 249; (c) J. M. Keith, J. F. Larrow and E. N. Jacobsen, *Adv. Synth. Catal.*, 2001, **343**, 5.
7. M. A. Beckett and R. B. Homer, *Inorg. Chim. Acta*, 1986, **122**, L5.

8. (a) K. Masutani, T. Uchida, R. Irie and T. Katsuki, *Tetrahedron Lett.*, 2000, **41**, 5119; (b) Y. Nakamura, H. Egami, K. Matsumoto, T. Uchida and T. Katsuki, *Tetrahedron*, 2007, **63**, 6383; (c) H. Mizoguchi, T. Uchida and T. Katsuki, *Angew. Chem. Int. Ed.*, 2014, **53**, 3178.
9. (a) D. R. Jensen, J. S. Pugsley and M. S. Sigman, *J. Am. Chem. Soc.*, 2001, **123**, 7475; (b) E. M. Ferreira and B. M. Stoltz, *J. Am. Chem. Soc.*, 2001, **123**, 7725.
10. J. T. Bagdanoff and B. M. Stoltz, *Angew. Chem. Int. Ed.*, 2004, **43**, 353.
11. (a) S.-S. Weng, M.-W. Shen, J.-Q. Kao, Y. S. Munot and C.-T. Chen, *Proc. Natl. Acad. Sci. U. S. A.*, 2006, **103**, 3522; (b) V. D. Pawar, S. Bettigeri, S.-S. Weng, J.-Q. Kao and C.-T. Chen, *J. Am. Chem. Soc.*, 2006, **128**, 6308; (c) C.-T. Chen, S. Bettigeri, S.-S. Weng, V. D. Pawar, Y.-H. Li, C.-Y. Liu and W.-Z. Lee, *J. Org. Chem.*, 2007, **72**, 8175; (d) C.-T. Chen, J.-Q. Kao, S. B. Salunke and Y.-H. Lin, *Org. Lett.*, 2011, **13**, 26.
12. A. T. Radosevich, C. Musich and F. D. Toste, *J. Am. Chem. Soc.*, 2005, **127**, 1090.
13. K. Murata, T. Ikariya and R. Noyori, *J. Org. Chem.*, 1999, **64**, 2186.
14. Z. M. Heiden and T. B. Rauchfuss, *J. Am. Chem. Soc.*, 2007, **129**, 14303.
15. (a) S. Arita, T. Koike, Y. Kayaki and T. Ikariya, *Angew. Chem. Int. Ed.*, 2008, **47**, 2447; (b) S. Arita, T. Koike, Y. Kayaki and T. Ikariya, *Chem. Asian J.*, 2008, **3**, 1479; (c) T. Ikariya, S. Kuwata and Y. Kayaki, *Pure Appl. Chem.*, 2010, **82**, 1471.
16. P. Muthupandi, S. K. Alamsetti and G. Sekar, *Chem. Commun.*, 2009, 3288.
17. (a) H. Egami and T. Katsuki, *J. Am. Chem. Soc.*, 2009, **131**, 6082; (b) K. Matsumoto, H. Egami, T. Oguma and T. Katsuki, *Chem. Commun.*, 2012, **48**, 5823; (c) H. Egami, K. Matsumoto, T. Oguma, T. Kunisu and T. Katsuki, *J. Am. Chem. Soc.*, 2010, **132**, 13633.
18. (a) J. P. Klinman, *Chem. Rev.*, 1996, **96**, 2541; (b) Y. Wang, J. L. DuBois, B. Hedman, K. O. Hodgson and T. D. P. Stack, *Science*, 1998, **279**, 537; (c) M. M. Whittaker and J. W. Whittaker, *Biochemistry*, 2001, **40**, 7140; (d) J. W. Whittaker, *Chem. Rev.*, 2003, **103**, 2347.
19. H. Shimizu, S, Onitsuka, H. Egami and T. Katsuki, *J. Am. Chem. Soc.*, 2005, **127**, 5396.
20. T. Kunisu, T. Oguma and T. Katsuki, *J. Am. Chem. Soc.*, 2011, **133**, 12937.
21. (a) S. K. Mandal, D. R. Jensen, J. S. Pugsley and M. S. Sigman, *J. Org. Chem.*, 2003, **68**, 4600; (b) S. K. Mandal and M. S. Sigman, *J. Org. Chem.*, 2003, **68**, 7535; (c) S. Krishnan, J. T. Bagdanoff, D. C. Ebner, Y. K. Ramtohul, U. K. Tambar and B. M. Stoltz, *J. Am. Chem. Soc.*, 2008, **130**, 13745; (d) D. C. Ebner, J. T. Bagdanoff, E. M. Ferreira, R. M. McFadden, D. D. Caspi, R. M. Trend and B. M. Stoltz, *Chem. Eur. J.*, 2009, **15**, 12978.
22. (a) A. Miyata, M. Murakami, R. Irie and T. Katsuki, *Tetrahedron Lett.*, 2001, **42**, 7076; (b) A. Miyata, M. Furukawa, R. Irie and T. Katsuki, *Tetrahedron Lett.*, 2002, **43**, 3481; (c) H. Egami, H. Shimizu and T. Katsuki, *Tetrahedron Lett.*, 2005, **46**, 783.

23. (a) H. Shimizu, K. Nakata and T. Katsuki, *Chem. Lett.*, 2002, **31**, 1080; (b) H. Shimizu and T. Katsuki, *Chem. Lett.*, 2003, **32**, 480.
24. (a) H. Akahori, K. Morita, A. Nishijima, T. Mitsuhashi, K. Ohkubo and S. Fukuzumi, *J. Soc. Photogr. Sci. Technol. Jpn.*, 2003, **66**, 491; (b) H. Akahori, K. Morita, A. Nishijima, T. Mitsuhashi, K. Ohkubo and S. Fukuzumi, *J. Imaging Sci. Technol.*, 2003, **47**, 124.
25. (a) M. McCarthy and P. J. Guiry, *Tetrahedron*, 2001, **57**, 3809; (b) Y. Chen, S. Yekta and A. K. Yudin, *Chem. Rev.*, 2003, **103**, 3155; (c) J. M. Brunel, *Chem. Rev.*, 2005, **105**, 857; (d) M. C. Kozlowski, B. J. Morgan and E. C. Linton, *Chem. Soc. Rev.*, 2009, **38**, 3193.
26. M. Nakajima, K. Kanayama, I. Miyoshi and S. Hashimoto, *Tetrahedron Lett.*, 1995, **36**, 9519.
27. (a) X. Li, J. Yang and M. C. Kozlowski, *Org. Lett.*, 2001, **3**, 1137; (b) X. Li, J. B. Hewgley, C. A. Mulrooney, J. Yang and M. C. Kozlowski, *J. Org. Chem.*, 2003, **68**, 5500; (c) A. Caselli, G. B. Giovenzana, G. Palmisano, M. Sisti and T. Pilati, *Tetrahedron: Asymmetry*, 2003, **14**, 1451; (d) K. H. Kim, D.-W. Lee, Y.-S. Lee, D.-H. Ko and D.-C. Ha, *Tetrahedron*, 2004, **60**, 9037; (e) J. Gao, J. H. Reibenspies and A. E. Martell, *Angew. Chem. Int. Ed.*, 2003, **42**, 6008.
28. (a) R. Irie, K. Masutani and T. Katsuki, *Synlett*, 2000, 1433; (b) K. Masutani, R. Irie and T. Katsuki, *Chem. Lett.*, 2002, **31**, 36.
29. (a) C.-Y. Chu, D.-R. Hwang, S.-K. Wang and B.-J. Uang, *Chem. Commun.*, 2001, 980; (b) Z. Luo, Q. Liu, L. Gong, X. Cui, A. Mi and Y. Jiang, *Chem. Commun.*, 2002, 914; (c) Z. Luo, Q. Liu, L. Gong, X. Cui, A. Mi and Y. Jiang, *Angew. Chem. Int. Ed.*, 2002, **41**, 4532.
30. Q.-X. Guo, Z.-J. Wu, Z.-B. Luo, Q.-Z. Liu, J.-L. Ye, S-W. Luo, L.-F. Cun and L.-Z. Gong, *J. Am. Chem. Soc.*, 2007, **129**, 13927.
31. (a) H. Somei, Y. Asano, T. Yoshida, S. Takizawa, H. Yamataka and H. Sasai, *Tetrahedron Lett.*, 2004, **45**, 1841; (b) S. Takizawa, T. Katayama, C. Kameyama, K. Onitsuka, T. Suzuki, T. Yanagida, T. Kawai and H. Sasai, *Chem. Commun.*, 2008, 1810; (c) S. Takizawa, T. Katayama and H. Sasai, *Chem. Commun.*, 2008, 4133; (d) S. Takizawa, T. Katayama, H. Somei, Y. Asano, T. Yoshida, C. Kameyama, D. Rajesh, K. Onitsuka, T. Suzuki, M. Mikami, H. Yamataka, D. Jayaprakash and H. Sasai, *Tetrahedron*, 2008, **64**, 3361; (e) S. Takizawa, *Chem. Pharm. Bull.*, 2009, **57**, 1179.
32. (a) T. Glaser, R. H. Pawelke and M. Heidemeier, *Z. Anorg. Allg. Chem.*, 2003, **629**, 2274; (b) J. B. H. Strautmann, S. D. George, E. Bothe, E. Bill, T. Weyhermüller, A. Stammier, H. Bögge and T. Glaser, *Inorg. Chem.*, 2008, **47**, 6804; (c) T. Kurahashi, Y. Kobayashi, S. Nagatomo, T. Tosha, T. Kitagawa and H. Fujii, *Inorg. Chem.*, 2005, **44**, 8156.
33. (a) D. Magdziak, S. J. Meek and T. R. R. Pettus, *Chem. Rev.*, 2004, **104**, 1383; (b) T. Dohi, *Chem. Pharm. Bull.*, 2010, **58**, 135; (c) L. Pouységu, D. Deffieux and S. Quideau, *Tetrahedron*, 2010, **66**, 2235; (d) S. P. Roche and J. A. Porco Jr, *Angew. Chem. Int. Ed.*, 2011, **50**, 4068; (e) C.-X. Zhuo, W. Zhang and S.-L. You, *Angew. Chem. Int. Ed.*, 2012, **51**, 12662.

34. (a) J. Zhu, N. P. Grigoriadis, J. P. Lee and J. A. Porco Jr, *J. Am. Chem. Soc.*, 2005, **127**, 9342; (b) S. Dong, J. Zhu and J. A. Porco Jr, *J. Am. Chem. Soc.*, 2008, **130**, 2738.
35. T. Oguma and T. Katsuki, *J. Am. Chem. Soc.*, 2012, **134**, 20017.
36. (a) L. Jurd, *Tetrahedron*, 1977, **33**, 163; (b) R. W. Van de Water and T. R. R. Pettus, *Tetrahedron*, 2002, **58**, 5367; (c) *Quinone Methide*, ed. S. E. Rokita, John Wiley, Hoboken, NJ, 2009; (d) T. P. Pathak and M. S. Sigman, *J. Org. Chem.*, 2011, **76**, 9210.
37. T. Oguma and T. Katsuki, *Chem. Commun.*, 2014, **50**, 5053.

CHAPTER 10

Overview: Representative Experimental Procedures, Comparative Tables and Conclusions

FRANCESCA CARDONA*[a] AND CAMILLA PARMEGGIANI*[b]

[a] Dipartimento di Chimica 'Ugo Schiff', Università degli Studi di Firenze, via della Lastruccia 3–13, 50019 Sesto Fiorentino (FI), Italy; [b] Consiglio Nazionale delle Ricerche, Istituto Nazionale di Ottica, c/o Laboratorio Europeo di Spettroscopia Non Lineare, via Nello Carrara 1, 50019 Sesto Fiorentino (FI), Italy
*Email: francesca.cardona@unifi.it; camilla.parmeggiani@unifi.it

10.1 Introduction

This chapter aims to give a brief overview of the most important and mature procedures developed for the aerobic oxidation of alcohols catalyzed by transition metals, which are covered in detail in the previous chapters. We focus on the possibility of performing the oxidation of one class of alcohol in the presence of another, by furnishing some representative procedures. Moreover, the opportunity to employ alternative solvents (aiming at the recovery and reuse of catalysts) is mentioned, with some interesting examples. Finally, some procedures that have been scaled up to a larger scale are mentioned.

RSC Green Chemistry No. 28
Transition Metal Catalysis in Aerobic Alcohol Oxidation
Edited by Francesca Cardona and Camilla Parmeggiani

Published by the Royal Society of Chemistry, www.rsc.org

10.2 Oxidation of Primary Alcohols to Aldehydes

The main problem with this reaction is the selectivity, since aldehydes are susceptible to over-oxidation to carboxylic acids. This is particularly evident with aliphatic aldehydes and when water is employed as solvent, since the hydrate form of aldehydes is more inclined to undergo oxidation. Indeed, most procedures were developed with activated allylic and benzylic substrates and not with the less reactive aliphatic alcohols.

Copper is perhaps the metal of choice for this transformation. After the pioneering work of Markò and co-workers,[1] many other catalytic systems have been developed. One of the most effective procedure for the conversion of a wide range of primary alcohols to aldehydes (including non-activated ones) is that developed by Hoover and Stahl, which uses a bipyridyl (bpy)-Cu(I)/TEMPO catalytic system in the presence of *N*-methylimidazole (NMI) as the base and CH_3CN as solvent (see also Chapter 2, Figure 2.6a).[2] Exquisite selectivities for primary over secondary alcohols were observed, allowing selective oxidation of diols without the need for protecting groups (Scheme 10.1).

Representative procedure for the oxidation of primary alcohols to aldehydes.[2] To a solution of alcohol (1 mmol) in dry CH_3CN (1 mL) in a 20 mm culture tube, the following solutions were added: (1) $[Cu(MeCN)_4]X^-$ ($X = OTf^-$, BF_4^- or PF_6^-, 0.05 mmol in 1 mL CH_3CN); (2) bpy (0.05 mmol in 1 mL CH_3CN); (3) TEMPO (0.05 mmol in 1 mL CH_3CN); (4) *N*-methylimidazole (0.1 mmol in 1 mL CH_3CN). The dark red–brown reaction mixture was stirred rapidly while open to air and monitored by TLC until no starting material remained (often accompanied by a change in reaction color to green–blue). Preliminary studies indicated that the reactions described here are not subject to mass transfer effects and the rate of mixing and stir bar shape do not have a significant impact on the outcome of the reaction. Larger scale reactions were run in oversized round-bottomed flasks: (i) 10 mmol scale reactions were carried out with 50 mL of CH_3CN in a 250 mL flask; (ii) 50 mmol scale reactions were carried out with 250 mL of CH_3CN in a 1 L flask. Reactions to form volatile aldehydes were carried out in a round-bottomed flask fitted with a reflux condenser, a septum and a balloon of house air (or O_2).

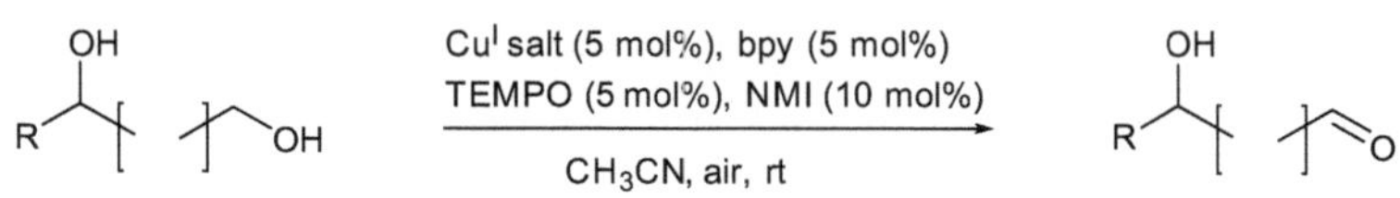

Scheme 10.1 Selective oxidation of primary over secondary alcohols in unprotected diols.

Representative procedure for the selective oxidation of unprotected diols.[2] To a solution of alcohol (1 mmol) in dry CH_3CN (1 mL) in a 20 mm culture tube were added the following solutions: (1) Cu salt (0.05 mmol in 1 mL CH_3CN); (2) bpy (0.05 mmol in 1 mL CH_3CN); (3) TEMPO (0.05 mmol in 1 mL CH_3CN); (4) base (0.1 mmol in 1 mL CH_3CN). The reaction mixture was stirred rapidly while open to air and monitored by TLC until no starting material remained. The product was then worked up according to one of the following two methods. *Workup Method A.* The reaction mixture was neutralized with 1 N HCl and diluted with water (~10 mL) and extracted with dichloromethane (3×20 mL). The combined organics were washed with brine, dried over Na_2SO_4 and concentrated. The crude product was purified by silica gel column chromatography (gradient elution of EtOAc in hexane). *Workup Method B.* The crude reaction mixture was concentrated and purified by silica gel column chromatography (gradient elution of EtOAc in hexane).

10.3 Oxidation of Secondary Alcohols to Ketones

Iron- and ruthenium-based catalysts were reported to achieve this reaction in a highly selective way. Oxidation of secondary alcohols in the presence of primary alcohols, which were inert under the reaction conditions, was reported using a binary Fe(III) compound (see also Chapter 6, Scheme 6.1),[3] while a biomimetic bimetallic catalytic system composed of $RuCl(OAc)(Ph_3)_3$ in combination with hydroquinone and a cobalt–Schiff base complex led to one of the fastest catalytic systems reported for the oxidation of secondary alcohols (see also Chapter 3, Figures 3.1 and 3.2, and Chapter 7, Equation 7.2).[4,5]

However, the best optimized homogeneous procedures for the oxidation of secondary alcohols employ palladium-based catalysts. Sigman and co-workers developed three different Pd(II) catalysts,[6] evaluated the substrate scope and the reaction conditions of each of them in comparison studies and concluded that the simple $Pd(OAc)_2$–TEA system represents the most convenient among the three,[7] as discussed in detail in Chapter 4 (Table 4.3).

Representative procedure for the $Pd(OAc)_2$–TEA-catalyzed oxidation of benzylic, secondary aliphatic and cyclic allylic alcohols.[7] To a 25 mL round-bottomed flask equipped with a stir bar were added 6.7 mg of $Pd(OAc)_2$ (0.03 mmol, 0.03 equiv.) and 200 mg of powdered, freshly activated 3 Å molecular sieves. To this were added 0.5 mL of THF, 2.83 mL of toluene and 8.4 μL of TEA (0.06 mmol, 0.06 equiv.). A balloon of oxygen was attached *via* a three-way joint. The flask was evacuated and refilled with oxygen three times followed by vigorous stirring for 30 min at room temperature under O_2. To this solution 1 mmol of alcohol was added and the mixture was stirred vigorously at room temperature under a balloon of O_2. The reaction progress was monitored by GC. After 12 h, the reaction mixture was placed directly on a plug of silica, washed with pentane to remove toluene and eluted with diethyl ether. The ether was removed *in vacuo* to yield the desired carbonyl product. For alcohols with incomplete oxidation, the desired carbonyl product was isolated *via* column chromatography with mixtures of diethyl ether–hexanes as the eluting solvent.

10.4 Oxidation of Primary Alcohols to Carboxylic Acids

The oxidation of primary alcohols typically affords the corresponding carboxylic acids when water is used as solvent, since this medium facilitates aldehyde hydration and therefore its easy oxidation.

Various metals supported on solid matrices have been investigated for this conversion using water as solvent or co-solvent. Ruthenium supported on carbon (10% Ru/C) (see also Chapter 3) as a catalyst in toluene (at 90 °C) under an oxygen atmosphere was able to convert primary aliphatic alcohols selectively to carboxylic acids when water was added as a co-solvent (Scheme 10.2).[8]

Also concerning heterogeneous catalysis, Hutchings and co-workers[9] showed that gold supported on graphite could oxidize glycerol to glycerate with 100% selectivity using dioxygen as the oxidant in water, with yields approaching 60% (see also Chapter 5). It was observed that the selectivity to glyceric acid and the glycerol conversion were strongly dependent on the glycerol to-NaOH ratio (Scheme 10.3).

Representative procedure for the oxidation of glycerol to glyceric acid.[9] The catalyst (1% Au/graphite) was suspended in a solution of glycerol (12 mmol) in deionized water (20 mL) containing NaOH (12 mmol) in a Parr autoclave (50 mL) and the autoclave was pressurized to the required pressure (3 bar) with oxygen and heated to 60 °C. The reaction mixture was stirred (1500 rpm) for 3 h, affording 56% glycerol conversion and 100% glyceric acid selectivity.

Finally, the selective oxidation of benzylic alcohols to carboxylic acids or esters by self-esterification was recently reported in water-soluble Pd and Au nanoparticles.[10]

The important role of water in the selective oxidation of primary alcohols to the corresponding acids was shown recently with gold nanoparticles

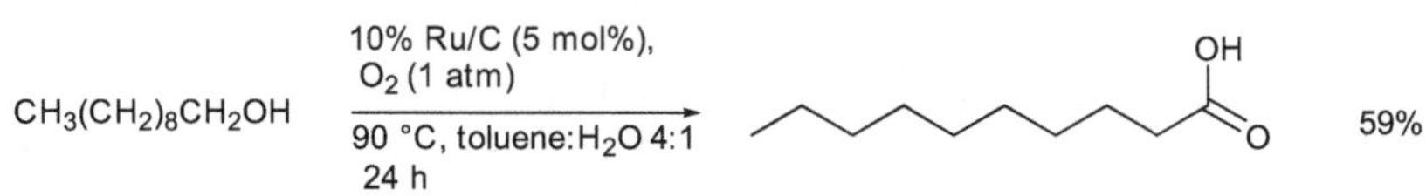

Scheme 10.2 Ru/C-catalyzed aerobic oxidation of primary aliphatic alcohols to carboxylic acids.

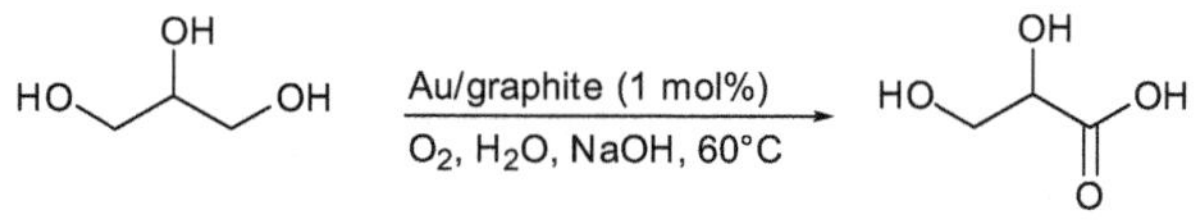

Scheme 10.3 Glycerol oxidation to glyceric acid.

deposited on carbon nanotubes. Such nanohybrid structures were able to convert benzylic alcohol selectively to benzoic acid when employing toluene–water (1 : 1) as solvent, while the reaction afforded benzaldehyde selectively when THF was employed. Upon addition of H_2O to the reaction mixture, benzaldehyde was further oxidized into benzoic acid in 99% yield.[11]

10.5 Oxidation of Propargyl Alcohols

Vanadium is certainly the metal of choice for oxidizing secondary propargyl alcohols to the corresponding ketones, as widely discussed in Chapter 6. The use of 1 mol% $VO(acac)_2$ in acetonitrile at 80 °C oxidized a variety of propargyl alcohols, including aryl, vinyl, alkynyl and aliphatic substrates (see also Chapter 6, Scheme 6.16).[12]

General procedure for the oxidation of propargylic alcohols.[12b] To a solution of $VO(acac)_2$ (2.65 mg, 0.01 mmol) in acetonitrile (1.5 mL) in a 10 mL two-necked round-bottomed flask was added MS 3Å (500 mg, powder). Next, a solution of propargylic alcohol (1 mmol) in acetonitrile (0.5 mL) was added and the resulting mixture was stirred. Oxygen gas was then introduced into the flask from an O_2 balloon under atmospheric pressure and the mixture was stirred vigorously for 3 h at 80 °C under oxygen. The mixture was then cooled to room temperature and MS 3Å was separated by filtration through a glass filter. The amount of the product was determined by GLC analysis. For isolation of the product, the solvent was evaporated and the residue was purified by column chromatography (Merck silica gel 60; hexane–ethyl acetate as eluent).

A practical synthetic procedure for the conversion of propargyl alcohol to propionaldehyde on a fairly large scale (56 g) was also reported recently by Liu and Ma employing $Fe(NO_3)_3 \cdot 9H_2O$, TEMPO and NaCl as catalysts, with oxygen at room temperature in *p*-xylene.[13] Further details can be found in Chapter 6.

10.6 Nanoparticles Preparation

Palladium and gold catalysts have been found particularly active in the form of nanoparticles. The general routes to nanocluster/nanoparticle synthesis involve chemical reduction of transition metal salts with a reducing agent in the presence of a stabilizer for the metal. The resulting stabilized metal nanoclusters dispersed in solution can be used as catalysts as such or subsequently heterogenized on solid supports using different techniques (*e.g.* surface adsorption, covalent anchoring, embedding by sol–gel techniques). This topic is discussed in Chapters 4 and 5, for palladium and gold, respectively. Here we present only some representative procedures.

Preparation of hydroxyapatite-supported Pd nanoclusters.[14] $(NH_4)_2HPO_4$ (40.0 mmol) was dissolved in deionized water (150 mL) and the pH was adjusted to 11 with aqueous NH_3 solution. To a solution of $Ca(NO_3)_2 \cdot 4H_2O$ (66.7 mmol) in deionized water (120 mL) adjusted to pH 11 with aqueous NH_3 solution was added dropwise over 30 min the above solution with vigorous stirring at room temperature and then the milky solution obtained was heated at 90 °C for 10 min. The precipitate was filtered, washed with deionized water and dried at 110 °C, giving stoichiometric hydroxyapatite, $Ca_{10}(PO_4)_6(OH)_2$ (HAP-0). HAP-0 (2.0 g) was stirred at 25 °C for 3 h in 150 mL of an acetone solution of $PdCl_2(PhCN)_2$ (2.67×10^{-4} M). The slurry obtained was filtered, washed with acetone and dried under vacuum, yielding 2.01 g of PdHAP-0 (Pd content 0.02 mmol g^{-1}).

Preparation of Pd–neocuproine nanoparticles.[15] Neocuproine (0.05 mmol, 10.6 mg) was added to a solution of $Pd(O_2CCF_3)_2$ (0.05 mmol, 16.6 mg) in water (12.5 mL). The mixture was stirred for 1 h and the cosolvent (ethylene carbonate, dipropylene glycol or PEG-600, 12.5 mL) was added. After 3 min, the mixture was reduced under H_2 with vigorous stirring for 15 min. Oxidation of 2-hexanol: standard catalytic experiments were carried out in a closed Hastelloy C autoclave (150 mL). 2-Hexanol (10 mmol) and NaOAc (0.5 mmol, 41 mg) were added to a solution of the palladium–neocuproine complex or a suspension of the Pd nanoparticles in 1:1 water–co-solvent mixture (25 mL) and pressurized with 8% O_2 in N_2 to 50 bar and heated to 100 °C, while stirring at 750 rpm. After the reaction, the autoclave was cooled to room temperature and depressurized. The product mixture was extracted with Et_2O and the organic layer was washed with water and dried over $MgSO_4$, *n*-dodecane was added as an external standard to the organic solutions and the latter were analyzed by GC.

Preparation of silica-supported gold nanoparticles.[16] Gold catalyst (1 wt%) supported on silica (Aerosil 200, Grace and Davison, surface area 200 m^2 g^{-1}, pore volume 1.7 mL g^{-1}) was prepared in the following manner: 8.5 mL of an aqueous solution of chloroauric acid containing 50 mg of Au was added to 5 g of silica while mixing with a glass rod for 10 min, then 8 mL of aqueous NH_3 (3%) was added followed by 10 mL of an aqueous solution of $NaBH_4$ (60 mg per 10 mL) and the suspension was stirred at 333 K for 10 min. The solid product was filtered and washed several times with water until the washings contained no chloride.

10.7 Oxidations in Alternative Solvents

10.7.1 Water

Water is in principle an ideal solvent from the environmental point of view. Moreover, it solves the problems of flammability in processes that employ oxygen. An improved safety profile and more acceptable scalability can be achieved by performing aerobic oxidations in this solvent. The use of water, in addition to being benign itself, can introduce many other benefits such as reaction and selectivity improvements, simplification of the work-up procedures and mild reaction conditions.[17]

As addressed in Chapter 5, gold based catalysts are the ideal choice in water, since they do not suffer from deactivation when employed in the presence of oxygen. Some oxidations of benzylic alcohols in water as solvent were also reported using copper-based catalysts (see Chapter 2 for further details).

A highly active catalyst in water is represented by the water-soluble palladium(II) complex of sulfonated bathophenanthroline introduced by Sheldon and co-workers.[18] This stable, recyclable catalyst allowed oxidation in a two-phase aqueous–organic medium in 5 h at 100 °C and 30 bar air with 0.25 mol% catalyst. No organic solvent was required (except for solid alcohols) and the carbonyl product was recovered easily by phase separation. Primary alcohols afforded the corresponding carboxylic acids *via* further oxidation of the aldehyde intermediate; otherwise, in the presence of 1 mol% of TEMPO, the aldehyde was obtained selectively in high yield. For further details, see Chapter 4.

The hydroxyapatite-supported palladium nanoclusters (Pd/HAP-0) described above[14] were also applied to the oxidation of wide variety of alcohols bearing heteroatoms, which were oxidized by this catalyst in water at 110 °C.

10.7.2 Ionic Liquids

Ionic liquids are peculiar alternative solvents for catalytic reactions[19] and have also been employed in aerobic alcohol oxidations catalyzed by transition metals. At the end of the reaction, products can be extracted with standard organic solvents and the ionic liquid containing the catalyst can be reused for the next run.

The use of a pyridyl-based ionic liquid, namely 1-butyl-4-methylpyridinium hexafluorophosphate, [bmpy][PF_6], was reported in a room-temperature aerobic oxidation of primary alcohols catalyzed by the three-component system acetamido-TEMPO–$Cu(ClO_4)_2$–DMAP (see also Chapter 2, Figure 2.21). This procedure allowed the recovery and reuse of catalyst for up to five runs without a decrease in activity.[20] The final aldehyde products were extracted with *n*-pentane, in which both acetamido-TEMPO and DMAP were insoluble. No oxidation was observed with secondary alcohols.

Tetramethylammonium hydroxide and Aliquat 336 (tricaprylmethylammonium chloride) were used as solvents for the rapid oxidation of

1-phenylethanol and benzyl alcohol in a ruthenium-catalyzed reaction (see also Chapter 3).[21] The tetramethylammonium hydroxide–$RuCl_2(PPh_3)_3$ catalyst could be reused after extraction of the product with diethyl ether.

10.7.3 Fluorinated Solvents

Fluorinated solvents offer an interesting alternative to standard organic solvents in aerobic alcohol oxidation. If the catalyst is suitably tailored with 'fluorous ponytails', it becomes soluble in the fluorous phase and this allows its recovery and reuse by simple extraction of the products with an organic solvent at the end of the reaction. Several examples of this technique with copper-based catalysts have been described.

Knochel and co-workers used a biphasic solvent system composed of chlorobenzene and perfluorooctyl bromide together with a pyridine ligand containing fluorinated ponytails for a CuBr–Me_2S–TEMPO catalytic system.[22] The biphasic reaction mixture was stirred at 90 °C while a gentle stream of oxygen was passing. At the end of the reaction, the mixture was cooled to 0 °C, the organic layer was decanted and the fluorous phase was washed with chlorobenzene. The fluorous phase was used directly up to eight times in further reaction runs with little decrease in activity (see also Chapter 2, Figure 2.19). Other fluorinated ligands for copper complexes have been reported in Chapter 2, (Figure 2.23).

10.7.4 Solvent-Free Systems

Chemists must take in account that a solvent is not always necessary, and several interesting examples of aerobic oxidations of alcohols that do not need the presence of a solvent have been reported.

The three-component system composed of acetamido-TEMPO–$Cu(ClO_4)_2$–DMAP described above was also applied under solvent-free conditions (see also Chapter 2). The three catalyst components were simply recovered by addition of a non-polar solvent (hexanes) that selectively dissolved the product aldehydes.[23] In the case of solid alcohols, PEG-200 (not oxidized under these reaction conditions) was used as solvent.

Gold-based catalysts usually show selectivity towards carboxylic acids when used in water; employing solvent-free conditions instead results in good selectivity towards carbonyl compounds (see also Chapter 5 for further details). For instance, Corma and co-workers showed that Au/CeO_2 catalyst was active in the selective oxidation of alcohols to aldehydes and ketones under solvent-free conditions, by using O_2 as oxidant without the requirement for the addition of NaOH to achieve high activity.[24]

Among solvent-free oxidations, of particular interest are those performed in the gaseous phase (see also Chapters 5 and 8), which proceed with the same selectivity and employ silica-supported gold nanoparticles as catalysts.[16]

A nice example of iron-based catalysis working in solvent-free conditions was recently reported by Zhang and co-workers.[25] A task-specific bimagnetic imidazolium salt, [Imim-TEMPO][$FeCl_4$], containing cooperative functionalities (see also Chapter 6, Scheme 6.8) was easily synthesized and applied to the selective aerobic oxidation of aromatic alcohols. In each run, [Imim-TEMPO][$FeCl_4$] could be readily recovered by extraction with diethyl ether, then subjected to a subsequent run by charging with fresh substrate and $NaNO_2$. The catalyst could be reused at least five times with retention of high activity and selectivity.

Bimetallic nanoclusters have also been found to work well in solvent-free conditions, as reported by Hutchings and co-workers using Au/TiO_2 alloyed with Pd[26] (see also Chapters 5 and 7).

10.7.5 Supercritical Carbon Dioxide

Supercritical carbon dioxide ($scCO_2$) has been demonstrated to offer interesting opportunities as a medium for performing various catalytic reactions. In addition to its benign nature, the good miscibility with gaseous reagents and organic compounds allows remarkable accelerations of many reactions due to increased diffusion rates.[27]

A representative example of aerobic oxidation in $scCO_2$ was reported by Leitner and co-workers.[28] They found that the giant palladium cluster [$Pd_{561}phen_{60}(OAc)_{180}$], dispersed in poly(ethylene glycol) (PEG), efficiently catalyzes the aerobic oxidation of alcohols in $scCO_2$ (Scheme 10.4).

The use of $scCO_2$ with Pd as catalyst is discussed in Chapter 4. In particular, the use of this alternative solvent in continuous-flow microreactors is described in Section 4.2.2, and an example of the oxidation of cinnamyl alcohol is reported in Section 4.2.5.

10.8 Catalyst Recovery

The main advantage of heterogeneous over homogeneous catalysts is the easy recovery by simple filtration of the catalyst and the opportunity to reuse it. This is probably one of the main reasons why industry prefers heterogeneous catalysts. However, there is another interesting approach, that offers promising opportunities. This consists in the immobilization of the homogeneous catalysts in multiphase operation (liquid–liquid biphasic catalysis).[29] Examples of this technique have already been provided in previous sections dealing with fluorinated solvents, ionic liquids and $scCO_2$. Some other interesting examples of catalyst recovery working in a homogeneous phase are briefly mentioned in this section.

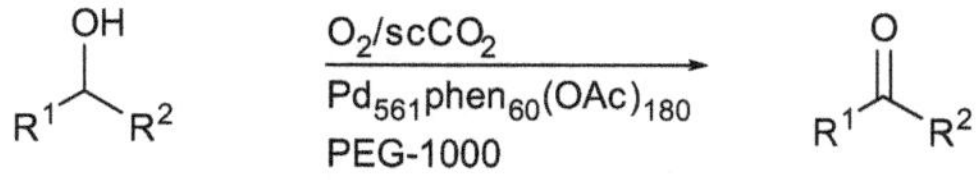

Scheme 10.4 Oxidation with palladium nanoparticles in $scCO_2$.

Jiang and Ragauskas (see Chapter 2, Figure 2.9a)[30] reported an oxidation system in which the catalyst (a modified TEMPO with an NHAc group) soluble in DMSO was recovered by simple extraction with *n*-pentane of the final products and the catalytic phase was reused by addition of fresh starting material. Another similar interesting example recovered the catalyst by antisolvent precipitation (addition of hexanes to the reaction mixture and decantation), as described in Section 10.7.4.[23]

A different and unusual approach for catalyst recovery was developed by Mizuno and co-workers, who showed that a ruthenium hydroxide species on magnetite [$Ru(OH)_x/Fe_3O_4$] performed very well and catalyst/product(s) separation was extremely simple. Indeed, after completion of the oxidation reaction, a permanent magnet was attached to the outside wall of the glass reactor to 'hold' the catalyst magnetically and the reaction solution including the product(s) was separated by simple decantation (see also Chapter 3, Figure 3.14).[31]

10.9 Some Scaled-Up Procedures

Large-scale applications of aerobic alcohol oxidation are often prevented by safety concerns associated with the combination of O_2 and organic solvents and reagents and by the frequent use of halogenated solvents.[32] Moreover, some oxidation methods have been optimized with non-standard solvents such as fluorinated compounds or ionic liquids, and some of the catalysts with the best reported activities are not commercially available or they have a limited lifetime. All these factors limit the widespread adoption of aerobic oxidation reactions in synthetic chemistry and in particular make the development of practical and scalable procedures very difficult.

However, some efficient and scaled-up procedures have been reported. The method of Stahl and co-workers employing (bpy)Cu(I) salts/TEMPO in CH_3CN as solvent with ambient air as the oxidant was applied on a 10 g scale to eight different alcohols, with reaction times varying from 20 min to 24 h, depending on the alcohol used and the purification methods, each taking about 2 h.[33] The total time necessary for the complete protocol ranged from 3 to 26 h.

The safety hazards typically associated with aerobic oxidations on a large scale could also be overcome by using dilute air (8% O_2 in N_2) as the oxidant in a continuous flow reactor, which allowed a $Pd(OAc)_2$/pyridine-catalyzed aerobic oxidation reaction to be tested on a 1 kg scale in a 7 L flow reactor.[34]

Representative experimental procedure for aerobic alcohol oxidation in a 7 L flow reactor.[34] An oven used tφ regulate the reaction-zone temperature of the flow reactor was set to 100 °C. The flow reactor was rinsed with dry toluene and dried by passing nitrogen gas through the tubing at 100 °C. The reactor was pressurized by applying a 500 psig nitrogen back-pressure from a high-pressure nitrogen cylinder connected to the vapor–liquid separator. The regulator for the diluted O_2 gas cylinder (8% O_2

in N_2) was set to 100 psi higher than the N_2 background pressure (600 psig). Two sequential metering valves connected to the O_2 outlet were adjusted to obtain the desired gas flow rate. The total gas flow out of the vapor-outlet valve was maintained at around 5 scf h^{-1}. The reactor was then purged with the dilute oxygen gas for 10–15 min (8% O_2 in N_2; 500 psig). The first syringe pump was charged with a $Pd(OAc)_2$ stock solution in toluene (10 mM) and the second with alcohol–pyridine stock solution in toluene (1.0 M/40 mM, respectively). The feed rates of both pumps were adjusted to achieve the desired liquid residence time in the reaction zone. The flow tube between the reaction zone and the vapor–liquid separator consisted of a jacketed stainless-steel tube and, prior to the start of the reaction, a flow of ethylene glycol cooling fluid (–10 to –20 °C) was passed through the outer jacket. Both syringe pumps were started to initiate the flow of liquid solution with the dilute oxygen gas through the reactor. After starting the pumps ($t = 0$), the time when the liquid started to accumulate in the liquid product tank was recorded as the actual liquid residence time.

Palladium(III) acetate in combination with neocuproine as ligand was also applied on a fairly large scale (20 mmol).[35]

Representative procedure for aerobic alcohol oxidation with Pd–neocuproine catalyst on a 20 mmol scale.[35] Standard catalytic experiments were carried out in a closed Hastelloy C autoclave (175 mL). The (neocuproine)$Pd(OAc)_2$ catalyst (0.02 mmol, 8.6 mg) and $NaOAc \cdot H_2O$ (1.0 mmol) were dissolved in 1 : 1 DMSO–water (50 mL). The autoclave was charged with the catalyst solution, alcohol (20 mmol) and internal standard (*n*-heptane, *n*-octane or *n*-dodecane) and was pressurized with 8% O_2 in N_2 and heated to 100 °C (30 ± 60 bar) while stirring at 750 rpm. After reaction, the autoclave was cooled to room temperature and depressurized. The product mixture was extracted with Et_2O and the organic layer was washed with water and dried over $MgSO_4$. For water–DMSO mixtures above 50 ± 55 vol.% water, virtually no DMSO was extracted with diethyl ether. Below this value, the organic layer could be washed with water to remove DMSO. A different external standard (*n*-dodecane or *n*-hexadecane) was added to the organic solutions and the latter were analyzed by GC-MS. Recoveries were always 98 ± 4% with this procedure. No blank reaction was observed. Isolated yields were generally above 90%.

A palladium catalyst with improved stability towards oxidative conditions was developed by Muldoon and co-workers.[36] They investigated the use of commercially available *N*,*O*-ligands to produce highly active Pd(II) complexes, which compared well with the previously reported *N,N*-ligands in the aerobic oxidation of 2-octanol on the gram scale.

10.10 Comparative Tables

Comparative tables are presented in Tables 10.1–10.5.

10.11 Conclusion

In the last 15–20 years, there has been substantial and still increasing interest in the area of metal-catalyzed aerobic alcohol oxidations. We can definitely affirm that the oxidation of alcohols represents a benchmark reaction for the development of new catalysts.

Regarding homogeneous alcohol oxidations, several procedures have been optimized. Based on copper as the active metal, Markò *et al.*'s Cu–phen and Stahl and co-workers' (bpy)Cu(I)–TEMPO–NMI systems are among the most mature procedures in the field, and with palladium as the active metal excellent reactions have been discovered, Sheldon and co-workers' Pd–(sulfonated bathophenanthroline) and Sigman and co-workers' Pd(OAc_2)–TEA systems being among the most optimized. Some of these catalytic systems were also tested on a larger scale to explore their potential utility in industrial processes.

Considerable efforts have also been made regarding sustainability, for instance, to replace common organic solvents with alternative ones such as ionic liquids, fluorinated solvents or $scCO_2$ or to perform the oxidation reactions in water or without the use of any solvent. Selective methods to obtain aldehydes or carboxylic acids from primary alcohols were also developed, in addition to selective oxidations in which one type of alcohol (*e.g.* primary *versus* secondary or the opposite) can be oxidized in the presence of the other.

The discovery that Pd and Au nanoparticles (or bimetallic systems) are effective catalysts for the oxidation of alcohol moieties has further expanded this research field in the search for new heterogeneous systems, which can allow the recovery and reuse of the metal catalyst and the formation of pure products. Although a tremendous amount of effort has been applied to the development of new catalysts, many improvements can still be envisioned. For instance, in order to use these methods in target synthesis, the scope of the individual catalytic systems must be broadened to include more complex alcohols that are synthetically relevant. Therefore, many challenges still need to be addressed in terms not only of their substrate scope, but also regarding their availability at low cost and robustness in the reaction mixture, with the aim of recycling and reuse in an industrial perspective.

Table 10.1 Representative selective oxidation methods.

Alcohol type	Catalyst	Solvent	Oxidant	Product	Ref.
Primary	Cu(I) salt, bpy, TEMPO, NMI	CH_3CN	Air	Aldehyde	2
Primary	Ru	Toluene–H_2O	O_2	Carboxylic acid	8
Secondary	Pd(II) salt, TEA	THF–toluene	O_2	Ketone	6,7
Propargyl	$VO(acac)_2$	CH_3CN	O_2	Ketone	12

Table 10.2 Selected homogeneous catalyzed oxidation of benzyl alcohol to benzaldehyde.

Catalyst	Oxidant	*T* (°C)	Time (h)	Solvent	Conversion (%) (selectivity, %)	Ref.
$[Cu(MeCN)_4]X^a$ (5 mol%) bpy (5 mol%) TEMPO (5 mol%) NMI (10 mol%)	Air	25	3	CH_3CN	>95 (<99)	2
$CuBr_2(bpy)$ (5 mol%) TEMPO (5 mol%) *t*BuOK (5 mol%)	Air	25	2.5	CH_3CN–H_2O (2:1)	100 (>99)	37
$RuCl_2(PPh_3)_3$ (0.5 mol%) TEMPO (1.5 mol%)	O_2–N_2 (8:92 v/v) 10 bar	100	2.5	PhCl	99 (>99)	38
$Pd(OAc)_2$ (3 mol%) TEA (6 mol%) 3 Å MS	O_2	25	12	THF–toluene	84 (>99)	6*a*
$(HQ)_2V^V(O)(O\textit{i}Pr)$ (2 mol%)^b NEt_3 (10 mol%)	Air	60	24	DCE	92 (>99)	39

^{a}X = PF_4^- or other anions.
^{b}HQ = 8-quinolinate.

Table 10.3 Selected homogeneous catalyzed oxidation of benzyl alcohol to benzaldehyde in alternative solvents or solvent-free conditions.

Catalyst	Oxidant	T (°C)	Time (h)	Solvent	Conversion (%) (selectivity, %)	Ref.
$Cu(ClO_4)_2$ (5 mol%) DMAP (10 mol%) Acetamido-TEMPO (5 mol %)	O_2	25	5	$[bmpy][PF_6]$	99 (>99)	20
Cu(II) salt (1 mol%)[a] TEMPO (5 mol%) K_2CO_3 (33 mol%)	Air	50	48	H_2O	99 (>99)	40
CuBr (1 mol%) Ligand (3 mol%) Acetamido-TEMPO (2 mol%)	O_2	25	0.6	None	100 (>95)	23
$Ru(PPh_3)_3Cl_2$ (1 mol%)	O_2	80	5	TMAH[b]	91 (>99)	21
$PhenS^*Pd(OAc)_2$[c] (0.5 mol%) NaOAc (10 mol%)	Air (30 bar)	100	10	H_2O	100 (>99)	18*a*
$[Imim\text{-}TEMPO][FeCl_4]$ (5 mol%) $NaNO_2$ (5 mol%)	O_2 (0.2 MPa)	30	1.5	H_2O	100 (>99)	25
$VO(acac)_2$ (5 mol%) DABCO (10 mol%)	O_2	80	6	$[bmim][PF_6]$	99 (>99)	41

[a]Salt: tetramer $[O \subset Cu_4(tea)_4(BOH)_4][BF_4]_2$.
[b]Tetramethylammonium hydroxide.
[c]PhenS* = bathophenanthroline disulfonate.

Table 10.4 Selected heterogeneous catalyzed oxidation of benzyl alcohol to benzaldehyde.

Catalyst	Oxidant	T (°C)	Time (h)	Solvent	Conversion (%) (selectivity, %)	Ref.
RuHAP	Air	80	3	Toluene	100 (95)	42
$Ru(OH)_x/Fe_3O_4$ (3.8 mol% Ru)	O_2	105	1	Toluene	>99 (>99)	31
Silica-supported Pd (3 mol% Pd)	O_2	90	2.5	Toluene	100 (>95)	43
Pd@SBA-15 (0.4 mol%)K_2CO_3 (1 equiv.)	Air	80	5.5	Toluene	>99 (<99)	44
Pd/MgO (1 mol%)	O_2	70–80	20	Trifluorotoluene	100 (92)	45
PdHAP-0 (0.2 mol%)	O_2	90	1	Trifluorotoluene	>99 (>99)	14
AuCNT (0.2 mol%); NaOH (3 equiv.)	Air	25	12	THF	100 (>99)	11
Au (2.7 wt%)/1%; CuO/3% MCM41	O_2	80	20	Toluene	73 (94)	46
Polyaniline-supported; $VO(acac)_2$ (2.3 mol% V)	O_2	100	9	Toluene	98 (>99)	47

Table 10.5 Selected heterogeneous catalyzed oxidation of benzyl alcohol to benzaldehyde in alternative solvents or solvent free conditions.

Catalyst	Oxidant	T (°C)	Time (h)	Solvent	Conversion (%) (selectivity, %)	Ref.
$[Cu(AcO)_2(py)]_2$/silica (0.75 mol%); TEMPO (3 mol%)	O_2 (1.5 bar)	80	12	$scCO_2$	100 (100)	48
Pd nanoparticles/PEG (5 mol% Pd)	O_2	80	13	$scCO_2$	83 (>99)	28*a*
ARP-Pd[a] (1 mol%)	O_2	100	1.5	H_2O	>97 (>99)	49
PdHAP-0	O_2	110	24	H_2O	>99 (>90)	14
Au nanoparticles (2.5% on TiO_2)	O_2 (2 atm)	100	–	None	77 (51)	50
$Pd/SiO_2–Al_2O_3–H_2$ (0.3 wt%)	O_2	50	24	None	99 (>99)	51

[a]ARP-Pd = amphiphilic resin dispersion of Pd nanoparticles'.

Abbreviations

bpy	2,2′-bipyridine
CNT	carbon nanotube
DABCO	1,4-diazabicyclo[2.2.2]octane
DMAP	*N*,*N*-dimethylaminopyridine
HAP	hydroxyapatite
Imim	imidazolium salt
MCM41	mesoporous silica-based support
MS	molecular sieves
NMI	*N*-methylimidazole
PEG	poly(ethylene glycol)
SBA-15	mesoporous silica-based support
TEA	triethylamine
TEMPO	2,2,6,6-tetramethylpiperidine-1-oxyl, free radical

References

1. I. E. Markò, P. R. Giles, M. Tsukazaki, S. M. Brown and C. J. Urch, *Science*, 1996, **274**, 2044.
2. J. M. Hoover and S. S. Stahl, *J. Am. Chem. Soc.*, 2011, **133**, 16901.
3. S. E. Martin and D. F. Suárez, *Tetrahedron Lett.*, 2002, **43**, 4475.
4. G.-Z. Wang, U. Andreasson and J.-E. Bäckvall, *J. Chem. Soc., Chem. Commun.*, 1994, 1037.
5. G. Csjernyik, A. H. Ell, L. Fadini, B. Pugin and J.-E. Bäckvall, *J. Org. Chem.*, 2002, **67**, 1657.
6. (a) M. J. Schultz, C. C. Park and M. S. Sigman, *Chem. Commun.*, 2002, 3034; (b) D. R. Jensen, M. J. Schultz, J. A. Mueller and M. S. Sigman, *Angew. Chem. Int. Ed.*, 2003, **42**, 3810.
7. M. J. Schultz, S. S. Hamilton, D. R. Jensen and M. S. Sigman, *J. Org. Chem.*, 2005, **70**, 3343.
8. S. Mori, M. Takubo, K. Makida, T. Yanase, S. Aoyagi, T. Maegawa, Y. Monguchi and H. Sajiki, *Chem. Commun.*, 2009, 5159.
9. S. Carrettin, P. McMorn, P. Johnston, K. Griffin and G. J. Hutchings, *Chem. Commun.*, 2002, 696.
10. L. Tang, X. Guo, Y. Li, S. Zhang, Z. Zha and Z. Wang, *Chem. Commun.*, 2013, **49**, 5213.
11. R. Kumar, E. Gravel, A. Hagege, H. Li, D. V. Jawale, D. Verma, I. N. N. Namboothiri and E. Doris, *Nanoscale*, 2013, **5**, 6491.
12. (a) Y. Maeda, N. Kakiuchi, S. Matsumura, T. Nishimura and S. Uemura, *Tetrahedron Lett.*, 2001, **42**, 8877; (b) Y. Maeda, N. Kakiuchi, S. Matsumura, T. Nishimura, T. Kawamura and S. Uemura, *J. Org. Chem.*, 2002, **67**, 6718.
13. J. Liu and S. Ma, *Synthesis*, 2013, 1624.
14. K. Mori, T. Hara, T. Mizugaki, K. Ebitani and K. Kaneda, *J. Am. Chem. Soc.*, 2004, **126**, 10657.

15. M. Mifsud, K. V. Parkhomenko, I. W. C. W. Arends and R. A. Sheldon, *Tetrahedron*, 2010, **66**, 1040.
16. S. Biella and M. Rossi, *Chem. Commun.*, 2003, 378.
17. M.-O. Simon and C.-J. Li, *Chem. Soc. Rev.*, 2012, **41**, 1415.
18. (a) G.-J. ten Brink, I. W. C. E. Arends and R. A. Sheldon, *Science*, 2000, **287**, 1636; (b) G.-J. ten Brink, I. W. C. E. Arends and R. A. Sheldon, *Adv. Synth. Catal.*, 2002, **344**, 355.
19. R. Sheldon, *Chem. Commun.*, 2001, 2399.
20. N. Jiang and A. J. Ragauskas, *Org. Lett.*, 2005, **7**, 3689.
21. A. Wolfson, S. Wuyts, D. E. De Vos, I. F. J. Vankelecom and P. A. Jacobs, *Tetrahedron Lett.*, 2002, **43**, 8107.
22. G. Ragagnin, B. Betzemeier, S. Quici and P. Knochel, *Tetrahedron*, 2002, **58**, 3985.
23. N. Jiang and A. J. Ragauskas, *ChemSusChem*, 2008, **1**, 823.
24. A. Abad, P. Concepción, A. Corma and H. García, *Angew. Chem. Int. Ed.*, 2005, **44**, 4066.
25. C.-X. Miao, J.-Q. Wang, B. Yu, W. G. Cheng, J. Sun, S. Chanfreau, L.-N. He and S.-J. Zhang, *Chem. Commun.*, 2011, **47**, 2697.
26. D. I. Enache, J. K. Edwards, P. Landon, B. Solsona-Espriu, A. F. Carley, A. A. Herzing, M. Watanabe, C. J. Kiely, D. W. Knight and G. J. Hutchings, *Science*, 2006, **311**, 362.
27. For a review on catalytic oxidations in dense carbon dioxide, see: T. Seki and A. Baiker, *Chem. Rev.*, 2009, **109**, 2409.
28. (a) Z. Hou, N. Theyssen, A. Brinkmann and W. Leitner, *Angew. Chem. Int. Ed.*, 2005, **44**, 1346; (b) Z. Hou, N. Theyssen and W. Leitner, *Green Chem.*, 2007, **9**, 127.
29. W. Keim, *Green Chem.*, 2003, **5**, 105.
30. N. Jiang and A. J. Ragauskas, *J. Org. Chem.*, 2006, **71**, 7087.
31. M. Kotani, T. Koike, K. Yamaguchi and N. Mizuno, *Green Chem.*, 2006, **8**, 735.
32. D. J. C. Constable, P. J. Dunn, J. D. Hayler, G. R. Humphrey, J. L. Leazer, R. J. Linderman, K. Lorenz, J. Manley, B. A. Pearlman, A. Wells, A. Zacks and T. Y. Zhang, *Green Chem.*, 2007, **9**, 411.
33. J. M. Hoover, J. E. Steves and S. S. Stahl, *Nat. Protoc.*, 2012, **7**, 1161.
34. X. Ye, M. D. Johnson, T. Diao, M. H. Yates and S. S. Stahl, *Green Chem.*, 2010, **12**, 1180.
35. G.-J. Ten Brink, I. W. C. E. Arends, M. Hoogenraad, G. Verspui and R. A. Sheldon, *Adv. Synth. Catal.*, 2003, **345**, 1341.
36. D. S. Bailie, G. M. A. Clendenning, L. McNamee and M. J. Muldoon, *Chem. Commun.*, 2010, **12**, 1180.
37. P. Gamez, I. W. C. E. Arends, J. Reedijk and R. A. Sheldon, *Chem. Commun.*, 2003, 2414.
38. A. Dijksman, A. Marino-Gonzalez, A. Mairata i Payeras, I. W. C. E. Arends and R. A. Sheldon, *J. Am. Chem. Soc.*, 2001, **123**, 6826.
39. S. K. Hanson, R. Wu and L. A. P. Silks, *Org. Lett.*, 2011, **13**, 1908.
40. P. J. Figiel, A. M. Kirillov, Y. Y. Karabach, M. N. Kopylovich and A. J. L. Pombeiro, *J. Mol. Catal. A: Chem.*, 2009, **305**, 178.

41. N. Jiang and A. J. Ragauskas, *Tetrahedron Lett.*, 2007, **48**, 273.
42. K. Yamaguchi, K. Mori, T. Mizugaki, K. Ebitani and K. Kaneda, *J. Am. Chem. Soc.*, 2000, **122**, 7144.
43. D. Choudhary, S. Paul, R. Gupta and J. H. Clark, *Green Chem.*, 2006, **8**, 479.
44. B. Karimi, S. Abedi, J. H. Clark and V. Budarin, *Angew. Chem. Int. Ed.*, 2006, **45**, 4776.
45. U. R. Pillai and E. Sahle-Demessie, *Green Chem.*, 2004, **6**, 161.
46. H. Wang, W. Fan, Y. He, J. Wang, J. N. Kondo and T. Tatsumi, *J. Catal.*, 2013, **299**, 10.
47. S. R. Reddy, D. Da and T. Punniyamurthy, *Tetrahedron Lett.*, 2004, **45**, 3561.
48. M. Herbert, F. Montilla and A. Galindo, *Dalton Trans.*, 2010, **39**, 900.
49. Y. Uozumi and R. Nakao, *Angew. Chem. Int. Ed.*, 2003, **42**, 194.
50. N. Zheng and G. D. Stucky, *Chem. Commun.*, 2007, 3862.
51. J. Chen, Q. Zhang, Y. Wang and H. Wan, *Adv. Synth. Catal.*, 2008, **350**, 453.

Subject Index

References to figures are given in *italic* type. References to tables as given in **bold** type.